621.3097
TURNING
1977
c1

Turning points in American electrical history / edited by James E. Brittain. -- New York : IEEE Press : distributor Wiley, c1977.
xi, 399 p. : ill. ; 28 cm. -- (IEEE Press selected reprint series)
Includes bibliographical references and indexes.
ISBN 0-87942-081-2. ISBN 0-87942-082-0 pbk.

1. Electric engineering--United States--History--Addresses, essays, lectures. 2. Telecommunication--United States--History--Addresses, essays, lectures. I. Brittain, James E., 1931-

TK23.T83 621.3'0973

Wa 76-18433

Turning Points in American Electrical History

OTHER IEEE PRESS BOOKS

Charge-Coupled Devices: Technology and Applications, *Edited by R. D. Melen and D. D. Buss*
Spread Spectrum Techniques, *Edited by R. C. Dixon*
Electronic Switching: Central Office Systems of the World, *Edited by A. E. Joel, Jr.*
Electromagnetic Horn Antennas, *Edited by A. W. Love*
Waveform Quantization and Coding, *Edited by N. S. Jayant*
Communication Satellite Systems: An Overview of the Technology, *Edited by R. G. Gould and Y. F. Lum*
Literature Survey of Communication Satellite Systems and Technology, *Edited by J. H. W. Unger*
Solar Cells, *Edited by C. E. Backus*
Computer Networking, *Edited by R. P. Blanc and I. W. Cotton*
Communications Channels: Characterization and Behavior, *Edited by B. Goldberg*
Large-Scale Networks: Theory and Design, *Edited by F. T. Boesch*
Optical Fiber Technology, *Edited by D. Gloge*
Selected Papers in Digital Signal Processing, II, *Edited by the Digital Signal Processing Committee*
A Guide for Better Technical Presentations, *Edited by R. M. Woelfle*
Career Management: A Guide to Combating Obsolescence, *Edited by H. G. Kaufman*
Energy and Man: Technical and Social Aspects of Energy, *Edited by M. G. Morgan*
Magnetic Bubble Technology: Integrated-Circuit Magnetics for Digital Storage and Processing, *Edited by H. Chang*
Frequency Synthesis: Techniques and Applications, *Edited by J. Gorski-Popiel*
Literature in Digital Signal Processing: Author and Permuted Title Index (Revised and Expanded Edition), *Edited by H. D. Helms, J. F. Kaiser, and L. R. Rabiner*
Data Communications via Fading Channels, *Edited by K. Brayer*
Nonlinear Networks: Theory and Analysis, *Edited by A. N. Willson, Jr.*
Computer Communications, *Edited by P. E. Green, Jr. and R. W. Lucky*
Stability of Large Electric Power Systems, *Edited by R. T. Byerly and E. W. Kimbark*
Automatic Test Equipment: Hardware, Software, and Management, *Edited by F. Liguori*
Key Papers in the Development of Coding Theory, *Edited by E. R. Berlekamp*
Technology and Social Institutions, *Edited by K. Chen*
Key Papers in the Development of Information Theory, *Edited by D. Slepian*
Computer-Aided Filter Design, *Edited by G. Szentirmai*
Laser Devices and Applications, *Edited by I. P. Kaminow and A. E. Siegman*
Integrated Optics, *Edited by D. Marcuse*
Laser Theory, *Edited by F. S. Barnes*
Digital Signal Processing, *Edited by L. R. Rabiner and C. M. Rader*
Minicomputers: Hardware, Software, and Applications, *Edited by J. D. Schoeffler and R. H. Temple*
Semiconductor Memories, *Edited by D. A. Hodges*
Power Semiconductor Applications, Volume II: Equipment and Systems, *Edited by J. D. Harnden, Jr. and F. B. Golden*
Power Semiconductor Applications, Volume I: General Considerations, *Edited by J. D. Harnden, Jr. and F. B. Golden*
A Practical Guide to Minicomputer Applications, *Edited by F. F. Coury*
Active Inductorless Filters, *Edited by S. K. Mitra*
Clearing the Air: The Impact of the Clean Air Act on Technology, *Edited by J. C. Redmond, J. C. Cook, and A. A. J. Hoffman*

Turning Points in American Electrical History

Edited by

James E. Brittain

Associate Professor, The History of Science and Technology
The Georgia Institute of Technology

A volume in the IEEE PRESS Selected Reprint Series, prepared under the sponsorship of the IEEE History Committee.

The Institute of Electrical and Electronics Engineers, Inc. New York

PRINTED IN THE UNITED STATES OF AMERICA

Library of Congress Catalog Card Number 76-18433

IEEE International Standard Book Numbers: Clothbound 0-87942-081-2
Paperbound 0-87942-082-0

Sole Worldwide Distributor (Exclusive of the IEEE):

JOHN WILEY & SONS, INC.
605 Third Ave.
New York, NY 10016

Wiley Order Numbers: Clothbound 0-471-02568-2
Paperbound 0-471-02569-0

Acknowledgments

THE initial stimulus that led ultimately to the publication of this collection of papers from the electrical history of America was provided by Reed Crone, Managing Editor of the IEEE Press. He proposed that the IEEE History Committee consider sponsoring such a project in a letter to Frederick Terman, then Chairman of the IEEE History Committee, dated February 12, 1974. Following discussions at the History Committee meeting in March 1974, Dr. Terman asked me to serve as Editor of the proposed book. I wish to extend special thanks to Mr. Crone and Dr. Terman for their enthusiastic and sustained interest, as well as their many helpful suggestions during the intervening months.

During the extended process of selecting papers, preliminary lists were circulated for comments among the members of an informal editorial advisory panel consisting of the following: Fred Terman; Jack Ryder, current Chairman of the IEEE History Committee; Bernard Finn, Curator of the Division of Electricity and Nuclear Energy at the Smithsonian Institution; Thomas Parke Hughes, Department of History and Sociology of Science at the University of Pennsylvania; Arthur L. Norberg, University of California, Berkeley and current Chairman of the Jovian Society. Philip L. Alger and Allan Schell also made suggestions of revisions of the lists. All of the suggestions were considered carefully in the process of arriving at a final choice, although it proved impossible to adopt all, and the final responsibility rests with me.

I am grateful for the efficient behind-the-scenes help provided by members of the staff of the IEEE Press. The facilities and staff assistance provided by the Library of Georgia Tech were essential to the success of the project.

Contents

Section II-D: Electronics

Section II-E: Feedback Control and Computers

Preface

THIS book is one of two projects undertaken as part of a cooperative effort by the IEEE History Committee and the Editors of the IEEE to commemorate the American bicentennial. The other project took the form of a special issue of the PROCEEDINGS OF THE IEEE, which contains original essays on electrical history written from an American perspective and which was published in September 1976. This book complements the special issue by providing a convenient compilation of primary source papers that report or summarize important electrical discoveries or innovations as they originated in or were diffused into America. As a byproduct, it proved possible to include papers by many of the great names in the history of electrical science and engineering in America.

Taken together, this book and the special issue of the PROCEEDINGS should constitute a valuable resource for the electrical engineering educator who wishes to integrate some historical material into his courses. They should prove especially useful in introductory courses in engineering and in specialized courses in engineering history where the bibliographic references included will facilitate access to the existing literature in electrical history. Finally, the book is intended for every contemplative electrical engineer with an interest in or curiosity about his professional heritage.

BIBLIOGRAPHIC NOTES

Each of the papers is accompanied by a brief biobibliographic note with references that help the reader gain access to further information about the subject of the paper and its author. Abbreviations have been used for a few standard biographical sources that are cited frequently. These include: *Dictionary of National Biography* (*DNB*); *Dictionary of American Biography* (*DAB*); *Dictionary of Scientific Biography* (*DSB*); *National Cyclopedia of American Biography* (*NCAB*); *National Academy of Sciences Biographical Memoirs* (*NAS*).

There are several specialized bibliographies that the serious student of electrical history will find quite useful. These include: George Shiers, *Bibliography of the History of Electronics* (Metuchen, NJ: Scarecrow Press, 1972); Thomas James Higgins, "A Biographical Bibliography of Electrial Engineers and Electrophysicists," *Technology and Culture*, vol. 2, pp. 28–32 and 146–165, 1961; Eugene S. Ferguson, *Bibliography of the History of Technology* (Cambridge, MA: M.I.T. Press, 1968); Jack Goodwin, "Current Bibliography in the History of Technology," *Technology and Culture* (annually). See also the references cited by Bernard S. Finn in a recent paper on the "History of Electrical Technology: The State of the Art," *Isis*, vol. 67, pp. 31–35, March 1976.

The special bicentennial issue of the PROCEEDINGS OF THE IEEE on electrical history, published in September 1976, contains both informed essays in electrical history and references to a wide range of sources. Anniversary issues of electrical engineering journals, such as the May 1962 issue of the PROCEEDINGS OF THE IRE and the May 1934 issue of *Electrical Engineering*, frequently contain comprehensive bibliographies in special fields of electrical engineering history.

Turning Points in American Electrical History

Introduction

AMERICANS have been fascinated by electricity at least since the time of Benjamin Franklin's spectacular lightning experiments of the mid 18th century. Franklin's success as a foreign diplomat was, in fact, based largely on his reputation as a natural philosopher gained from his early electrical investigations. The electrical history of the United States, in contrast to the comparative political stability that followed the American Revolution, has been one of repeated revolutions as established systems or devices were supplanted. The history of electricity reflects many of the characteristic features and long-range trends that have been observed in more general interpretations of American history. The explosive growth of electrical industries and the electrical engineering profession during the past century suggest an extraordinary American willingness to adopt and adapt to a continual sequence of electrical innovations having revolutionary social and economic implications. Some idea of the perplexity that the American enthusiasm for electrical novelties aroused in outside observers may be seen in a curious comment by a British electrical engineer, William Ayrton, in 1900. Ayrton stated that the Americans "are as everybody knows, an extremely radical people; but on the other hand they are extremely conservative" [1]. The statement was in the context of a discussion at a meeting of the Institution of Electrical Engineers on how the American electrical industry had been able to achieve the standardization necessary for mass production, while at the same time avoiding stagnation. It was by then becoming apparent to perceptive foreign observers that American electrical systems could not be transplanted easily in the absence of the same blend of conservative radicalism.

Other engineering visitors to America sought to explain the American infatuation with electrical systems. After a tour of the United States in 1889, G. L. Addenbrook argued before a meeting of British engineers that the British seemed to find it necessary to educate the public to accept new technology, but that in America electricity was regarded as a virtual necessity rather than a luxury [2]. William Preece, a leading British engineer–administrator, made several visits to the United States to evaluate communications innovations and methods during the late 19th century. After attending the International Electrical Exhibition in Chicago in 1893, Preece stated that he had found that the most distinguishing feature of the American electrical industry was its great scale and the rapidity with which innovations were adopted and developed [3]. Another engineer alluded to the American preference for novelty as opposed to durability. He stated that the Americans only wanted a product that would last until a new one appeared, while the Englishman wanted something "to last his day" [4].

American engineers also sought to explain the distinctive qualities of electrical engineering in the United States. In a paper presented at the International Electrical Congress in St. Louis in 1904, F. A. C. Perrine pointed out that American engineers had tended to standardize not only machines and ratings, but also methods of production, installation, and operation. He explained that this was due in part to the size of the country and the scale of the electrical industry. As a result, practices representing the ideas of the best engineers were followed throughout the country, thus improving greatly the quality of engineering work [5]. Charles Scott spoke at the same Congress and mentioned the importance of the absence of political and language barriers in the U.S. in comparison to Europe. Scott also noted that most American electrical engineers had been under similar influences, having graduated from the same schools, and that the influence of a few large companies had been decisive. He concluded with a story about a French engineer who had told him that the typical engineer in France regarded himself as something of an artist who wanted any plant he designed to be an original creation reflecting his own individuality [6].

Parallels between the American system of government and American electrical systems were sometimes suggested. For example, Theodore Vail, President of the American Telephone and Telegraph Company, wrote in 1910 that the Bell System was "not unlike the United States, each local company occupying its own territory and performing local functions, the A.T.&T. Company binding them all together with its long distance lines and looking after all the relations between the local companies and other companies" [7]. A radio doctrine analogous to the Monroe Doctrine was invoked shortly after World War I when the negotiations were underway that led to the creation of the Radio Corporation of America.

These examples and others that might be cited suggest that the character and style of American electrical engineering, like the American national character, have exhibited distinctive features. Identifying and interpreting these and tracing the patterns of international and interregional diffusion of electrical know-how will provide a continuing challenge for electrical historians.

Notes

[1] See the discussion in *J. Inst. Elec. Eng.*, vol. 29, p. 333, 1899–1900.
[2] G. L. Adenbrook, "Electrical Engineering in America," *J. Inst. Elec. Eng.*, vol. 18, pp. 770–798, 1889.
[3] W. H. Preece, "Notes on a trip to the United States and to Chicago," *J. Inst. Elec. Eng.*, vol. 23, pp. 40–86, 1894.
[4] See *Engineer*, vol. 89, pp. 148–149, 1900.
[5] F. A. C. Perrine, "American practice in high-tension line construction and operation," *Trans. Int. Elec. Congr.*, vol. 2, pp. 279–293, 1904.
[6] See the discussion in *ibid.*, pp. 790–795.
[7] *Reprints of Statements of the Structure of the Bell System and Some of its Fundamental Principles.* New York, 1922, pp. 9–10.

PART I
COLONIAL TO 1876

Philada Augt. 14. 1747.

If there is no other Use discover'd of Electricity, this, however, is something considerable, that it may help to make a vain Man humble.

B Franklin

Portion of a letter from Franklin to Collinson.
Courtesy, Burndy Library and Pierpont Morgan Library.

Section I-A
Franklin's Electrical Theory and Lightning Rod

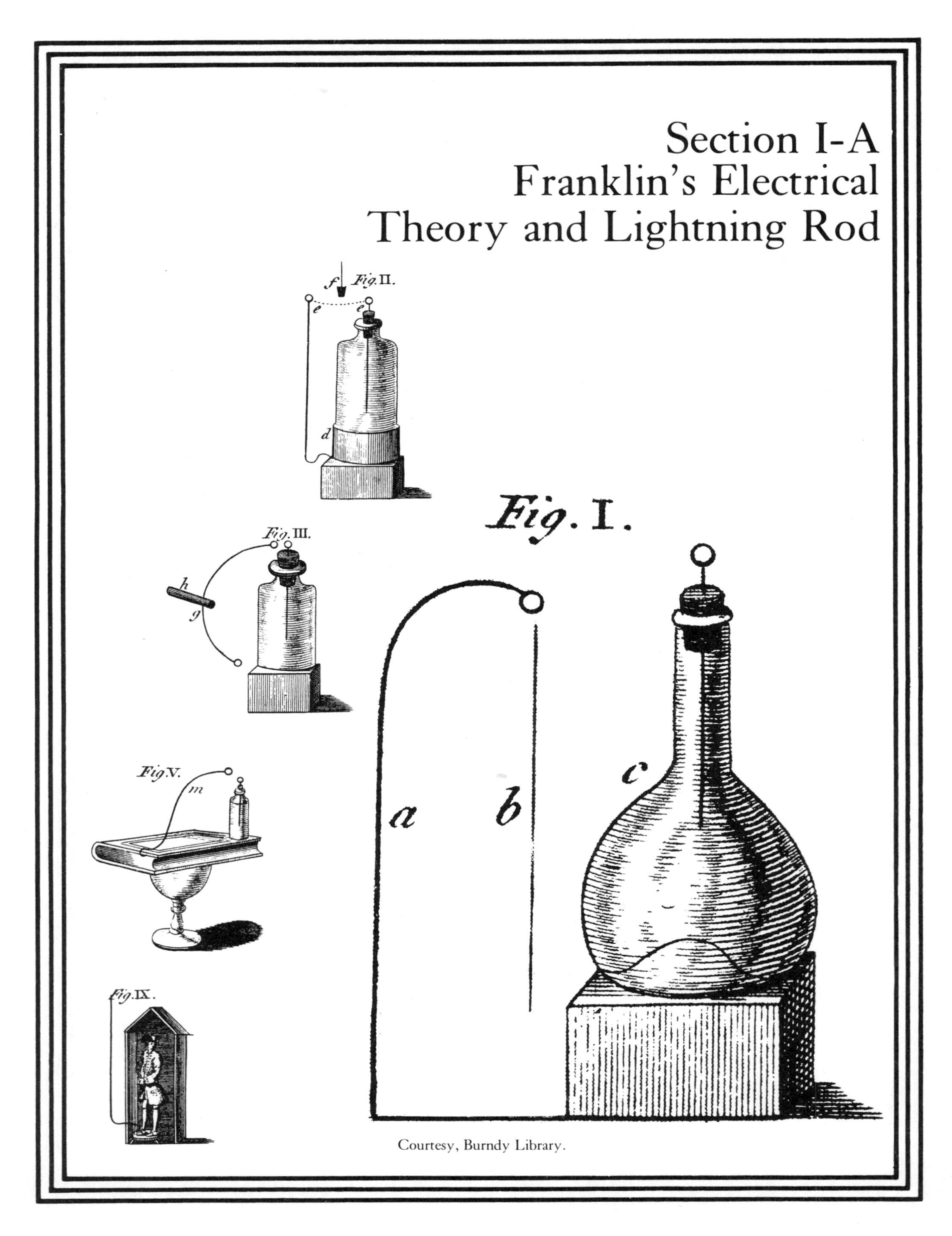

Courtesy, Burndy Library.

William Watson's Review of Franklin's Experiments and Observations on Electricity

WHEN this review of Benjamin Franklin's book on electricity appeared, the Royal Society of London had been in existence for almost a century. Its journal, *Philosophical Transactions*, had become the world's oldest and most prestigious scientific periodical. The same volume in which this review appeared carried reports of the dramatic success of the "sentry-box experiment" in France that verified Franklin's hypothesis that lightning was a large-scale electrical phenomenon. Although the spectacular lightning experiments tended to overshadow Franklin's conceptual contributions, he was responsible for the single-fluid theory of electricity that remained the dominant paradigm in electrical science until it was supplanted by Faraday's electromagnetic field theory. Watson was obviously impressed by Franklin's ability to combine theoretical speculation and experimental execution.

William Watson (1715–1787) was a leading British authority on electricity, and had been awarded the prestigious Copley Medal of the Royal Society in 1745 for his own electrical researches. See the *DNB*, vol. 20, pp. 956–958.

Benjamin Franklin (1706–1790) was born in Boston and served an apprenticeship as a printer before moving to Philadelphia where his electrical investigations were conducted during the late 1740's. The importance of his contribution was recognized by the award of the Copley Medal in 1753 and he was elected a Fellow of the Royal Society in 1756. The international reputation he achieved as a scientist proved very important to his later activities as a diplomat in England and France. See the *DSB*, vol. 5, pp. 129–139. For an authoritative account of Franklin's scientific contributions, see I. Bernard Cohen, *Franklin and Newton, An Inquiry into Speculative Newtonian Experimental Science and Franklin's Work in Electricity as an Example Thereof* (Philadelphia, 1956). See also *Benjamin Franklin's Experiments*, edited with a critical and historical introduction by I. Bernard Cohen (Cambridge, MA, 1941). A paper on Franklin's electrical researches by Bernard S. Finn is included in a special issue of the PROCEEDINGS OF THE IEEE, published in September 1976. See also Bernard S. Finn, "An Appraisal of the Origins of Franklin's Electrical Theory," *Isis*, vol. 60, pp. 362–369, 1969. Several papers related to Franklin's lightning rod appeared in the *Journal of the Franklin Institute*, vol. 253, 1952.

XXXI. *An Account of Mr.* Benjamin Franklin's *Treatise, lately published, intituled,* Experiments and Observations on Electricity, made at Philadelphia in America; *by* Wm. Watson, *F. R. S.*

Read June 6. 1751.

MR. Franklin's treatise, lately presented to the Royal Society, consists of four letters to his correspondent in England, and of another

Reprinted from *Phil. Trans. Royal Soc. London*, vol. 47, pp. 202–211, 1751–1752.

other part intituled " Opinions and conjectures con-
" cerning the properties and effects of the electrical
" matter arising from experiments and observations."

The four letters, the last of which contains a new hypothesis for explaining the several phænomena of thunder-gusts, have either in the whole or in part been before communicated to the Royal Society. It remains therefore, that I now only lay before the Society an account of the latter part of this treatise, as well as that of a letter intended to be added thereto by the author, but which arrived too late for publication with it, and was therefore communicated to the Society by our worthy brother Mr. Peter Collinson.

This ingenious author, from a great variety of curious and well-adapted experiments, is of opinion, that the electrical matter consists of particles extremely subtil; since it can permeate common matter, even the densest metals, with such ease and freedom, as not to receive any perceptible resistance: and that if any one should doubt, whether the electrical matter passes through the substance of bodies, or only over and along their surfaces, a shock from an electrified large glass jar, taken through his own body, will probably convince him.

Electrical matter, according to our author, differs from common matter in this, that the parts of the latter mutually attract, and those of the former mutually repel, each other; hence the divergency in a stream of electrified effluvia §: but that, tho' the particles

§ As the electric stream is observed to diverge very little, when the experiment is made *in vacuo*, this appearance is more owing to

particles of electrical matter do repel each other, they are ſtrongly attracted by all other matter.

From theſe three things, *viz.* the extreme ſubtilty of the electrical matter, the mutual repulſion of its parts, and the ſtrong attraction between them and other matter, ariſes this effect, that when a quantity of electrical matter is applied to a maſs of common matter of any bigneſs or length within our obſervation (which has not already got its quantity) it is immediately and equally diffuſed thro' the whole.

Thus common matter is a kind of ſponge to the electrical fluid; and as a ſponge would receive no water, if the parts of water were not ſmaller than the pores of the ſponge; and even then but ſlowly, if there was not a mutual attraction between thoſe parts and the parts of the ſponge; and would ſtill imbibe it faſter, if the mutual attraction among the parts of the water did not impede, ſome force being required to ſeparate them; and faſteſt, if, inſtead of attraction, there were a mutual repulſion among thoſe parts, which would act in conjunction with the attraction of the ſponge: ſo is the caſe between the electrical and common matter. In common matter indeed there is generally as much of the electrical as it will contain within its ſubſtance: if more is added, it lies without upon the ſurface ||, and forms what we call an

---

to the reſiſtance of the atmoſphere, than to any natural tendency in the electricity itſelf. *W. W.*

|| The author of this account is of opinion, that what is here added, lies not only without upon the ſurface, but penetrates with the ſame degree of denſity the whole maſs of common matter, upon which it is directed.

an electrical atmosphere; and then the body is said to be electrified.

'Tis supposed, that all kinds of common matter do not attract and retain the electrical with equal force, for reasons to be given hereafter; and that those called electrics *per se*, as glass, *&c.* attract and retain it the strongest, and contain the greatest quantity.

We know, that the electrical fluid is in common matter, because we can pump it out by the globe or tube; and that common matter has near as much as it can contain; because, when we add a little more to any portion of it, the additional quantity does not enter, but forms an electrical atmosphere: and we know, that common matter has not (generally) more than it can contain; otherwise all loose portions of it would repel each other, as they constantly do when they have electric atmospheres.

The form of the electrical atmosphere is that of the body, which it surrounds. This shape may be render'd visible in a still air, by raising a smoke from dry resin dropp'd into a hot tea spoon under the electrised body, which will be attracted and spread itself equally on all sides, covering and concealing the body. And this form it takes, because it is attracted by all parts of the surface of the body, though it cannot enter the substance already replete. Without this attraction it would not remain round the body, but be dissipated in the air.

The atmosphere of electrical particles surrounding an electrified sphere is not more disposed to leave it, or more easily drawn off from any one part of the sphere than from another, because it is equally attracted by every part. But that is not the case with bodies

bodies of any other figure. From a cube it is more eaſily drawn at the corners than at the plane ſides, and ſo from the angles of a body of any other form, and ſtill moſt eaſily from the angle that is moſt acute; and for this reaſon points have a property of drawing on, as well as throwing off the electrical fluid, at greater diſtances than blunt bodies can.

From various experiments recited in our author's treatiſe, to which the curious may have recourſe, the preceding obſervations are deduced. You will obſerve how much they coincide with and ſupport thoſe which I ſome time ſince communicated to the Society upon the ſame ſubject.

To give even the ſhorteſt account of all the experiments contained in Mr. Franklin's book, would exceed greatly the time allowed for theſe purpoſes by the Royal Society: I ſhall content myſelf therefore with laying a few of the moſt ſingular ones before you.

The effects of lightning, and thoſe of electricity, appear very ſimilar. Lightning has often been known to ſtrike people blind. A pigeon, ſtruck dead to appearance by the electrical ſhock, recovering life, drooped ſeveral days, eat nothing, tho' crumbs were thrown to it, but declined and died. Mr. Franklin did not think of its being deprived of ſight; but afterwards a pullet, ſtruck dead in like manner, being recovered by repeatedly blowing into its lungs, when ſet down on the floor, ran headlong againſt the wall, and on examination appeared perfectly blind: hence he concluded, that the pigeon alſo had been abſolutely blinded by the ſhock. From this obſervation we ſhould be extremely cautious, how in electriſing we draw

draw the ſtrokes, eſpecially in making the experiment of Leyden, from the eyes, or even from the parts near them.

Some time ſince it was imagined, that deafneſs had been relieved by electriſing the patient, by drawing the ſnaps from the ears, and by making him undergo the electrical commotion in the ſame manner. If hereafter this remedy ſhould be fantaſtically applied to the eyes in this manner to reſtore dimneſs of ſight, I ſhould not wonder, if perfect blindneſs were the conſequence of the experiment.

By a very ingenious experiment our author endeavours to evince the impoſſibility of ſucceſs, in the experiments propoſed by others of drawing forth the effluvia of non-electrics, cinamon, for inſtance, and by mixing them with the electrical fluid, to convey them with that into a perſon electrified: and our author thinks, that tho' the effluvia of cinamon and the electrical fluid ſhould mix within the globe, they would never come out together through the pores of the glaſs, and thus be conveyed to the prime conductor; for he thinks, that the electrical fluid itſelf cannot come through, and that the prime conductor is always ſupplied from the cuſhion, and this laſt from the floor. Beſides, when the globe is filled with cinamon, or other non-electrics, no electricity can be obtained from its outer ſurface, for the reaſons before laid down. He has tried another way, which he thought more likely to obtain a mixture of the electrical and other effluvia together, if ſuch a mixture had been poſſible. He placed a glaſs plate under his cuſhion, to cut off the communication between the cuſhion and the floor: he then brought a

ſmall

ſmall chain from the cuſhion into a glaſs of oil of turpentine, and carried another chain from the oil of turpentine to the floor, taking care, that the chain from the cuſhion to the glaſs touched no part of the frame of the machine. Another chain was fixed to the prime conductor, and held in the hand of a perſon to be electrified. The ends of the two chains in the glaſs were near an inch from each other, the oil of turpentine between. Now the globe being turned could draw no fire from the floor through the machine, the communication that way being cut off by the thick glaſs plate under the cuſhion: it muſt then draw it through the chains, whoſe ends were dipp'd in the oil of turpentine. And as the oil of turpentine being in ſome degree an electric *per ſe*, would not conduct what came up from the floor, the electricity was obliged to jump from the end of one chain to the end of the other, which he could ſee in large ſparks; and thus it had a fair opportunity of ſeizing of the fineſt particles of the oil in its paſſage, and carrying them off with it: but no ſuch effect followed, nor could he perceive the leaſt difference in the ſmell of the electrical effluvia thus collected, from what it had when collected otherwiſe; nor does it otherwiſe affect the body of the perſon electrified. He likewiſe put into a phial, inſtead of water, a ſtrong purging liquid, and then charged the phial, and took repeated ſhocks from it; in which caſe every particle of the electrical fluid muſt, before it went through his body, have firſt gone thro' the liquid, when the phial is charging, and returned through it when diſcharging; yet no other effect followed than if the phial had been charged with water.

He

He has alſo ſmelt the electrical fire, when drawn thro' gold, ſilver, copper, lead, iron, wood, and the human body, and could perceive no difference; the odour being always the ſame, where the ſpark does not burn what it ſtrikes; and therefore he imagines, that it does not take that ſmell from any quality of the bodies it paſſes through. There was no abridging this experiment, which I think very well conceived, and as well conducted, in a manner to make it intelligible; and therefore I have laid the author's words nearly before you.

As Mr. Franklin, in a letter to Mr. Collinſon ſome time ſince, mentioned his intending to try the power of a very ſtrong electrical ſhock upon a turkey, I deſired Mr. Collinſon to let Mr. Franklin know, that I ſhould be glad to be acquainted with the reſult of that experiment. He accordingly has been ſo very obliging as to ſend an account of it, which is to the following purpoſe. He made firſt ſeveral experiments on fowls, and found, that two large thin glaſs jars gilt, holding each about 6 gallons, and ſuch as I mentioned I had employed in the laſt paper I laid before you upon this ſubject, were ſufficient, when fully charged, to kill common hens outright; but the turkeys, though thrown into violent convulſions, and then, lying as dead for ſome minutes, would recover in leſs than a quarter of an hour. However, having added three other ſuch to the former two, though not fully charged, he killed a turkey of about ten pounds weight, and believes that they would have killed a much larger. He conceited, as himſelf ſays, that the birds kill'd in this manner eat uncommonly tender.

In

In making theſe experiments, he found, that a man could, without great detriment, bear a much greater ſhock than he imagined: for he inadvertently received the ſtroke of two of theſe jars through his arms and body, when they were very near fully charged. It ſeemed to him an univerſal blow throughout the body from head to foot, and was followed by a violent quick trembling in the trunk, which went gradually off in a few ſeconds. It was ſome minutes before he could recollect his thoughts, ſo as to know what was the matter; for he did not ſee the flaſh, tho' his eye was on the ſpot of the prime conductor, from whence it ſtruck the back of his hand; nor did he hear the crack, tho' the byſtanders ſaid it was a loud one; nor did he particularly feel the ſtroke on his hand, tho' he afterwards found it had raiſed a ſwelling there of the bigneſs of half a ſwan-ſhot, or piſtol-bullet. His arms and the back of his neck felt ſomewhat numbed the remainder of the evening, and his breaſt was ſore for a week after, as if it had been bruiſed. From this experiment may be ſeen the danger, even under the greateſt caution, to the operator, when making theſe experiments with large jars; for it is not to be doubted, but that ſeveral of theſe fully charged would as certainly, by increaſing them, in proportion to the ſize, kill a man, as they before did the turkey.

Upon the whole, Mr. Franklin appears in the work before us to be a very able and ingenious man; that he has a head to conceive, and a hand to carry into execution, whatever he thinks may conduce to enlighten the ſubject-matter, of which he is treating: and altho' there are in this work ſome few opinions, in in which I cannot perfectly agree with him, I think ſcarce any body is better acquainted with the ſubject of electricity than himſelf.

Section I-B
The Electrochemical Battery and Electromagnetism

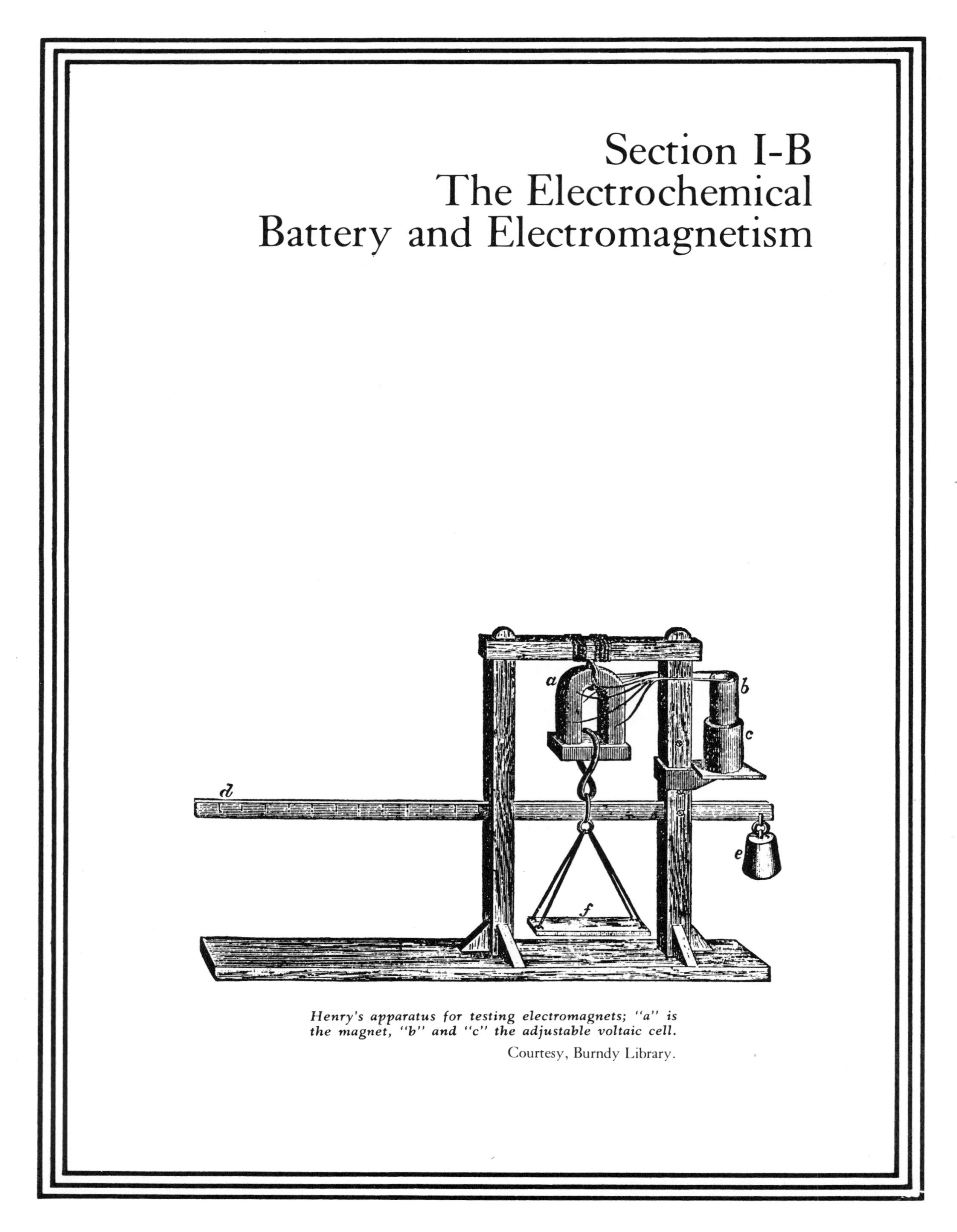

Henry's apparatus for testing electromagnets; "a" is the magnet, "b" and "c" the adjustable voltaic cell.

Courtesy, Burndy Library.

J. F. Dana's Electrical Battery

THIS paper appeared in the first volume of the *American Journal of Science*, popularly known as *Silliman's Journal* in honor of Benjamin Silliman (1779–1864) who founded the journal and served as Editor until his death. It served the emerging American scientific community as the *Philosophical Transactions* had served leading colonial scientists such as Franklin, and it published numerous papers on electrical science and technology during the 19th century. The term "battery" in the title illustrates the evolutionary change in meaning of familiar terms over time. Although the electrochemical or voltaic battery had been in use since 1800, the term "Leyden battery" continued to be used to describe devices used to store an electric charge. For example, both the Leyden and voltaic batteries were discussed in Faraday's *Experimental Researches in Electricity* published in 1839. Dana is describing an improved design of the Leyden battery. The utilitarian and practical orientation of many early American scientists is evidenced by Dana's effort to design a more convenient, inexpensive, and efficient battery.

James Freeman Dana (1793–1827) was born in New Hampshire and graduated from Harvard College in 1813. He studied at Cambridge University before returning to the United States to be awarded the M.D. degree from Harvard in 1817. He soon gave up his medical practice to join the faculty at Dartmouth where he taught from 1820 to 1826. Both Samuel Morse and Joseph Henry credited Dana with introducing them to electromagnetism shortly before his untimely death. See the *DAB*, vol. 5, p. 56 and the *NCAB*, vol. 10, p. 390.

Cambridge, January 25, 1819.

PROFESSOR SILLIMAN.

Dear Sir,

IF the following observations are worthy of a place in your valuable Journal, please to insert them, and oblige yours, with real esteem, J. F. DANA.

ART. XIV. *On a New Form of the Electrical Battery, by* J. F. DANA, M. D. *Chemical Assistant in Harvard University, and Lecturer on Chemistry and Pharmacy in Dartmouth College.*

THE Electrical Battery in its common form is an unmanageable and inconvenient apparatus. When the coated surface is

Reprinted from *Amer. J. Sci.*, vol. I, pp. 292–294, 1819.

comparatively small, the instrument occupies a large space, and it cannot be readily removed from place to place without much trouble and risk; the apparatus is, moreover, very expensive, and when one of the jars is broken, another of the same dimensions cannot readily be found to supply its place.

It occurred to me, that a Battery might be constructed of plates of glass and sheets of tinfoil, in which the same extent of coated surface should occupy a much smaller space, and consequently that the apparatus would be more convenient and more portable. I selected several panes of glass, the surfaces of which coincided closely with each other, and then arranged them with sheets of tinfoil in this order, viz. pane of glass, sheet of tinfoil, then another pane of glass, then a second sheet of tinfoil, and so on; the sheets of foil being smaller than the plates of glass by two inches all around; the glass being 10 by 12, and the foil 6 by 8. This apparatus contained six plates of tinfoil, and the lowest plate being numbered *one*, was connected with the ground, and by slips of tinfoil passing over the edges, with the *third* plate, and this, in like manner with the *fifth*. The *second* plate was connected with the *fourth*, and this with the *sixth*, which communicated with the conductor of the machine; in this manner each plate positively electrified will be opposed by one negatively electrified, and vice versa; the 6th, 4th, and 2d plates positive, and the 5th, 3d, and 1st, negative. Into this apparatus I could introduce a powerful charge, but not possessing a battery of the common form, could not make comparative experiments. The annexed figures will explain the construction of this apparatus.

(See Plate.)

Fig. 1. *a* 1, *a* 2, &c. the tinfoil.
b b b, plates of glass.
c, the intermediate slips connecting the plates 6, 4, and 2.
d, the slips connecting 5, 3, 1, and the ground.

Fig. 2. *a*, the intermediate slips passing over the edges of the glass and connecting plates, 1, 3, and 5.
b, the slip which connects the upper sheet of foil with the 4th, &c.

(Figures not included)

In a battery of the ordinary form, it is evident that a much less surface is coated than in one of the above construction;

in a battery of the common form, two feet long, one foot wide, and ten inches high, and containing 18 coated jars, there will be no more than 3500 square inches of coated surface, while in a battery of the same dimensions on the proposed construction, there will be no less than 8000 square inches covered with tinfoil, allowing the sheet of glass and of foil to be $\frac{1}{4}$ inch thick.

When plate glass is employed for making this battery, the ring of glass exterior to the tinfoil may be covered with varnish, and then the next plate laid over it; the tinfoil will then be shut out for ever from the access of moisture, and the insulation will remain perfect. This form of the Electrical Battery is very portable, may be packed in a case with the machine, and indeed a powerful battery occupies no greater space than a quarto volume. It is cheap and easily constructed.

George T. Bowen on Hare's Calorimotor

BOWEN'S paper appeared in the *American Journal of Science* in 1822. It well illustrates the diffusion of scientific knowledge from Europe and the unsettled state of the theory of galvanic electricity during the interval between Oersted's discovery of electromagnetism in 1820 and the publication of Ohm's theory in 1827. The calorimotor was actually a powerful electrochemical battery, but its name was chosen by its inventor, Robert Hare, to conform to his "new theory of galvanism." Hare opposed the idea that heat was a manifestation of molecular motion, and instead contended that caloric and electricity were material fluids. (See "Professor Hare on new Galvanic Apparatus and Theory," *American Journal of Science*, vol. 3, pp. 105–117, 1821.) Hare argued that the galvanic fluid was a mixture of caloric and electric fluids with one or the other being predominant, depending on the design of the source.

George T. Bowen (1803–1828) was born in Providence, R.I., and graduated from Yale College in 1822. His investigation of Hare's calorimotor was done under the direction of Professor Silliman. Bowen studied medicine in Philadelphia before accepting a position as Professor of Chemistry at the University of Nashville in 1825. See the *NCAB*, vol. 12, p. 108.

Art. XXI.—*Notice of Magnetic effects produced* by Dr. Hare's Calorimotor; by* George T. Bowen, *of Providence.*

In Vol. V. p. 352 of the Edin. Phil. Journal, is a description of an electro-magnetic apparatus by Prof. Moll of Utrecht. After having given a description of the instrument, that gentleman observes, "A remarkable feature in the effect of this spiral voltaic apparatus, is the strong adhesion of iron filings to the conductive wire. If the zinc plate be new, or well cleaned, the acid strong, and of course the galvanic process going on with energy, then if iron filings, on a paper, are brought backward and forward, under, and near to the horizontal conductive wire of copper, the iron filings will begin to stand erect as if in the vicinity of a loadstone, and they will even adhere strongly to the copper wire when brought into contact with it." These observations appearing to me to be interesting, I was desirous of seeing what would be the effect produced by presenting iron filings to the wire which connects the opposite poles of the calorimotor of Doct. Hare. The results of the experiments which were performed with this instrument are as follows :—A large copper wire, about one foot in length, was bent in the form of a semi-circle, and its ends connected by means of small vices to the opposite poles of the calorimotor. The instrument was then immersed in the weak acid solution. On bringing a paper containing iron filings into the vicinity of the copper, which had already become hot, the filings began to stand erect, and when brought into contact with the wire they were powerfully attracted by it; adhering to it, and forming a fringe upon its surface. The calorimotor was then raised out of the diluted acid, and it was observed that the filings dropped from the connecting wire, the instant the instrument left the surface of the fluid. This experiment was often repeated; (the size of the connecting wire being varied,) and always with the same results. A platina wire was then bent as in the above experiment, and its ends connected with the opposite poles of the calorimotor. Upon immersing the plates, and bringing the iron filings on a paper, near to, and into contact with the platina, they were powerfully attracted, and

* In the laboratory of Yale College.

Reprinted from *Amer. J. Sci.*, vol. V, pp. 357–360, 1822.

adhered to the wire as in the preceding experiment. The instrument being then raised from the diluted acid, the filings fell from the wire. These experiments were repeated, and varied by employing different metals to complete the communication between the two galvanic poles. Wires of iron, copper, brass, lead, zinc, silver, and platina were used; the length of the wire employed being in all cases so regulated, that although it became hot, it was not ignited. In every instance the same magnetic properties were exhibited by the connecting wire; the iron filings being strongly attracted by it, so long as the calorimotor was immersed in the acid solution, and immediately falling from it when the instrument was raised from the fluid.

In Vol. VI. p. 83 of the work above quoted, are detailed some experiments of Prof. Moll, in which he succeeded in imparting magnetism to steel needles by inclosing them in a glass tube, about which was wound a spiral of brass wire, and passing *strong electrical discharges* through the spiral. These experiments were repeated in the following manner: Around a glass tube one quarter of an inch in diameter and four inches in length, was wound spirally, a brass wire, from *left to right*, forming ten spirals. The ends of the wires were then connected with the opposite galvanic poles and a needle, which had been previously ascertained to be free from magnetism was placed within the tube. The instrument was then immersed and remained in the fluid for thirty seconds. Upon examining the needle after the plates had been raised from the acid solution, it was found to have become powerfully magnetic, having a north and south pole; one of which was attracted and the other repelled by the poles of a magnetic needle suspended in the usual manner—they also took up iron filings abundantly. The end of the needle which had been placed nearest to the copper or negative side of the calorimotor had acquired north polarity. while that which had been next to tne zinc or positive side had acquired south polarity. This experiment was often repeated, and always with the same results; the end of the needle placed nearest to the copper plates constantly acquiring a north polarity.

One of the needles which had been magnetized in this manner, was again enclosed within the glass tube, its *north pole* being placed next to the copper side of the apparatus. The plates were immersed, and again raised from the fluid.

Upon removing the needle, its poles were found to have been unaffected, the end which had been nearest the copper, still retaining its north polarity. The same needle was again submitted to the galvanic action, its *north pole* being now placed nearest to the *zinc plates* of the instrument; upon examination, its poles were found to be reversed; its south pole which had been placed nearest the *copper* plates, had acquired north polarity, while its *north* pole which was next to the *zinc* plates had acquired south polarity.

A *common magnetic needle* was then enclosed in the tube, its south pole being placed next to the copper side of the apparatus. The plates having been immersed the usual time, the needle was examined. The end which had previously been its south pole, and which was placed next the copper plates, had now acquired north polarity, and in every instance that end of a needle which was connected with the negative side of the calorimotor became its north pole, so long as the spiral brass wire upon the glass tube was wound from *left to right.* I then took the same glass tube and wound a brass wire spirally around it, the spirals however, being now wound from *right to left.* A needle was placed within the tube, and the ends of the spirals connected with the opposite poles of the calorimotor. After the immersion of the plates, the needle was removed from the glass tube and was found to have become magnetic—its *north pole* being that end of the needle which had been connected with *zinc plates* of the instrument, and vice versa. The needle was then again enclosed in the tube and the plates immersed; its acquired south pole being placed in connection with the zinc plates. When examined, its poles were found to be reversed; its former *south* pole which had been connected with the *zinc* plates having now acquired *north polarity;* and in all cases, that end of a needle which was connected with the *zinc* plates of the instrument, acquired *north* polarity, when the spirals about the glass tube were wound from *right to left.*

A steel needle free from magnetism, was then enclosed within a tube of glass four inches long and an eighth of an inch in diameter. This glass tube was then placed within a tube of lead, and the lead tube again enclosed in one of glass, around which was placed a spiral of brass wire, wound from *left to right;* the ends of the spirals being connected with the opposite galvanic poles—The plates were immersed and suffered to remain in the fluid during half a minute

The needle on examination, was found to have become magnetic; its *north pole* being that end of the needle which had been connected with the copper or negative side of the calorimotor. It was again enclosed in the tube as before; the end which had acquired *north* polarity being now placed next to the *zinc* plates. After the immersion of the instrument, the poles of the needle were found to be reversed--the former south pole having now acquired north polarity, and vice versa—the results obtained in this method of operating were always the same; the needle acquiring *north* polarity at the end which was placed nearest the *copper* plates, while the spirals of brass around the glass tube, passed from *left to right.* When the direction of the spirals was changed, and the brass wire wound about the glass tube from *right to left;* then, that end of the needle which was connected with the *zinc* plates always acquired north polarity. Being desirous of ascertaining how long it was necessary the plates should be immersed in order to produce these effects, a needle was inclosed in the tube as in the former experiments, and the plates were then immersed, and *immediately* withdrawn from the fluid. On examination, the needle was found magnetic. Another needle having been placed within the tube, the calorimotor was lowered until the plates had descended into the fluid *one quarter of an inch,* when it was instantly raised. Even in this instance, when the plates had descended only *one quarter of an inch* into the acid solution, and had remained there only *one second,* the needle was found to have become powerfully magnetic, and readily took up iron filings. This experiment was often repeated, and with the same results. The preceding experiments lead to the conclusion, that when a needle is subjected to the galvanic action in the manner above described, it instantly becomes magnetic, and that end of the needle which is connected with the *copper or negative* side of the calorimotor, always acquires *north* polarity, when the turns of the spiral about the glass tube pass from *left to right;* and that end connected with the copper plates always acquires a *south polarity* when the turns of the spiral pass from *right to left.*

From these experiments it appears that the same magnetic effects are produced by Dr. Hare's calorimotor, as by powerful electrical batteries—although he justly considers his instrument, as producing a great flow of caloric almost without electricity.

Henry and Ten Eyck on a Large Electromagnet

IN this paper from the *American Journal of Science* of 1831, the Editor, Benjamin Silliman, credits Joseph Henry with having constructed far more powerful electromagnets than any yet made in Europe. This suggests an element of cultural nationalism, as Americans took great pride in any evidence of technical or scientific achievements that compared favorably to those in Europe. On another level, Silliman's note reflects a behind-the-scenes discussion over allocation of credit appropriate to the relative contributions of Henry and Ten Eyck to the project. Neither man was entirely satisfied by the editorial note. Henry alluded to the dispute in a later letter to Silliman dated February 28, 1832. He concluded that "I am now willing to forget and forgive but at the same time shall be somewhat tenacious of my rights in reference to magnetism." (For this letter and additional details on the dispute, see *The Papers of Joseph Henry*, vol. I, edited by Nathan Reingold, p. 406 and passim.)

Joseph Henry (1797–1878) was born in Albany, N.Y. and graduated from the Albany Academy in 1822. He worked as a tutor and surveyor before accepting a teaching position at Albany Academy where he began his electromagnetic experiments in 1827. He later taught at Princeton University from 1832 until 1846 when he became first Secretary of the Smithsonian Institution in Washington, a position he retained for the rest of his life. See the *DAB*, vol. 6, pp. 277–281. See also the discussion of Henry in *Science in Nineteenth Century America. A Documentary History*, edited by Nathan Reingold (New York, 1964), pp. 62–65. Nathan Reingold is directing a project currently underway to publish selected manuscripts from the Joseph Henry Papers at the Smithsonian. The first volume has already been published, and its introduction and editorial notes provide an excellent summary of the context of Henry's work to 1832. A paper by Arthur P. Molella dealing with Henry's views on technological progress and its relation to societal needs as evidenced by the electric motor and telegraph is included in the September 1976 issue of the PROCEEDINGS OF THE IEEE.

Philip Ten Eyck (1802–1892) was also a native of Albany, N.Y. His M.D. degree was received in 1825 and he became Henry's successor at the Albany Academy in 1832. Apparently his collaboration with Henry on the electromagnet was his only significant contribution to science. See *The Papers of Joseph Henry*, vol. I, p. 214.

An account of a large Electro-Magnet, made for the Laboratory of Yale College*; by JOSEPH HENRY and DR. TEN EYCK.

(Extract of a letter to Prof. Silliman, accompanying the Magnet.)

THE magnet is constructed on precisely the same principles as that described in the last number of the Journal. It weighs 59½ lbs. avoirdupois, (exclusive of the copper wire which surrounds it,) and

---

* This magnet is now arranged in its frame, in the laboratory of Yale College. Being myself out of town when the instrument arrived, the necessary experiments and fixtures were satisfactorily made by Mr. C. U. Shepard, (Chem. Assis.) and Dr. Titus W. Powers, of Albany, who was so obliging as to bring the magnet to New Haven. There has not been time (as the magnet came just as this No. was finishing) to do any thing more than make a few trials, which have however fully substantiated the statements of Prof. Henry.† He has the honor of having constructed by far, the most powerful magnets that have ever been known, and his last, weighing, armature and all, but 82½ lbs., sustains over a ton. It is eight times more powerful than any magnet hitherto known in Europe, and between six and seven times more powerful than the great magnet in Philadelphia. We understand that the experiments described in the last No. of this Journal, (except those ascribed to Dr. Ten Eyck) were devised by Professor Henry alone, who (except forging the iron) constructed the magnet with his own hand. The plan of the frame, and the fixtures, and the drawing in the last No., were done by Dr. Ten Eyck. In the Yale College magnet, the plan was drawn by Professor Henry, and the iron forged under his direction. The length of the wires being agreed upon, the winding was done by Dr. Ten Eyck, and the experiments were mutually performed.—*Ed.*

† It may be worth while to state a single experiment, which I made with a view to learn the chemical effects of this instrument. As its magnetic flow was so powerful, I had strong hopes of being able to accomplish the decomposition of water by its means. My experiment, however, which was made as follows, proved unsuccessful. The battery being immersed, to the extremities of the magnet were applied two broad, polished plates of iron, terminating in flattened wires, which were united with the wires of the ordinary apparatus for decomposing water, and the contact heightened by the use of cups of mercury: not the slightest decomposition was, however, observable. Aware, that had any chemical effect been produced, this arrangement could have decided nothing, (except perhaps from the degree of energy in the decomposition) as respects the point whether simple magnetism is adequate to decompose water, since it might under these circumstances be attributed to the electricity from the battery, I had determined in a second experiment, had the first proved successful, to have interrupted the galvanic flow by a non-conductor; in which case, had the decomposition ensued, pure magnetism might have been considered as the decomposing agent. But as my preliminary experiment was unsuccessful, I proceeded no farther; I hope, however, to resume the research hereafter, under more favorable circumstances. C. U. SHEPARD.

Reprinted from *Amer. J. Sci.*, vol. XX, pp. 201–203, 1831.

was formed from a bar of Swede's iron three inches square and thirty inches long. Before bending the bar into the shape of a horse-shoe, it was flattened on the edges, so as to form an octagonal prism, having a perimeter of $10\frac{3}{4}$ inches. The other dimensions of the magnet, as measured before winding it with wire, are as follows:—perpendicular height of the exterior arch of the horse-shoe $11\frac{3}{4}$ inches—around the outside from one pole to the other $29\frac{9}{10}$ inches—internal distance between the poles $3\frac{1}{2}$ inches.

The armature or lifter is formed from a piece of iron from the same bar, not flattened on the edges; it is nearly 3 inches square, $9\frac{1}{2}$ inches long, and weighs 23 lbs. The upper surface is made perfectly flat, except about an inch in the middle where the angles are rounded off so as to form a groove, into which the upper part of a strong iron stirrup, surrounding the armature, fits somewhat loosely. The weight to be supported is fastened to the lower part of the stirrup, and by means of the groove is made to bear directly on the center of the armature.

For the purpose of suspending the magnet, a piece of round iron with an eye on one end, is firmly screwed into the crown of the arch and is attached to the cross beam of a frame, similar to that figured in the last number of the Journal.

The magnet is wound with 26 strands of copper bell wire, covered with cotton thread 31 feet long; about 18 inches of the ends are left projecting, so that only 28 feet actually surround the iron; the aggregate length of the coils is therefore 728 feet. Each strand is wound on a little less than an inch; in the middle of the horse-shoe it forms three thicknesses of wire, and on the ends or near the poles it is wound so as to form six thicknesses.

Two small galvanic batteries are soldered to the wires of the magnet, one on each side of the supporting frame, in such a manner as to cause the poles to be instantaneously reversed, by merely dipping the batteries alternately into acid. To render these as compact as possible, they are formed of concentric copper cylinders with cylinders of zinc plates interposed and so united as to form but one galvanic pair. Each of these batteries presents to the action of the acid, measuring both surfaces of the plate, $4\frac{7}{9}$ square feet—they are 12 inches high and about 5 inches in diameter.

In experimenting with this magnet, a battery containing $\frac{2}{5}$ of a square foot of zinc surface was first attached to the wires; with this the magnet could not be made to support more than 500 lbs. Another battery was then substituted for the above, containing about three times the same quantity of zinc surface; with this, at the first instant of immersion, the magnet sustained 1600 lbs.; after the acid was removed, it continued to support, for a few minutes, 450 lbs.; and in one experiment, three days after the battery had been excited, more than 150 lbs. were added to the armature* before it fell. It was evident from these experiments, that this magnet required a considerably larger quantity of zinc surface in proportion to its weight, to magnetize it to saturation, than that described in the former paper. Accordingly the two batteries, before mentioned as containing $4\frac{7}{9}$ square feet, were prepared. With one of them, at the first immersion, the magnet readily supported 2000 lbs. A sliding weight was then attached to the bar; the battery was suffered to become perfectly dry, and on immersing it again, the magnet supported 2063 lbs. The effect of a larger battery was not tried.

To test its power of inducing magnetism on soft iron, two pieces of round iron $1\frac{1}{4}$ inches in diameter and 12 inches long, were interposed between the extremities of the magnet and the armature—with this arrangement, when one of the batteries was immersed, the pieces of iron became so powerfully magnetic as to support 155 lbs.

To exhibit the effects produced by instantaneously reversing the poles, the armature was loaded with 56 lbs. which added to its own weight made 89 lbs.; one of the batteries was then dipped into the acid and immediately withdrawn, when the weight of course continued to adhere to the magnet; the other battery was then suddenly immersed, when the poles were changed so instantaneously, that the weight did not fall. That the poles were actually reversed in this experiment, was clearly shown by a change in the position of a large needle placed at a small distance from the side of one extremity of the horse-shoe.

P. S. Last autumn, I commenced a series of observations on the magnetic intensity of the earth at Albany, and intend to begin a new series next month; the apparatus used was that sent by Capt. Sabine to Prof. Renwick, and was mentioned in the Journal, Vol. XVII, p. 145. I have constructed a similar apparatus for myself, and intend to pay considerable attention to the subject.

---

* The armature of 23 lbs. applied when the battery is immersed, only for an inch and an instant, remains, day after day, without falling, although the galvanic coils are perfectly dry.—*Ed.*

Joseph Henry's Electric Motor

In this paper from the 1831 volume of *Silliman's Journal*, Henry described his battery-powered motor as "a philosophical toy" although it might eventually "be applied to some useful purpose." He was already aware of the potential improvement if electromagnets were to be substituted for the permanent steel magnets. He gradually became more pessimistic about the usefulness of electric motors, while at the same time becoming very sanguine about the electric telegraph. His doubts about the battery-powered motor were reinforced by Charles G. Page's abortive efforts to perfect an electric locomotive in the early 1850's.

For a biobibliographic note on Henry, see Paper 4. For an authoritative account of Page's electric locomotive project, see Robert C. Post, "The Page Locomotive: Federal Sponsorship of Invention in Mid-19th-Century America," *Technology and Culture*, vol. 13, pp. 140–169, 1972. Post has also contributed an article on Page to the special issue of the PROCEEDINGS OF THE IEEE in September 1976.

Art. XVII.—*On a Reciprocating motion produced by Magnetic Attraction and Repulsion; by* Prof. Joseph Henry.

TO THE EDITOR.

Sir,—I have lately succeeded in producing motion in a little machine by a power, which, I believe, has never before been applied in mechanics—by magnetic attraction and repulsion.

Not much importance, however, is attached to the invention, since the atricle, in its present state, can only be considered a philosophical toy; although, in the progress of discovery and invention, it is not impossible that the same principle, or some modification of it on a more extended scale, may hereafter be applied to some useful purpose. But without reference to its practical utility, and only viewed

Reprinted from *Amer. J. Sci.*. vol. XX, pp. 340–343, 1831.

as a new effect produced by one of the most mysterious agents of nature, you will not, perhaps, think the following account of it unworthy of a place in the Journal of Science.

It is well known that an attractive or repulsive force is exerted between two magnets, according as poles of different names, or poles of the same name, are presented to each other.

In order to understand how this principle can be applied to produce a reciprocating motion, let us suppose a bar magnet to be supported horizontally on an axis passing through the center of gravity, in precisely the same manner as a dipping needle is poised; and suppose two other magnets to be placed perpendicularly, one under each pole of the horizontal magnet, and a little below it, with their north poles uppermost; then it is evident that the south pole of the horizontal magnet will be attracted by the north pole of one of the perpendicular magnets, and its north pole repelled by the north pole of the other: in this state it will remain at rest, but if, by any means, we reverse the polarity of the horizontal magnet, its position will be changed and the extremity, which was before attracted, will now be repelled; if the polarity be again reversed, the position will again be changed, and so on indefinitely: to produce, therefore, a continued vibration, it is only necessary to introduce, into this arrangement, some means by which the polarity of the horizontal magnet can be instantaneously changed, and that too by a cause which shall be put in operation by the motion of the magnet itself; how this can be effected, will not be difficult to conceive, when I mention that, instead of a permanent steel magnet, in the moveable part of the apparatus, a soft iron galvanic magnet is used.*

The change of polarity is produced simply by soldering to the extremities of the wires which surround the galvanic magnet, two small galvanic batteries in such a manner that the vibrations of the magnet itself may immerse these alternately into vessels of diluted acid; care being taken that the batteries are so attached that the current of galvanism from each shall pass around the magnet in an opposite direction.

Instead of soldering the batteries to the ends of the wires, and thus causing them at each vibration to be lifted from the acid by the power of the machine; they may be permanently fixed in the vessels,

---

* For a method of constructing the galvanic magnet on an improved plan, see my paper in Vol. XIX, p. 329 of this Journal.

and the galvanic communication formed by the amalgamated ends of the wires dipping into cups of mercury.

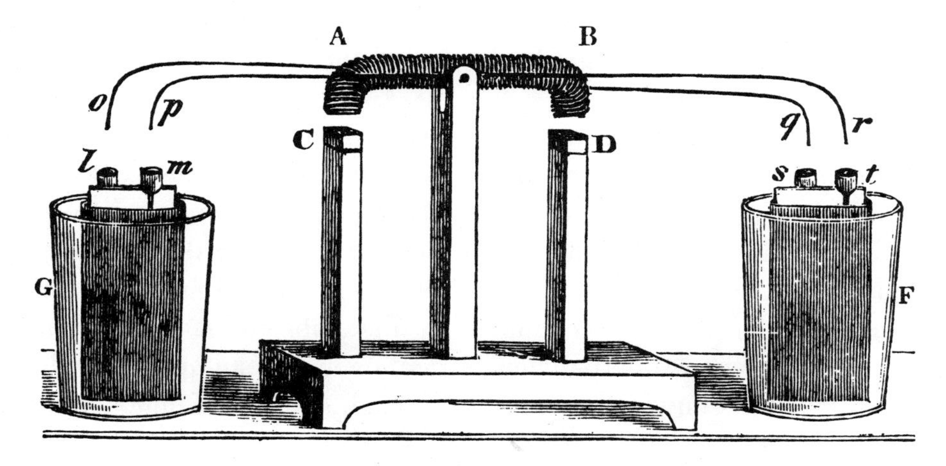

The whole will be more readily understood by a reference to the annexed drawing; A B is the horizontal magnet, about seven inches long, and moveable on an axis at the center: its two extremities when placed in a horizontal line, are about one inch from the north poles of the upright magnets C and D. G and F are two large tumblers containing diluted acid, in each of which is immersed a plate of zinc surrounded with copper. *l*, *m*, *s*, *t*, are four brass thimbles soldered to the zinc and copper of the batteries and filled with mercury.

The galvanic magnet AB is wound with three strands of copper bell wire, each about twenty five feet long; the similar ends of these are twisted together so as to form two stiff wires, which project beyond the extremity B, and dip into the thimbles *s*, *t*.

To the wires *q*, *r*, two other wires are soldered so as to project in an opposite direction, and dip into the thimbles *l*, *m*. The wires of the galvanic magnet have thus, as it were, four projecting ends; and by inspecting the figure it will be seen that the extremity *m*, which dips into the cup attached to the copper of the battery in G corresponds to the extremity *r* connecting with the zinc F.

When the batteries are in action, if the end B is depressed until *q*, *r* dips into the cups *s*, *t*, AB instantly becomes a powerful magnet, having its north pole at B; this of course is repelled by the north pole D, while at the same time it is attracted by C, the position is consequently changed, and *o*, *p* comes in contact with the mercury in *l*, *m*; as soon as the communication is formed, the poles are reversed, and the position again changed. If the tumblers be filled with strong diluted acid, the motion is at first very rapid and powerful, but it soon almost entirely ceases. By partially filling the tumblers with weak acid, and occasionally adding a small quantity of fresh acid, a uniform motion, at the rate of seventy five vibrations in a minute, has been kept up for more than an hour: with a large battery and very weak acid, the motion might be continued for an indefinite length of time.

The motion, here described, is entirely distinct from that produced by the electro-magnetic combination of wires and magnets; it results directly from the mechanical action of ordinary magnetism: galvanism being only introduced for the purpose of changing the poles.

My friend, Prof. Green, of Philadelphia, to whom I first exhibited this machine in motion, recommended the substitution of galvanic magnets for the two perpendicular steel ones. If an article of this kind was to be constructed on a large scale, this would undoubtedly be the better plan, as magnets of that kind can be made of any required power, but for a small apparatus, intended merely to exhibit the motion, the plan here described is perhaps the most convenient.

Robert Hare on Faraday's Electromagnetic Field Theory

IN this paper from the *American Journal of Science* of 1845, Hare revealed that he was still a determined advocate of material fluid theories of heat and electricity. (See the editorial note on Paper 3.) Hare had in fact been a persistent critic of Faraday during the latter's evolution of his field theory. According to Faraday's biographer, L. Pearce Williams, Hare's criticism was an important factor in causing Faraday to refine his theory and present it publically in 1844. This paper illustrates very well "the incommensurability of competing paradigms" described by Thomas Kuhn in *The Structure of Scientific Revolutions.*

Robert Hare (1781–1858) was born in Philadelphia. In addition to practical experience gained from working in a brewery owned by his family. Hare learned chemistry by self study and by auditing lectures at the University of Pennsylvania by James Woodhouse. He was a Professor of Chemistry at the University for about thirty years and a frequent contributor to *Silliman's Journal.* He was the inventor of an oxyhydrogen blow pipe and two types of electrochemical batteries that he called the calorimotor and the deflagrator. (See Paper 3.) Late in life, Hare took up the study of spiritualism. He claimed to have achieved communication with the spirit of Benjamin Franklin by means of a "spiritscope" of his own invention. See the *DSB*, vol. 6, pp. 114–115 and the *DAB*, vol. 8, pp. 263–264. For details of his dispute with Faraday, see L. Pearce Williams, *Michael Faraday* (New York, 1965), pp. 372–376.

Art. II.—*Remarks made by* Dr. Hare, *at a late meeting of the American Philosophical Society, on a recent speculation by Faraday on electric conduction and the nature of matter;*—communicated by the Author.

Philadelphia, Nov. 30, 1844.

Messrs. Editors—At the last meeting of the American Philosophical Society, I made some verbal remarks on a recent "speculation" of the celebrated Faraday, published in the London and Edinburgh Philosophical Magazine for February last. Of course a brief notice will be given of those remarks in the bulletin of the Proceedings. I send you for publication a statement of my reasoning on the questions at issue, hoping that it will not be found unworthy of the attention of philosophical chemists.

Your friend, Robert Hare.

Faraday objects to the Newtonian idea of an atom, being associated with combining ratios. These he conceives to have been more advantageously designated as chemical equivalents.*

This sagacious investigator adverts to the fact that after each atom in a mass of the metal potassium, has combined with an atom of oxygen and an atom of water, forming thus a hydrated oxide, the resulting aggregate occupies much less space than its metallic ingredient previously occupied; so that taking equal bulks of the hydrate and of potassium, there will be in the metal only four hundred and thirty metallic atoms, while in the hydrate there will be seven hundred such atoms. And in the latter, besides the seven hundred atoms, there will be an equal number of aqueous and oxygenous atoms, in all two thousand eight hundred ponderable atoms. It follows that if the atoms of potassium are to be considered as minute impenetrable particles, kept at certain distances by an equilibrium of forces, there must be, in a mass of potassium, vastly more space than matter. Moreover, it is the space alone that can be continuous. The non-contiguous material atoms cannot form a continuous mass. Consequently the well known power of potassium to conduct electricity must be a quality of the continuous empty space, which

* See his speculations touching electric conduction and the nature of matter, Vol. 24, 3d series, Philosophical Magazine and Journal, February, 1844.

Reprinted from *Amer. J. Sci.*, vol. XLVIII, pp. 247–252, 1845.

it comprises, not of the discontinuous particles of matter with which that space is regularly interspersed. It is in the next place urged that while, agreeably to these considerations, space is shown to be a conductor, there are considerations equally tending to prove it to be a non-conductor; since in certain non-conducting bodies, such as resins, there must be nearly as much vacant space as in potassium. Hence the supposition that atoms are minute impenetrable particles, involves the necessity of considering empty space as a conductor in metals and as a non-conductor in resins, and of course in sulphur and other electrics. This is considered as a *reductio ad absurdum.* To avoid this contradiction, Faraday supposes that atoms are not minute impenetrable bodies, but, existing throughout the whole space in which their properties are observed, may penetrate each other. Consistently, although the atoms of potassium pervade the whole space which they apparently occupy, the entrance into that space of an equivalent number of atoms of oxygen and water, in consequence of some reciprocal reaction, causes a contraction in the boundaries by which the combination thus formed is inclosed. This is an original and interesting view of this subject, well worthy of the contemplation of chemical philosophers.

But upon these premises Faraday has ventured on some inferences which, upon various accounts, appear to me unwarrantable. I agree that "*a*" representing a particle of matter and "*m*" representing its properties, it is only with "*m*" that we have any acquaintance, the existence of *a* resting merely on an inference. Heretofore I have often appealed to this fact, in order to show that the evidence both of ponderable and imponderable matter is of the same kind precisely: the existence of properties which can only be accounted for by inferring the existence of an appropriate matter to which those properties appertain. Yet I cannot concur in the idea that because it is only with "*m*" that we are acquainted, the existence of *a* must not be inferred; so that bodies are to be considered as constituted of their materialized powers. I use the word materialized, because it is fully admitted by Faraday, that by dispensing with an impenetrable atom "*a*," we do not get rid of the idea of matter, but have to imagine each atom as existing throughout the whole sphere of its force, instead of being condensed about the centre. This seems to follow from the following language.

Dr. Hare, on a recent "Speculation" by Faraday.

"The view now stated of the constitution of matter, would seem to involve necessarily the conclusion that matter fills all space, or at least the space to which gravitation extends, including the sun and its system, for gravitation is a property of matter, dependent on a certain force, and it is this force which constitutes matter."

Literally this paragraph seems to convey the impression, that agreeably to the new idea of matter, the sun and his planets are not distinct bodies, but consist of certain material powers reciprocally penetrating each other, and pervading a space larger than that comprised within the orbit of Uranus. We do not live upon, but within the matter of which the earth is constituted, or rather within a mixture of all the solar and planetary matter belonging to our solar system. I cannot conceive that the sagacious author seriously intended to sanction any notion involving these consequences. I shall assume, therefore, that, excepting the case of gravitation, his new idea of matter was intended to be restricted to those powers which display themselves within masses at insensible distances, and shall proceed to state the objections which seem to exist against the new idea as associated with those powers.

Evidently the arguments of Faraday against the existence, in potassium and other masses of matter, of impenetrable atoms endowed with cohesion, chemical affinity, momentum, and gravitation, rest upon the inference that in metals there is nothing to perform the part of an electrical conductor besides continuous empty space. This illustrious philosopher has heretofore appeared to be disinclined to admit the existence of any matter devoid of ponderability. The main object of certain letters which I addressed to him, was to prove that the phenomena of induction could not, as he had represented, be an "*action*" of ponderable atoms, but, on the contrary, must be considered as an *affection* of them consequent to the intervention of an imponderable matter, without which the phenomena of electricity would be inexplicable. This disinclination to the admission of an imponderable electrical cause, has been the more remarkable, as his researches have not only proved the existence of prodigious electrical power in metals, but likewise, that it is evolved during chemico-electric reaction, in equivalent proportion to the quantity of ponderable matter decomposed or combined.

According to his researches, a grain of water by electrolytic reaction with four grains of zinc, evolves as much electricity as would charge fifteen millions of square feet of coated glass. But in addition to the proofs of the existence of electrical powers in metals thus furnished, it is demonstrated that this power must be inseparably associated with metals, by the well known fact, that in the magneto-electric machine, an apparatus which we owe to his genius and the mechanical ingenuity of Pixii and Saxton, a coil of wire being subjected to the inductive influence of a magnet, is capable of furnishing, within the circuit which it forms, all the phenomena of an electrical current, whether of ignition, shock, or electrolysis.

The existence in metals of an enormous calorific power must be evident from the heat evolved by mere hammering. It is well known, that by a skillful application of the hammer, a piece of iron may be ignited. To what other cause than their inherent calorific power can the ignition of metals by a discharge of statical electricity be ascribed?

It follows that the existence of an immense calorific and electrical power is undeniable. The materiality of these powers, or of their cause, is all that has been questionable. But, according to the speculations of Faraday, all the powers of matter are material; not only the calorific and electrical powers are thus to be considered, but likewise the powers of cohesion, chemical affinity, inertia and gravitation, while *of all these material powers only the latter can be ponderable!!!*

Thus a disinclination on the part of this distinguished investigator to admit the existence of one or two imponderable principles, has led him into speculations involving the existence of a much greater number. But if the calorific and electrical powers of matter be material, and if such enormous quantities exist in potassium, as well as in zinc and all other metals, so much of the reasoning in question as is founded on the vacuity of the space between the metallic atoms, is really groundless.

Although the space occupied by the hydrated oxide of potassium comprises two thousand eight hundred ponderable atoms, while that occupied by an equal mass of the metal, comprises only four hundred and thirty, there may be in the latter proportionably as much more of the material powers of heat and electricity, as there is less of matter endowed with ponderability.

Thus while assuming the existence of fewer imponderable causes than the celebrated author of the speculation has himself proposed, we explain the conducting power of metals, without being under the necessity of attributing to void space the property of electrical conduction. Moreover, I consider it quite consistent to suppose that the presence of the material power of electricity is indispensable to electrical conduction, and that diversities in this faculty are due to the proportion of that material power present, and the mode of its association with other matter. The immense superiority of metals, as conductors, will be explained by referring it to their being peculiarly replete with the material powers of heat and electricity.

Hence Faraday's suggestions respecting the materiality of what has heretofore been designated as the properties of bodies, furnish the means of refuting his arguments against the existence of ponderable impenetrable atoms as the basis of cohesion, chemical affinity, momentum and gravitation.

But I will in the next place prove, that his suggestions not only furnish an answer to his objections to the views in this respect heretofore entertained, but are likewise pregnant with consequences directly inconsistent with the view of the subject which he has recently presented.

I have said that of all the powers of matter which are, according to Faraday's speculations, to be deemed material, gravitation alone can be ponderable. Since gravitation, in common with every power heretofore attributed to impenetrable particles, must be a matter independently pervading the space throughout which it is perceived, by what tie is it indissolubly attached to the rest? It cannot be pretended that either of the powers is the property of another. Each of them is an *m*, and cannot play the part of an *a*, not only because an *m* cannot be an *a*, but because no *a* can exist. Nor can it be advanced that they are the same power, since chemical affinity and cohesion act only at insensible distances, while gravitation acts at any and every distance, with forces inversely as their squares: and, moreover, the power of chemical affinity is not commensurate with that of gravitation. One part by weight of hydrogen has a greater affinity universally for any other element, than two hundred parts of gold. By what means then are cohesion, chemical affinity, and gravitation, inseparably associated, in all the ponderable elements of matter? Is it not fatal to the validity of the highly ingenious and interesting deductions of Faraday, that they are thus shown to be utterly incompetent to explain the inseparable association of cohesion, chemical affinity and inertia with gravitation; while the existence of a vacuity between Newtonian atoms, mainly relied upon as the basis of an argument against their existence, is shown to be inconsistent both with the ingenious speculation, which has called forth these remarks, and those Herculean "researches" which must perpetuate his fame.

Section I-C
Telegraphy and Telephony

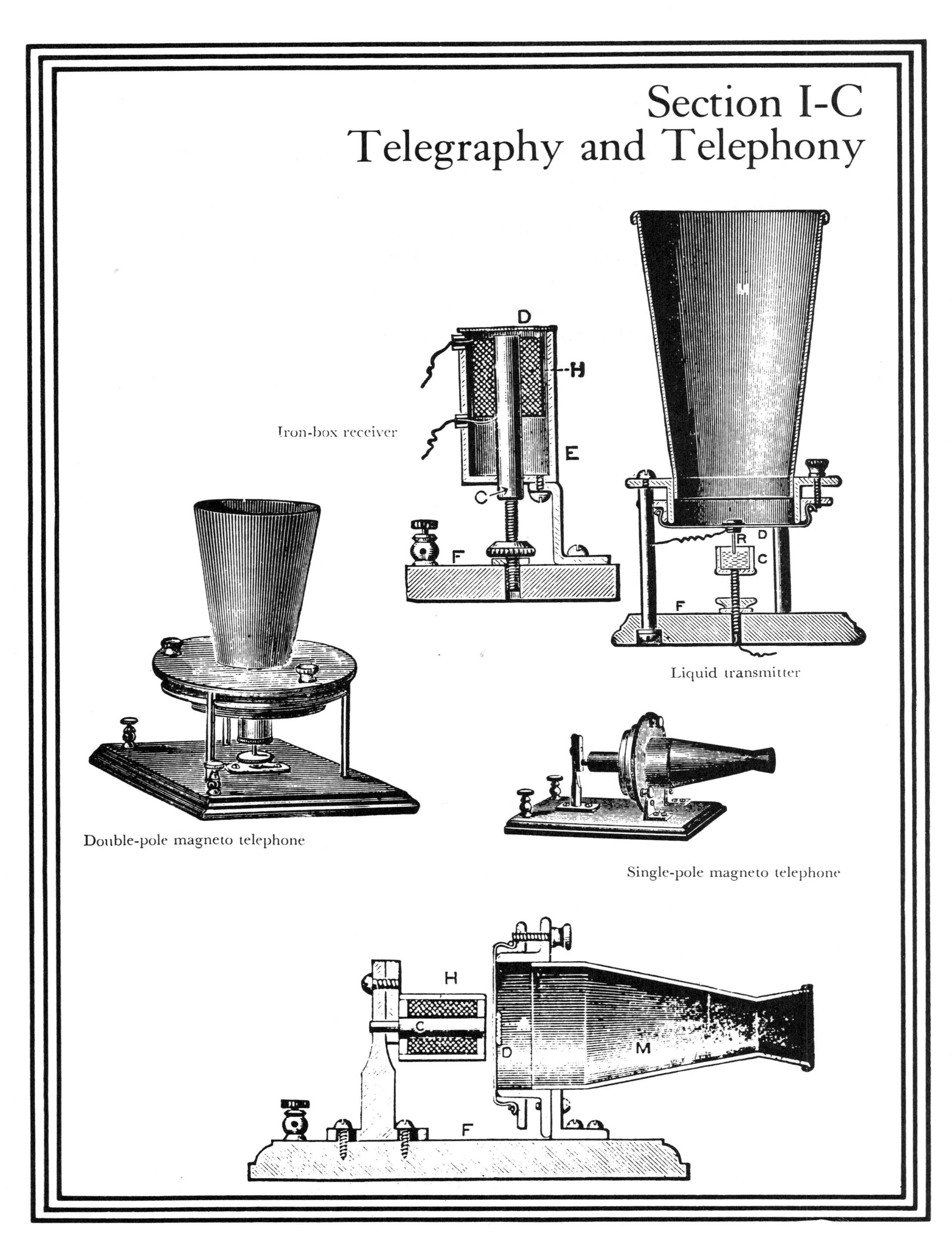

Iron-box receiver

Liquid transmitter

Double-pole magneto telephone

Single-pole magneto telephone

Morse and Draper on Electricity in Long Wires

As Morse mentions in this paper from the *American Journal of Science* of 1843, he was in the process of constructing an experimental telegraph line for the government. (He received a $30,000 appropriation for construction of a line from Washington to Baltimore that was completed in 1844.) These two papers were intended to provide proof that long-distance telegraphy was feasible by verification of a law that Morse and Draper attributed to Lenz. This was not Lenz's Law now familiar to students of electrical engineering. Emil Lenz (1804–1865) had published a paper "On the Laws of the Conducting Powers of Wires of Different Lengths and Diameters for Electricity" in 1835. Lenz himself had not laid claim to discovery of a new law, but regarded his experiments as confirmation of Ohm's theory of the galvanic circuit. (For a biographical essay on Lenz, see the *DSB*, vol. 8, pp. 187–191.) Morse and Draper were evidently not familiar with Ohm's theory, although it had been mentioned briefly by Joseph Henry in a paper in the *American Journal of Science* in 1841. Also, a complete English translation of Ohm's treatise of 1827 was published in vol. 2 of Taylor's *Scientific Memoirs* in 1841. The Grove battery used by Morse was invented by the English electrochemist William R. Grove (1811–1896) and described in articles in the *Philosophical Magazine* in 1838 and 1839. (For a biographical essay on Grove, see the *DSB*, vol. 5, pp. 559–561.) In a report on these experiments submitted to the Secretary of the Treasury in August 1843, Morse stated that his findings implied that telegraphy across the Atlantic was possible and that "the time will come when this project will be realized."

Samuel F. B. Morse (1791–1872) was born in Massachusetts and graduated from Yale in 1810. He studied art for four years in England before opening a studio in Boston in 1815. He became fairly successful as a portrait painter and received an appointment as a Professor of Painting and Sculpture at the University of the City of N.Y. in 1832. His concept of an electromagnetic telegraph system dates from 1832. Morse received important technical assistance during the developmental stage from Leonard D. Gale (1800–1883), a Professor of Chemistry at the University who was familiar with Joseph Henry's work, and from Alfred Vail (1807–1859), who witnessed a demonstration by Morse in 1837 and became an enthusiastic partner in the enterprise. Henry actively encouraged Morse and wrote to Morse in 1842 that "the science is now fully ripe for such an application of its principles." Following completion of the experimental line from Washington to Baltimore in 1844, Morse soon decided to leave further development of the industry to others. He was disappointed when Congress declined to accept his offer to make the telegraph a public utility. See the *DAB*, vol. 13, pp. 247–251 and the *NCAB*, vol. 4, pp. 449–450. For an interesting account of Joseph Henry's role in the creation of the telegraph, see a paper by Arthur P. Molella in the special issue of the PROCEEDINGS OF THE IEEE for September 1976.

John W. Draper (1811–1882) was born in England and studied at London University before coming to the United States in 1832. He received a medical degree from the University of Pennsylvania in 1836 and was a student of Robert Hare. (See Paper 6.) He became a Professor of Chemistry at the University of the City of N.Y. in 1839 and became acquainted with Morse at that time. His primary research interest was in the chemical effects of radiant energy and he was a pioneer in photographic science. See the *DSB*, vol. 4, pp. 181–183 and the *DAB*, vol. 5, pp. 438–441.

Art. XVI.—*Experiments made with one hundred pairs of Grove's Battery, passing through one hundred and sixty miles of insulated wire*;—in a letter from Prof. S. F. B. Morse, to the Editors, dated New York, Sept. 4th, 1843.

Dear Sirs,—On the 8th of August having completed my preparations of one hundred and sixty miles of copper wire for the electro-magnetic telegraph which I am constructing for the government, I invited several scientific friends to witness some experiments in verification of the law of Lenz, of the action of galvanic electricity through wires of great lengths.

I put in action a cup battery of one hundred pairs, which I had constructed, based on the excellent plan of Prof. Grove, but with some modifications of my own, economizing the platinum.

The wire was reeled upon eighty reels, containing two miles upon each reel, so that any length from two to one hundred and sixty miles could be made at pleasure to constitute the circuit.

My first trial of the battery was through the entire length of one hundred and sixty miles, making of course a circuit of eighty miles, and the magnetism induced in my electro-magnet, which formed a part of the circuit, was sufficient to move with great strength my telegraphic lever. Even forty-eight cups produced action in the lever, but not so promptly or surely.

We then commenced a series of experiments upon decomposition at various distances. The battery alone (one hundred pairs) gave in the measuring guage in one minute, 5.20 inches of gas. When four miles of wire were interposed, the result was 1.20 inches—ten miles of wire, .57 inch—twenty miles, .30 inch—fifty miles, .094.

The results obtained from a battery of one hundred pairs are projected in the following curve.

Reprinted from *Amer. J. Sci.*, vol. XLV, pp. 390–392, 1843.

Experiments with Grove's Battery.

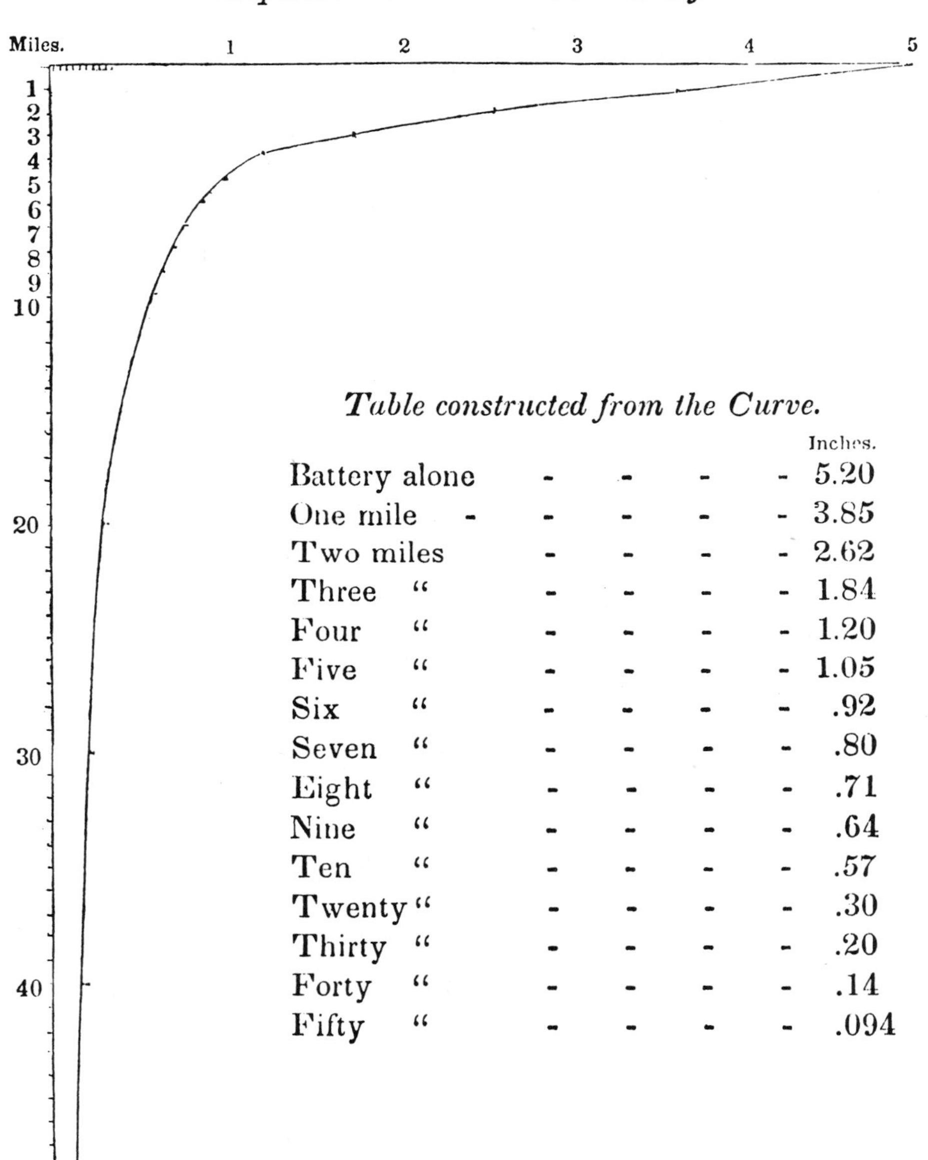

Table constructed from the Curve.

	Inches.
Battery alone - - - -	5.20
One mile - - - - -	3.85
Two miles - - - -	2.62
Three " - - - -	1.84
Four " - - - -	1.20
Five " - - - -	1.05
Six " - - - -	.92
Seven " - - - -	.80
Eight " - - - -	.71
Nine " - - - -	.64
Ten " - - - -	.57
Twenty " - - - -	.30
Thirty " - - - -	.20
Forty " - - - -	.14
Fifty " - - - -	.094

During the previous summer I made the following experiments upon a line of thirty-three miles, of number 17 copper wire, with a battery of fifty pairs. In this case, I used a small steelyard with weights, with which I was enabled to weigh with a good degree of accuracy the greater magnetic forces, but not the lesser, yet sufficiently approximating the recent results to confirm the law in question.

Experiments with Grove's Battery.

Table of Results.

Fifty pairs	through	2	miles	attracted and	raised	9 ozs.
"	"	4	"	"	"	4 "
"	"	6	"	"	"	3 "
"	"	8	"	"	"	$2\frac{1}{2}$ "
"	"	10	"	"	"	$2\frac{1}{4}$ "
"	"	12	"	"	"	$\frac{1}{8}$ "
"	"	14	"	"	"	$\frac{1}{8}$ "

and each successive addition of two miles up to thirty-three, still gave an attractive and lifting power of one-eighth of an ounce.

Curve from the Results.

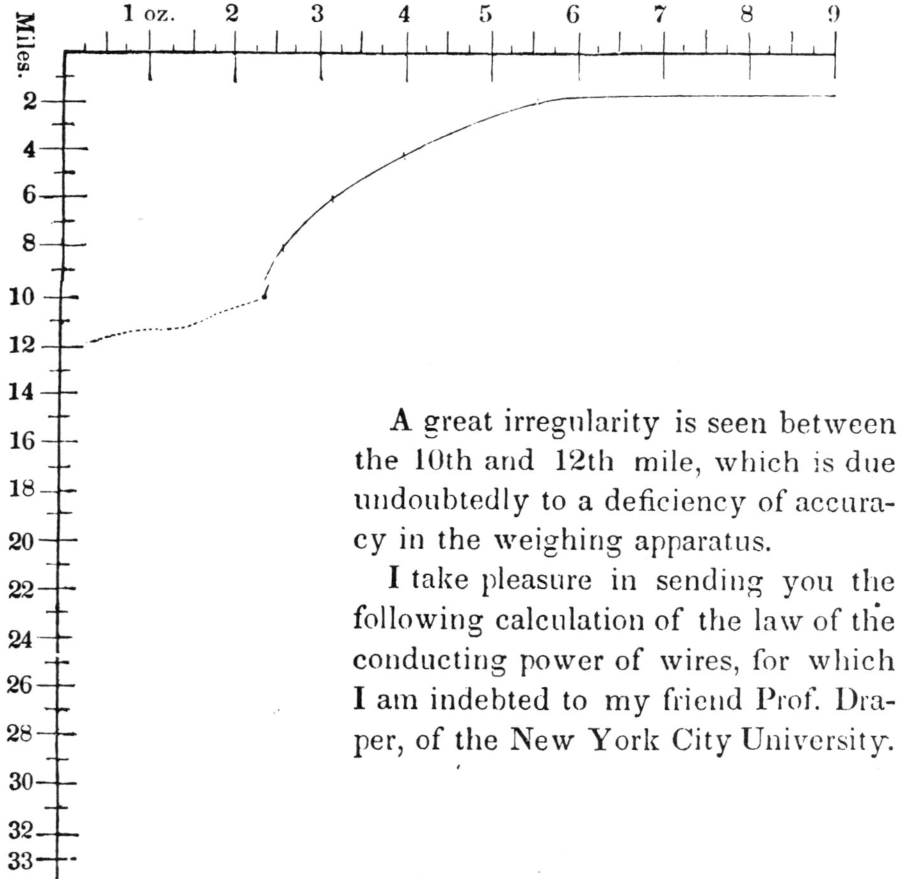

A great irregularity is seen between the 10th and 12th mile, which is due undoubtedly to a deficiency of accuracy in the weighing apparatus.

I take pleasure in sending you the following calculation of the law of the conducting power of wires, for which I am indebted to my friend Prof. Draper, of the New York City University.

On the law of the conducting power of wires; by JOHN W. DRAPER, M. D., &c. &c.

It has often been objected, that if the conducting power of wires for electricity was inversely as their length, and directly as as their section, the transmission of telegraphic signals through

Reprinted from *Amer. J. Sci.*, vol. XLV, pp. 392–394, 1843.

long wires, could not be carried into effect, and even the galvanic multiplier, which consists essentially of a wire making several convolutions round a needle, could have no existence.

This last objection was first brought forward by Prof. Ritchie, of the University of London, as an absolute proof that the law referred to is incorrect. There is, however, an exceedingly simple method of proving that signals may be despatched through very long wires, and that the galvanic multiplier, so far from controverting the law in question, depends for its very existence upon it.

Assuming the truth of the law of Lenz, the *quantities* of electricity which can be urged by a constant electromotoric source through a series of wires, the lengths of which constitute an arithmetical ratio, will always be in a geometrical ratio. Now the curve whose ordinates and abscissas bear this relation to each other is the logarithmic curve whose equation is $a^y = x$.

1st. If we suppose the base of the system which the curve under discussion represents be greater than unity, the values of y taken between $x=0$ and $x=1$, must be all negative.

2nd. By taking $y=0$ we find that the curve will intersect the axis of the x's at a distance from the origin equal to unity.

3rd. By making $x=0$ we find y to be infinite and negative.

Now these are the properties of the logarithmic curve which furnish an explanation of the case in hand. Assuming that the x's represent the quantities of electricity, and the y's the lengths of the wires, we perceive at once that those parts of the curve which we have to consider lie wholly in the fourth quadrant, where the abscissas are positive and the ordinates negative.

When, therefore, the battery current passes without the intervention of any obstructing wire, its value is equal to unity.

But as successive lengths of wire are continually added, the quantities of electricity passing, undergo a diminution at first rapid and then more and more slow. And it is not until the wire becomes infinitely long that it ceases to conduct at all; for the ordinate $-y$, when $x=0$, is an asymptote to the curve.

In point of practice, therefore, when a certain limit is reached the diminution of the intensity of the forces becomes *very small*, whilst the increase in the lengths of the wire is vastly great. It is, therefore, possible to conceive a wire to be a million times as long as another, and yet the two shall transmit quantities of electricity not perceptibly different, when measured by a delicate galvanometer.

But under these circumstances if the long wire be coiled so as to act as a multiplier, its influence on the needle will be inexpressibly greater than the one so much shorter than it.

Further, from this we gather that for telegraphic despatches, with a battery of given electromotoric power, when a certain distance is reached the diminution of effect for an increased distance becomes inappreciable.

The Atlantic Cable of 1858

THE Atlantic telegraph cable was among the most remarkable technological achievements of the 19th century. As indicated in this report from the *American Journal of Science*, the completion of the first cable in 1858 after several failures was greeted by tremendous public enthusiasm. Extracts from the journal of Cyrus Field, the man most responsible for the success of the project, are included. (Note that one of his distances is incorrect.) The 1858 cable became inoperative after a few weeks, and direct electrical communication between the United States and Europe was not restored until 1866. For background reading, see Bern Dibner, *The Atlantic Cable* (Norwalk, Conn., 1959) and Bernard S. Finn, *Submarine Telegraphy: The Grand Victorian Technology* (British Science Museum, London, 1973). Also see Finn's paper on life at the cable station at Heart's Content, Newfoundland in the PROCEEDINGS OF THE IEEE for September 1976.

Cyrus W. Field (1819–1892) was a native of Massachusetts who became wealthy in the paper manufacturing industry before becoming interested in the challenging problem of laying a trans-Atlantic cable from Ireland to Newfoundland. He devoted about 12 years of his life and much of his personal wealth to the endeavour before final success was achieved in 1866. He later became involved in efforts to build an elevated railroad system in New York City. See the *DAB*, vol. 6, pp. 357–359 and the *NCAB*, vol. 4, pp. 451–453.

V. MISCELLANEOUS SCIENTIFIC INTELLIGENCE.

1. *The Atlantic Cable.*—While all hoped earnestly for the successful issue of the last telegraphic expedition which left Queenstown, England, July 17, for mid ocean, we believe there were few whose fears of a third failure were not greater than their hopes of success. When, therefore, about noon of August 5th, the news of the safe arrival of the Niagara and Gorgon at Trinity Bay, with the cable unbroken, was flashed over the length and breadth of the land, there was at first, almost universal skepticism as to the authenticity of the news. This gave way, however, in a few hours before the accumulating evidence, and a wild gush of joy broke out from every quarter, such as no event in modern history has ever occasioned.

The great fact that the cable is laid without accident, and that its insulation and transmitting power are reported perfect, render the question of plateau or no plateau, discussed on pp. 157 and 219 of this volume of secondary importance to the public; and yet as a geographical and geological problem, it is still one of great interest. We may be permit-

Reprinted from *Amer. J. Sci.*, 2nd ser., vol. XXVI, pp. 285–288, Sept. 1858.

ted to say that while we believe, on what seem to us ample geological grounds, in the pretty uniform and rather moderate depth of that portion of the Atlantic where the cable now rests, it has never appeared to be a point of much practical moment whether the depth of water in which the cable was to be laid was 1000, 2000, or 3000 fathoms. In any case it was very sure to find bottom and rest there; and the distance across is no greater at the greater depth. The discussions of Professor Trowbridge afford valuable suggestions with regard to all deep sea soundings, and we trust that his method when perfected may be the means of giving us reliable results respecting the whole ocean.

In the absence of any official documents, we quote from the daily journals some facts relating to this great event, drawn from the private journal of CYRUS W. FIELD, Esq., to whose dauntless energy, skill, and hearty faith in a successful issue, the world will ever be eager to award the highest praise, since to him more than to any other person belongs, beyond doubt, the credit of conceiving and carrying through this great enterprise.

Saturday, July 17, 1858. The telegraphic fleet sailed from Queenstown, Ireland, as follows: the Valorous and Gorgon at 11 A. M.; the Niagara at 7½ P. M., and the Agamemnon a few hours later. All the steamers were to use as little coal as possible in getting to the place of rendezvous.

Friday, July 23. The Niagara arrived at the rendezvous, latitude 52° 5′, longitude 32° 40′ at 8.30 P. M., the Valorous followed at 4 A. M. of July 25, the Gorgon at 5 P. M. July 27, and the Agamemnon at 5 P. M. of July 28. The weather had been cloudy and squally until the 21st, but appears on the whole not to have been unfavorable.

Thursday, July 29. Latitude 52° 59′, longitude 32° 27′ W., the telegraphic fleet all in sight, with a smooth sea, cloudy sky, and a light wind from the S.E. to S. At 1 P. M. of this day the "splice" was successfully made between the two ends of the cable, and electrical signals passed perfectly through the whole length on board both ships. Depth of water 1550 fathoms. The distance from the entrance of Valentia harbor was eight hundred and thirteen nautical miles; to the entrance of Trinity Bay, N. F., eighteen hundred and twenty-two nautical miles, and from there sixty miles to the telegraph house at the head of the Bay of Bulls, equal in all to eight hundred and eighty-two nautical miles. The Niagara had sixty-nine miles farther to run than the Agamemnon. Each ship had on board about eleven hundred nautical miles of cable. The following table presents a condensed view of the Niagara's voyage:

Date.	Lat. N.	Long. W.	Dist. sailed in last 24 hours.	Miles and fathoms of cable paid out	Excess per cent.	Depth of water in fathoms.
July 30, Friday,	51° 50′	34° 49′	89	131*m.*	48	1550—1985
" 31, Saturday,	51° 50′	38° 28′	137	159*m.* 853*f.*	13	
Aug. 1, Sunday,	50° 32′	41° 55′	145	164*m.* 683*f.*	14	1950—2424
" 2, Monday,	49° 52′	45° 37′	154	177*m.* 150*f.*	15	1600—2385
" 3, Tuesday,	49° 17′	49° 23′	147	161*m.* 763*f.*	10	1742(?)—1827
" 4, Wedn'y,	48° 17′	52° 43′	146	154*m.* 360*f.*	6	200
" 5, Thursday,	Niagara anchored.		64	66*m.* 382*f.*	4	

It will be observed that the distances run during the five full days were remarkably uniform, 149 miles per day, and the excess of cable paid out was about 15 per cent more than the ship's record. Twice during the progress of the ship, from some unexplained cause, signals failed to pass between them, viz: at $7\frac{1}{2}$ P. M. July 29, for an hour, and August 2, from 12^h 38^m A. M. to 5.40 A. M. But during all the remainder of the time signals were constantly received; and at the last, the Agamemnon, August 5, 2.45 A. M., signalized the Niagara that they had paid out 1010 miles of cable. The cable was landed at Valentia Bay on Thursday morning, August 5, at 5^h 15^m; and at 6 A. M. the shore end was carried into the telegraph house, and a strong current of electricity received through the whole cable from the other side of the Atlantic.

On the 17th of August the Queen of Great Britain sent a congratulatory message to the President of the United States, expressing in courtly terms her joy in the completion of this great international bond, to which President Buchanan responded in the same spirit. It has been suggested elsewhere that the Boston fire-bells, which are now rung by electricity, should be placed in communication with London and rung by the current from the company's office there.*

In the absence of any accurate scientific report of the details of the electrical arrangements, we are able at present to state only a few facts, most of which are probably already familiar to our readers.

The cable is only about six-tenths of an inch in diameter outside of the encasing wires. The figure annexed is a correct view of the exterior form and cross section of the cable. The electrical conductor is formed of seven No. 22 copper wires, the centre wire being straight, while its six surrounding wires form a spiral about it. The insulation is of refined gutta percha, laid on by machinery in three distinct coatings the more perfectly to insure insulation. Finally, seventeen strands of iron wire, No. 20, composed each of seven separate wires, laid up like the electrical conductor, encase the whole into a strong but very flexible rope, as easily handled as a rope of hemp. All the mechanical work and materials used in the preparation of the cable have been of the best description. The whole length of separate wires in this cable amounts to 332,500 miles, or about fourteen times the earth's circumference. The weight of the cable is almost exactly one ton (2000 lbs.) to the mile, and it bears a direct minimum strain of four tons, and much more has been put on it without fracture.

The magneto-electrical current designed to be employed is derived from a powerful combination of inducing coils and soft iron magnets. The cable is regarded as a Leyden jar inductively charged, and Mr. Whitehouse, the electrician of the company, states that by alternating impulses of positive and negative electricity, transmitted in rapid succession to the copper conductor, the evil effects of the reflex induced current in the iron wire cover is entirely overcome. We look with much interest for the memoir, which we suppose of course will be published by the company, giving the physical details, and the constants established by this elaborate research. It will also be very interesting to see on record an official description of the manufacture and paying out of the cable, with full plates, and a statement of the chief causes of delay and embarrassment, and their cure. But the

* Boston Transcript, August 13, 1858.

portion most interesting to science will be the electrical history of this great enterprise, the details of which have hitherto been only partially revealed, and that not in a scientific form. The only official account of the undertaking which we have seen is a pamphlet issued in July, 1857, by the company in London, entitled "The Atlantic Telegraph, a history of preliminary experimental proceedings, and a descriptive account of the present state and prospects of the undertaking." This pamphlet contains some novel and very interesting statements, but they are wanting in that accuracy and fullness of detail indispensable to a scientific statement, and the results are generally given only in rather vague terms.

A. G. Bell on Telephony

BELL delivered this paper at a meeting of the American Academy of Arts and Sciences in Boston two months after his first successful voice transmission over wire. David Hounshell has referred to this paper as a "*tour de force*" in electroacoustics. The paper includes a well-documented history of relevant prior investigations, as well as the report of Bell's famous conversation with his assistant, Thomas Watson, of March 10, 1876. Bell's practice of informing the scientific community of his discovery was not followed by his major competitor, Elisha Gray. This proved decisive in Bell's defense in litigation to establish priority of invention. See David A. Hounshell, "Elisha Gray and the Telephone: On the Disadvantages of Being an Expert," *Technology and Culture*, vol. 16, pp. 133–161, 1975. Also see Hounshell's paper on Bell and Gray in the PROCEEDINGS OF THE IEEE of September 1976. For an interesting account of the initial failure of the Western Union Company to appreciate the commercial potential of the telephone, see Michael F. Wolff, "The Marriage that Almost Was," *IEEE Spectrum*, pp. 41–51, February 1976.

Alexander Graham Bell (1847–1922) was born in Edinburgh, Scotland and continued in the family's tradition by becoming a teacher of the hearing impaired. Following the death of his two brothers, the family moved to Canada in 1870, and Bell soon accepted a teaching position at a school for the deaf in Boston. He became enthusiastic about the possibilities of multiplex telegraphy after reading about a duplex telegraph invention in 1872. With the financial support of his future father-in-law, Gardiner G. Hubbard, Bell attempted to perfect a frequency-multiplex telegraph. By the summer of 1875, he realized that electrical transmission of speech was possible, an invention reduced to practice and patented the following year. As in the case of Morse and the telegraph, Bell was content to let others develop his new communication system commercially. He became interested in aeronautics and financially supported Langley's efforts to achieve powered flight during the 1890's. Bell was instrumental in founding the journal *Science* and subsidized it for several years. See *NAS*, vol. 23, 1–29 and the *DSB*, vol. 1, pp. 582–583. The definitive biography of Bell is Robert V. Bruce, *Bell: Alexander Graham Bell and the Conquest of Solitude* (Boston, 1973).

PROCEEDINGS

OF THE

AMERICAN ACADEMY

OF

ARTS AND SCIENCES.

VOL. XII.

PAPERS READ BEFORE THE ACADEMY.

I.

RESEARCHES IN TELEPHONY.

BY A. GRAHAM BELL.

Presented May 10, 1876, by the Corresponding Secretary.

1. IT has long been known that an electro-magnet gives forth a decided sound when it is suddenly magnetized or demagnetized. When the circuit upon which it is placed is rapidly made and broken, a succession of explosive noises proceeds from the magnet. These sounds produce upon the ear the effect of a musical note, when the current is interrupted a sufficient number of times per second. The discovery of "Galvanic Music," by Page,* in 1837, led inquirers in different parts of the world almost simultaneously to enter into the field of telephonic research; and the acoustical effects produced by magnetization were carefully studied by Marrian,† Beatson,‡ Gassiot,§ De la Rive,‖

---

* *C. G. Page.* "The Production of Galvanic Music." Silliman's Journ., 1837, XXXII., p. 396; Silliman's Journ., July, 1837, p. 354; Silliman's Journ., 1838, XXXIII., p. 118; Bibl. Univ. (new series), 1839, II., p. 398.

† *J. P. Marrian.* Phil. Mag., XXV., p. 382; Inst., 1845, p. 20; Arch. de l'Électr., V., p. 195.

‡ *W. Beatson.* Arch. de l'Électr., V., p. 197; Arch. de Sc. Phys. et Nat. (2d series), II., p. 113.

§ *Gassiot.* See "Treatise on Electricity," by De la Rive, I., p. 300.

‖ *De la Rive.* Treatise on Electricity, I., p. 300; Phil. Mag., XXXV., p. 422; Arch. de l'Electr., V., p 200; Inst., 1846, p. 83; Comptes Rendus, XX., p. 1287; Comp. Rend., XXII., p. 432; Pogg. Ann., LXXVI., p. 637; Ann. de Chim. et de Phys., XXVI., p. 158.

Reprinted from *Proc. Amer. Acad. Arts and Sci.*, vol. XII, pp. 1–10, 1876.

Matteucci,* Guillemin,† Wertheim,‡ Wartmann,§ Janniar,‖ Joule,¶ Laborde,** Legat,†† Reis,‡‡ Poggendorff,§§ Du Moncel,‖‖ Delezenne,¶¶ and others.***

2. In the autumn of 1874, I discovered that the sounds emitted by an electro-magnet under the influence of a discontinuous current of electricity are not due wholly to sudden changes in the magnetic condition of the iron core (as heretofore supposed), but that a portion of the effect results from vibrations in the insulated copper-wires composing the coils. An electro-magnet was arranged upon circuit with an instrument for interrupting the current, — the rheotome being placed in a distant room, so as to avoid interference with the experiment. Upon applying the ear to the magnet, a musical note was clearly perceived, and the sound persisted after the iron core had been removed. It was then much feebler in intensity, but was otherwise unchanged, — the curious crackling noise accompanying the sound being well marked.

The effect may probably be explained by the attraction of the coils of the wire for one another during the passage of the galvanic current,

---

* *Matteucci.* Inst., 1845, p. 315; Arch. de l'Électr., V., 389.

† *Guillemin.* Comp. Rend., XXII., p. 264; Inst., 1846, p. 30; Arch. d. Sc. Phys. (2d series), I., p. 191.

‡ *G. Wertheim.* Comp. Rend., XXII., pp. 336, 544; Inst., 1846, pp. 65, 100; Pogg. Ann., LXVIII., p. 140; Comp. Rend., XXVI., p. 505; Inst., 1848, p. 142; Ann. de Chim. et de Phys., XXIII., p. 302; Arch. d. Sc. Phys. et Nat., VIII., p. 206; Pogg. Ann., LXXVII., p. 43; Berl. Ber., IV., p. 121.

§ *Elie Wartmann.* Comp. Rend., XXII., p. 544; Phil. Mag. (3d series), XXVIII., p. 544; Arch. d. Sc. Phys. et Nat. (2d series), I., p. 419; Inst., 1846, p. 290; Monatscher. d. Berl. Akad., 1846, p. 111.

‖ *Janniar.* Comp. Rend., XXIII., p. 319; Inst., 1846, p. 269; Arch. d. Sc. Phys. et Nat. (2d series), II., p. 394.

¶ *J. P. Joule.* Phil. Mag., XXV., pp. 76, 225; Berl. Ber., III., p. 489.

** *Laborde.* Comp. Rend., L., p. 692; Cosmos, XVII., p. 514.

†† *Legat.* Brix. Z. S., IX., p. 125.

‡‡ *Reis.* "Téléphonie." Polytechnic Journ., CLXVIII., p. 185; Böttger's Notizbl., 1863, No. 6.

§§ *J. C. Poggendorff.* Pogg. Ann., XCVIII., p. 192; Berliner Monatsber., 1856, p. 133; Cosmos, IX., p. 49; Berl. Ber., XII., p. 241; Pogg. Ann., LXXXVII., p. 139.

‖‖ *Du Moncel.* Exposé, II., p. 125; also, III., p. 83.

¶¶ *Delezenne.* "Sound produced by Magnetization," Bibl. Univ. (new series), 1841, XVI., p. 406.

*** See London Journ., XXXII., p. 402; Polytechnic Journ., CX., p. 16; Cosmos, IV., p. 43; Glösener —— Traité général, &c., p. 350; Dove.-Repert., VI., p. 58; Pogg. Ann., XLIII., p. 411; Berl. Ber., I., p. 144; Arch. d. Sc. Phys. et Nat., XVI., p. 406; Kuhn's Encyclopædia der Physik, pp. 1014–1021.

and the sudden cessation of such attraction when the current is interrupted. When a spiral of fine wire is made to dip into a cup of mercury, so as thereby to close a galvanic circuit, it is well known that the spiral coils up and shortens. Ferguson * constructed a rheotome upon this principle. The shortening of the spiral lifted the end of the wire out of the mercury, thus opening the circuit, and the weight of the wire sufficed to bring the end down again, — so that the spiral was thrown into continuous vibration. I conceive that a somewhat similar motion is occasioned in a helix of wire by the passage of a discontinuous current, although further research has convinced me that other causes also conspire to produce the effect noted above. The extra currents occasioned by the induction of the voltaic current upon itself in the coils of the helix no doubt play an important part in the production of the sound, as very curious audible effects are produced by electrical impulses of high tension. It is probable, too, that a molecular vibration is occasioned in the conducting wire, as sounds are emitted by many substances when a discontinuous current is passed through them. Very distinct sounds proceed from straight pieces of iron, steel, retort-carbon, and plumbago. I believe that I have also obtained audible effects from thin platinum and German-silver wires, and from mercury contained in a narrow groove about four feet long. In these cases, however, the sounds were so faint and outside noises so loud that the experiments require verification. Well-marked sounds proceed from conductors of all kinds when formed into spirals or helices. I find that De la Rive had noticed the production of sound from iron and steel during the passage of an intermittent current, although he failed to obtain audible results from other substances. In order that such effects should be observed, extreme quietness is necessary. The rheotome itself is a great source of annoyance, as it always produces a sound of similar pitch to the one which it is desired to hear. It is absolutely requisite that it should be placed out of earshot of the observer, and at such a distance as to exclude the possibility of sounds being mechanically conducted along the wire.

3. Very striking audible effects can be produced upon a short circuit by means of two Grove elements. I had a helix of insulated copper-wire (No. 23) constructed, having a resistance of about twelve ohms. It was placed in circuit with a rheotome which interrupted the current one hundred times per second. Upon placing the helix to my ear I

---

* *Ferguson.* Proceedings of Royal Scottish Soc. of Arts, April 9, 1866; Paper on "A New Current Interrupter."

could hear the unison of the note produced by the rheotome. The intensity of the sound was much increased by placing a wrought-iron nail inside the helix. In both these cases, a crackling effect accompanied the sound. When the nail was held in the fingers so that no portion of it touched the helix, the crackling effect disappeared, and a pure musical note resulted.

When the nail was placed inside the helix, between two cylindrical pieces of iron, a loud sound resulted that could be heard all over a large room. The nail seemed to vibrate bodily, striking the cylindrical pieces of metal alternately, and the iron cylinders themselves were violently agitated.

4. Loud sounds are emitted by pieces of iron and steel when subjected to the attraction of an electro-magnet which is placed in circuit with a rheotome. Under such circumstances, the armatures of Morse-sounders and Relays produce sonorous effects. I have succeeded in rendering the sounds audible to large audiences by interposing a tense membrane between the electro-magnet and its armature. The armature in this case consisted of a piece of clock-spring glued to the membrane. This form of apparatus I have found invaluable in all my experiments. The instrument was connected with a parlor organ, the reeds of which were so arranged as to open and close the circuit during their vibration. When the organ was played the music was loudly reproduced by the telephonic receiver in a distant room. When chords were played upon the organ, the various notes composing the chords were emitted simultaneously by the armature of the receiver.

5. The simultaneous production of musical notes of different pitch by the electric current, was foreseen by me as early as 1870, and demonstrated during the year 1873. Elisha Gray,* of Chicago, and Paul La Cour,† of Copenhagen, lay claim to the same discovery. The fact that sounds of different pitch can be simultaneously produced upon any part of a telegraphic circuit is of great practical importance; for the duration of a musical note can be made to signify the dot or dash of the Morse alphabet, and thus a number of telegraphic messages may be sent simultaneously over the same wire without confusion by making signals of a definite pitch for each message.

6. If the armature of an electro-magnet has a definite rate of oscillation of its own, it is thrown bodily into vibration when the interrup-

* *Elisha Gray.* Eng. Pat. Spec., No. 974. See "Engineer," March 26, 1875.

† *Paul la Cour.* Telegraphic Journal, Nov. 1, 1875.

tions of the current are timed to its movements. For instance, present an electro-magnet to the strings of a piano. It will be found that the string which is in unison with the rheotome included in the circuit will be thrown into vibration by the attraction of the magnet.

Helmholtz,* in his experiments upon the synthesis of vowel sounds caused continuous vibration in tuning-forks which were used as the armatures of electro-magnets. One of the forks was employed as a rheotome. Platinum wires attached to the prongs dipped into mercury.

The intermittent current occasioned by the vibration of the fork traversed a circuit containing a number of electro-magnets between the poles of which were placed tuning-forks whose normal rates of vibration were multiples of that of the transmitting fork. All the forks were kept in continuous vibration by the passage of the interrupted current. By re-enforcing the tones of the forks in different degrees by means of resonators, Helmholtz succeeded in reproducing artificially certain vowel sounds.

I have caused intense vibration in a steel strip, one extremity of which was firmly clamped to the pole of a U-shaped electro-magnet, the free end overhanging the other pole. The amplitude of the vibration was greatest when the coil was removed from the leg of the magnet to which the armature was attached.

7. All the effects noted above result from rapid interruptions of a voltaic current, but sounds may be produced electrically in many other ways.

The Canon Gottoin de Coma,† in 1785, observed that noises were emitted by iron rods placed in the open air during certain electrical conditions of the atmosphere; Beatson ‡ produced a sound from an iron wire by the discharge of a Leyden jar; Gore § obtained loud musical notes from mercury, accompanied by singularly beautiful crispations of the surface during the course of experiments in electrolysis; and Page ‖ produced musical tones from Trevelyan's bars by the action of the galvanic current.

8. When an intermittent current is passed through the thick wires of a Ruhmkorff's coil, very curious audible effects are produced by the

---

* *Helmholtz.* Die Lehre von dem Tonempfindungen.

† See "Treatise on Electricity," by De la Rive, I., p. 300.

‡ Ibid.

§ *Gore.* Proceedings of Royal Society, XII., p. 217.

‖ *Page.* "Vibration of Trevelyan's bars by the galvanic current." Silliman's Journal, 1850, IX., pp. 105–108.

currents induced in the secondary wires. A rheotome was placed in circuit with the thick wires of a Ruhmkorff's coil, and the fine wires were connected with two strips of brass (A and B), insulated from one another by means of a sheet of paper. Upon placing the ear against one of the strips of brass, a sound was perceived like that described above as proceeding from an empty helix of wire during the passage of an intermittent voltaic current. A similar sound, only much more intense, was emitted by a tin-foil condenser when connected with the fine wires of the coil.

One of the strips of brass, A (mentioned above), was held closely against the ear. A loud sound came from A whenever the slip B was touched with the other hand. It is doubtful in all these cases whether the sounds proceeded from the metals or from the imperfect conductors interposed between them. Further experiments seem to favor the latter supposition. The strips of brass A and B were held one in each hand. The induced currents occasioned a muscular tremor in the fingers. Upon placing my forefinger to my ear a loud crackling noise was audible, seemingly proceeding from the finger itself. A friend who was present placed my finger to his ear, but heard nothing. I requested him to hold the strips A and B himself. He was then distinctly conscious of a noise (which I was unable to perceive) proceeding from his finger. In these cases a portion of the induced currents passed through the head of the observer when he placed his ear against his own finger; and it is possible that the sound was occasioned by a vibration of the surfaces of the ear and finger in contact.

When two persons receive a shock from a Ruhmkorff's coil by clasping hands, each taking hold of one wire of the coil with the free hand, a sound proceeds from the clasped hands. The effect is not produced when the hands are moist. When either of the two touches the body of the other a loud sound comes from the parts in contact. When the arm of one is placed against the arm of the other, the noise produced can be heard at a distance of several feet. In all these cases a slight shock is experienced so long as the contact is preserved. The introduction of a piece of paper between the parts in contact does not materially interfere with the production of the sounds, while the unpleasant effects of the shock are avoided.

When a powerful current is passed through the body, a musical note can be perceived when the ear is closely applied to the arm of the person experimented upon. The sound seems to proceed from the muscles of the fore-arm and from the biceps muscle. The musical note is the unison of the rheotome employed to interrupt the primary circuit. I failed to obtain audible effects in this way when the pitch of the rheotome was high. Elisha Gray* has also produced audible effects by the passage of induced electricity through the human body. A musical note is occasioned by the spark of a Ruhmkorff's coil when the primary circuit is made and broken sufficiently rapidly. When two rheotomes of different pitch are caused simultaneously to open and close the primary circuit, a double tone proceeds from the spark.

9. When a voltaic battery is common to two closed circuits, the current is divided between them. If one of the circuits is rapidly opened and closed, a pulsatory action of the current is occasioned upon the other.

All the audible effects resulting from the passage of an intermittent current can also be produced, though in less degree, by means of a pulsatory current.

10. When a permanent magnet is caused to vibrate in front of the pole of an electro-magnet, an undulatory or oscillatory current of electricity is induced in the coils of the electro-magnet, and sounds proceed from the armatures of other electro-magnets placed upon the circuit. The telephonic receiver referred to above (par. 4), was connected in circuit with a single-pole electro-magnet, no battery being used. A steel tuning-fork which had been previously magnetized was caused to vibrate in front of the pole of the electro-magnet. A musical note similar in pitch to that produced by the tuning-fork proceeded from the telephonic receiver in a distant room.

11. The effect was much increased when a battery was included in the circuit. In this case, the vibration of the permanent magnet threw the battery-current into waves. A similar effect was produced by the vibration of an unmagnetized tuning-fork in front of the electro-magnet. The vibration of a soft iron armature, or of a small piece of steel spring no larger than the pole of the electro-magnet in front of which it was placed, sufficed to produce audible effects in the distant room.

12. Two single-pole electro-magnets, each having a resistance of ten ohms, were arranged upon a circuit with a battery of five carbon elements. The total resistance of the circuit, exclusive of the battery, was about twenty-five ohms. A drum-head of gold-beater's skin, seven centimetres in diameter, was placed in front of each electro-magnet, and a circular piece of clock-spring, one centimetre in diameter, was glued to the middle of each membrane. The telephones so constructed were placed in different rooms. One was retained in

* *Elisha Gray.* Eng. Pat. Spec., No. 2646, see "Engineer," Aug. 14, 1874.

the experimental room, and the other taken to the basement of an adjoining house.

Upon singing into the telephone, the tones of the voice were reproduced by the instrument in the distant room. When two persons sang simultaneously into the instrument, two notes were emitted simultaneously by the telephone in the other house. A friend was sent into the adjoining building to note the effect produced by articulate speech. I placed the membrane of the telephone near my mouth, and uttered the sentence, "Do you understand what I say?" Presently an answer was returned through the instrument in my hand. Articulate words proceeded from the clock-spring attached to the membrane, and I heard the sentence: "Yes; I understand you perfectly."

The articulation was somewhat muffled and indistinct, although in this case it was intelligible. Familiar quotations, such as, "To be, or not to be; that is the question." "A horse, a horse, my kingdom for a horse." "What hath God wrought," &c., were generally understood after a few repetitions. The effects were not sufficiently distinct to admit of sustained conversation through the wire. Indeed, as a general rule, the articulation was unintelligible, excepting when familiar sentences were employed. Occasionally, however, a sentence would come out with such startling distinctness as to render it difficult to believe that the speaker was not close at hand. No sound was audible when the clock-spring was removed from the membrane.

The elementary sounds of the English language were uttered successively into one of the telephones and the effects noted at the other. Consonantal sounds, with the exception of L and M, were unrecognizable. Vowel-sounds in most cases were distinct. Diphthongal vowels, such as *a* (in ale), *o* (in old), *i* (in isle), *ow* (in now), *oy* (in boy), *oor* (in poor), *oor* (in door), *ere* (in here), *ere* (in there), were well marked.

Triphthongal vowels, such as *ire* (in fire), *our* (in flour), *ower* (in mower), *ayer* (in player), were also distinct. Of the elementary vowel-sounds, the most distinct were those which had the largest oral apertures. Such were *a* (in far), *aw* (in law), *a* (in man), and *e* (in men).

13. Electrical undulations can be produced directly in the voltaic current by vibrating the conducting wire in a liquid of high resistance included in the circuit.

The stem of a tuning-fork was connected with a wire leading to one of the telephones described in the preceding paragraph. While the tuning-fork was in vibration, the end of one of the prongs was dipped into water included in the circuit. A sound proceeded from the distant telephone. When two tuning-forks of different pitch were connected together, and simultaneously caused to vibrate in the water, two musical notes (the unisons respectively of those produced by the forks) were emitted simultaneously by the telephone.

A platinum wire attached to a stretched membrane, completed a voltaic circuit by dipping into water. Upon speaking to the membrane, articulate sounds proceeded from the telephone in the distant room. The sounds produced by the telephone became louder when dilute sulphuric acid, or a saturated solution of salt, was substituted for the water. Audible effects were also produced by the vibration of plumbago in mercury, in a solution of bichromate of potash, in salt and water, in dilute sulphuric acid, and in pure water.

14. Sullivan * discovered that a current of electricity is generated by the vibration of a wire composed partly of one metal and partly of another; and it is probable that electrical undulations were caused by the vibration. The current was produced so long as the wire emitted a musical note, but stopped immediately upon the cessation of the sound.

15. Although sounds proceed from the armatures of electro-magnets under the influence of undulatory currents of electricity, I have been unable to detect any audible effects due to the electro-magnets themselves. An undulatory current was passed through the coils of an electro-magnet which was held closely against the ear. No sound was perceived until a piece of iron or steel was presented to the pole of the magnet. No sounds either were observed when the undulatory current was passed through iron, steel, retort-carbon, or plumbago. In these respects an undulatory current is curiously different from an intermittent one. (See par. 2.)

16. The telephonic effects described above are produced by three distinct varieties of currents, which I term respectively intermittent, pulsatory, and undulatory. *Intermittent currents* are characterized by the alternate presence and absence of electricity upon the circuit; *Pulsatory currents* result from sudden or instantaneous changes in the intensity of a continuous current; and *undulatory currents* are produced by gradual changes in the intensity of a current analogous to the changes in the density of air occasioned by simple pendulous vibrations. The varying intensity of an undulatory current can be

---

* *Sullivan.* "Currents of Electricity produced by the vibration of Metals." Phil. Mag., 1845, p. 261; Arch. de l'Électr., X., p. 480.

represented by a sinusoidal curve, or by the resultant of several sinusoidal curves.

Intermittent, pulsatory, and undulatory currents may be of two kinds, — *voltaic*, or *induced;* and these varieties may be still further discriminated into *direct* and *reversed* currents; or those in which the electrical impulses are all positive or negative, and those in which they are alternately positive and negative.

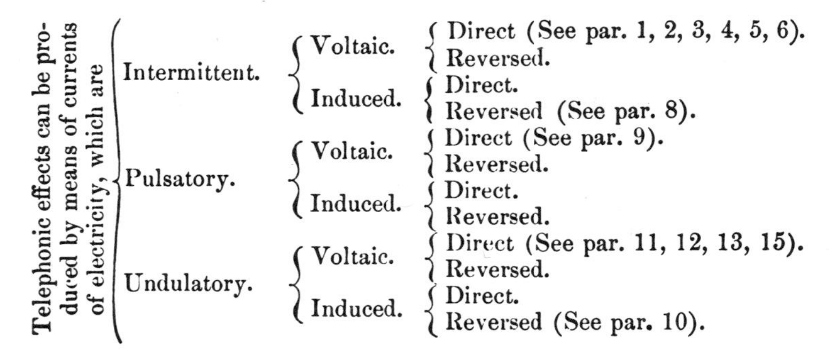

Telephonic effects can be produced by means of currents of electricity, which are			
	Intermittent.	Voltaic.	Direct (See par. 1, 2, 3, 4, 5, 6).
			Reversed.
		Induced.	Direct.
			Reversed (See par. 8).
	Pulsatory.	Voltaic.	Direct (See par. 9).
			Reversed.
		Induced.	Direct.
			Reversed.
	Undulatory.	Voltaic.	Direct (See par. 11, 12, 13, 15).
			Reversed.
		Induced.	Direct.
			Reversed (See par. 10).

17. In conclusion, I would say that the different kinds of currents described above may be studied optically by means of König's manometric capsule.* The instrument, as I have employed it, consists simply of a gas-chamber closed by a membrane to which is attached a piece of clock-spring. When the spring is subjected to the attraction of an electro-magnet, through the coils of which a "telephonic" current of electricity is passed, the flame is thrown into vibration.

I find the instrument invaluable as a rheometer, for an ordinary galvanometer is of little or no use when "telephonic" currents are to be tested. For instance, the galvanometer needle is insensitive to the most powerful undulatory current when the impulses are reversed, and is only slightly deflected when they are direct. The manometric capsule, on the other hand, affords a means of testing the amplitude of the electrical undulations; that is, of deciding the difference between the maximum and minimum intensity of the current.

---

* *König.* "Upon Manometric Flames," Phil. Mag., 1873, XLV., No. 297, 298.

Section I-D
Electric Lighting

Moses Farmer's Analysis of Electric Lighting

FARMER'S evaluation of the economics of electric versus gas lighting appeared in the *American Journal of Science* in 1868. By coincidence, the same issue carried a description of a self-exciting dynamo, a machine that Thomas Edison would utilize as the basis for a successful electric-lighting industry more than a decade later. Farmer was one of America's most highly regarded electrical experts of the mid 19th century. According to A. G. Bell, Farmer was as celebrated in the U.S. as Charles Wheatstone was in England.

Moses G. Farmer (1820–1893) was born in New Hampshire and attended Dartmouth College, although he failed to graduate. He was the inventor of a machine to manufacture paper window shades, and he used his income from this to finance his electrical experiments that began around 1845. He experimented with a model electric train and worked for a time as a telegraph operator. Farmer was one of the first to propose a multiplex telegraph and was the inventor of an electric fire-alarm system used in Boston. His experiments with incandescent electric lights date from as early as 1858. He was among the electrical experts consulted by Bell prior to the invention of the telephone, and he served as technical consultant to the U.S. Electric Lighting Company during the early 1880's. See the *DAB*, vol. 6, pp. 279–281 and the *NCAB*, vol. 7, pp. 361–362.

On the cost of the Electric Light; by MOSES G. FARMER.—The time appears to be near at hand when the electric light will be used for a variety of purposes. Is is worth our while to inquire as to its cost. The expense and inconvenience attendant upon the production of electricity upon a large scale has hitherto been an obstacle in the way of using the electric light, except for lecture rooms and a few other purposes. But the recent improvements in the construction of magneto-electric machines and thermo-electric batteries have put in our power to command the services of this beautiful illuminating agent on any desirable scale of magnitude.

In order to examine the question of cost intelligently, let us refer both electrical and illuminating effects to the common measure of power, viz: the foot-pound per minute. The experiments of Mr. Julius Thomson, of Copenhagen, have shown that the power to maintain the light to that of a standard candle for one minute is equal to the raising of a weight not exceeding thirteen pounds, one foot high in that time. I have arrived at a similar result from a reduction of recorded experiments made by Müller, Ritchie, myself, and others. I am satisfied that, where an electric light of not less than eight hundred to one thousand candles is produced, under proper management, the power required will not greatly exceed 15 foot-pounds per minute per candle. For smaller amounts of light the power required will be greater.

Reprinted from *Amer. J. Sci.*, 2nd ser., vol. XLV, pp. 112–115, 1868.

Now let us inquire what amount of electricity is the equivalent of, or is represented by 15 foot-pounds per minute. If 100 feet of No. 18 pure copper wire be coiled into a helix and immersed in a pound of water, and if the ends of this wire be connected to the poles of one cell of the Grove battery (pint cup size as used in telegraphing), the temperature of the water will begin to rise at the rate of 1° F. in 9½ minutes, or 0·105° per minute. Now if the temperature of one pound of water be raised one degree (Fah.) per minute, this effect will be the thermal equivalent of 772 pounds raised one foot high in space per minute; the heating effect then, of our Grove cell upon the water is the equivalent of $0{\cdot}105 \times 772 = 81$ (call it 80) foot-pounds per minute.

It is well known that a galvanic battery will perform its maximum work when the external resistance which it encounters is equal to the internal resistance of the battery. I have found the internal resistance of the pint cup Grove cell to be equal, on the average, to that of 100 feet of pure copper wire, No. 18 size. Hence the maximum external effect of the ordinary Grove cell may be set down as the equivalent of 80 foot-pounds per minute, equal to the production of $80 \div 15 = 5\frac{1}{3}$ candle lights. I would not be understood as saying that this amount of light can be produced by a single Grove cell, but that 1,000 cells, if properly arranged, would be capable of evolving somewhat more than 5,000 candle lights from a single lamp.

With sulphuric acid costing 2½ cents, nitric acid 10 cents, zinc 8 cents, and mercury 50 cents per pound, the cost of running 1,000 Grove cells one hour, while doing their maximum work, would be $27.65. This would give for 5,000 candles a cost of about 5½ mills per hour per candle.

The cost of gas light per candle per hour would be about one mill, if gas costs $3.25 per thousand cubic feet, and if one cubic foot per hour gives the light of three candles.

With the Smee battery, carefully managed, the cost of 5,000 candle lights would be about the same as with gas.

Let us now look at the cost of electricity as developed by the magneto-electric machine. The power expended on the machine is consumed in friction, in heating the wires, magnets, etc. On a well built machine which I examined in 1861, 1,100 foot-pounds per minute were required to keep the machine in motion when the circuit was open, and the machine doing no work. But when the circuit was closed 3,200 foot-pounds per minute were required to maintain the same velocity of rotation; nearly all this excess of power (viz., 2,100 foot-pounds) was measured as electricity, about two-thirds (say 1,300 foot-pounds) being expended internally, heating the coils and magnets, etc., and the balance, 800 foot-pounds, measured as external useful effect. Had the external resistance been larger, a greater proportion of the expended power would have appeared as useful effect. Suppose, however, that only 800 foot-pounds per minute could be utilized by this machine and used for illuminating purposes. This would be the equivalent of $800 \div 15 = 53{\cdot}33$ candles, and the total power required (including friction, etc.), would be $3{,}200 \div 53{\cdot}33 = 60$, about sixty foot-pounds per minute per candle.

In the vicinity of Boston, power is furnished, per horse power, at the rate of \$180 per year of 313 days of ten hours each, or at the rate of $\frac{\$180}{313 \times 10} = 0{\cdot}0575$ ($5\frac{3}{4}$ cents) per hour. If only one-fourth of this power could be utilized as light $\frac{33{,}000}{4 \times 15} = 550$ candles would be the equivalent of one horse power, and would cost $\$0{\cdot}0575 \div 550 = \$0.0001046$, about one-tenth of a mill per hour per candle, being about one-tenth the cost of gas light.

Let us for a moment take another view of the matter. The average hourly consumption of coal by a good steam engine may be set down at four pounds per hour per horse-power, $=(33{,}000 \times 60) \div 4 = 495{,}000$ foot-pounds from one pound of coal. Utilizing as electricity, and thence light, one-fourth part of this we get $495{,}000 \div 4 = 123{,}750$ foot-pounds, or as light, $\frac{123{,}750}{15 \times 60} = 137.5$ hour candle lights from one pound of coal, through the agency of steam engine and the magneto-electric machine.

With the thermo-electric battery I have been able to develop 130,000 foot-pounds of electricity from one pound of coal $= \frac{130{,}000}{15 \times 60}$ $= 144{\cdot}4 =$ to about 144 candle lights.

There is still another point of view worthy our attention. Common gas coal will yield about ten thousand cubic feet of gas per ton. This, at three hour candle lights per cubic foot, would give $(3 \times 10{,}000) \div 2{,}000 = 15$ hour candle-lights per pound of coal. About twenty-five cubic feet of illuminating gas weigh one pound. Hence one pound of gas, after it is made from the coal, will yield a light equal to that of a candle for seventy-five hours. One pound of pure carbon, wholly burned to carbonic acid gas, yields 14,500 units of heat, equal to $772 \times 14{,}500 = 11{,}200{,}000$, or $11\frac{1}{5}$ millions of foot-pounds of work: Hence, were the total energy of one pound of pure carbon converted into light, it would be equivalent to one candle light for the time of

$$\frac{11{,}200{,}000}{15 \times 365 \times 24 \times 60} = 1\tfrac{5}{12};$$ or one year and five months.

To recapitulate: the gas made from one pound of coal would yield a candle light for fifteen hours; one pound of the gas would yield a light equal to one candle for seventy-five hours; but could all the energy in a pound of carbon be converted into light, it would be equivalent to the burning of a candle for 12,410 hours.

Thus it will appear that by our ordinary methods of gas lighting we utilize much less than one per cent of the energy stored in the coal. I think we may reasonably expect that electricity, as developed by the thermo-electric battery, the magneto-electric machine, or some still more efficient apparatus, will help us in some way to bridge the chasm between fifteen and twelve thousand hour candle lights from a pound of coal.—*Scientific American.*

PART II
1877–1976

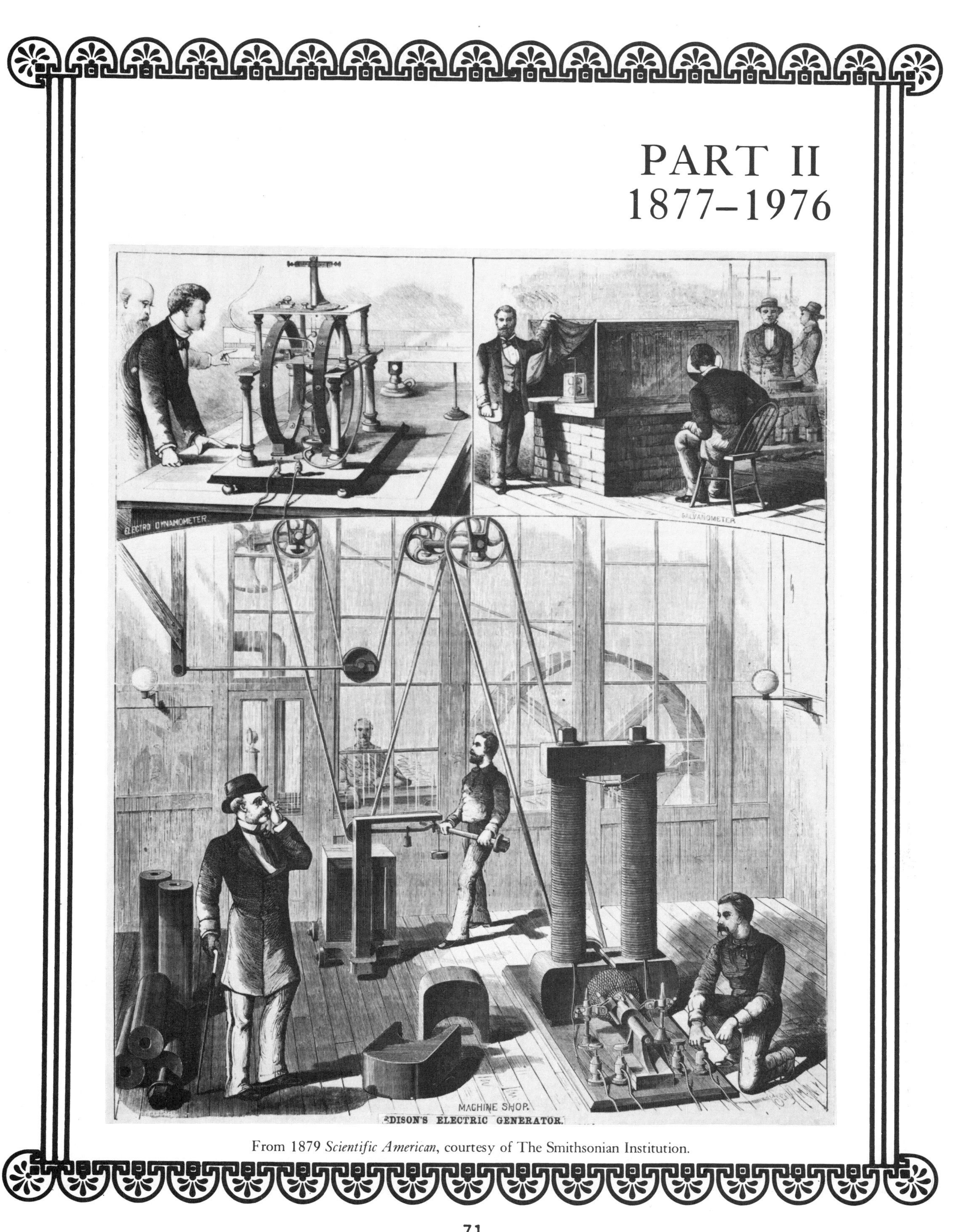

From 1879 *Scientific American*, courtesy of The Smithsonian Institution.

Section II-A
Professionalization

Stephen D. Field - Electro Com. Tel. Co.
S. D. Schuyler Prest. S. E. L. Co.
L. E. Curtis Secy U. S. E. L. Co
Wm. A. Hovey Pres. Merchant E. L. & P. Co
Norvin Green Prest. W. Union Tel Co
A. H. Bates, Prest. B. & O. T.
Thomas A. Edison Inventor
George B. Prescott Electrician
A. A. Hayes President Brush-Swan
J. H. Frost Supt. A. D. T. Co
Jos. P. Davis V. Prest. Met. Tel & T
Frank L. Pope Solicitor of Patents
George A. Hamilton Electrician W. U. Tel
S. B. Eaton Prest. Edison E. L. Co
W. H. McGrath Supt. Brush Elect. L. Co.
F. A. P. Barnard, Prest. Columbia Coll.
[illegible] . B. & M. Tel Co
Edwin J. Houston Electrician
Thomson-Houston Electric Light Co.
N. S. Keith Electrical Engineer.
Thos D. Lockwood Electrician & Engineer Am Bell Telephone
W. D. Sargent Engr. Tel Co
Richard R. Hazard P. [illegible]
Chas L. Buckingham Counsel W. U. Tel. Co.

Some well-known signatures on the 1884 call for an AIEE organization meeting.

H. J. Ryan on the First Forty Years of the AIEE

THE organization of the American Institute of Electrical Engineers in 1884 was a major event in the social history of electrical engineering in the United States. In his presidential address on the 40th anniversary of the AIEE, Harris J. Ryan engaged in a contemplative review of the circumstances that led to the founding of the AIEE and of the first papers presented. Note his prediction that future electrical engineers will need to assume a major responsibility for locating "water, ore and fuel." Ryan also suggests a distinction between the scientist and the engineer in that the latter studies "natural phenomena for the immediate purpose of promulgating practical results through the economic expenditure of capital." See Michael McMahon's paper on the social origins of the AIEE in the special issue of the PROCEEDINGS OF THE IEEE of September 1976.

Harris J. Ryan (1866–1934) was born on a farm in Pennsylvania, and was among the first students to enroll in the electrical engineering program at Cornell where he graduated in 1887. Ryan later stated that his life-long interest in high-voltage transmission was stimulated by a conversation with Frank Sprague when Ryan's class visited Sprague's factory. Ryan became a partner with J. G. White and D. C. Jackson in an engineering company in Lincoln, Nebr., but returned to Cornell a year later to teach. He gave a pioneering paper on transformers at an AIEE meeting in 1890 and was the first American engineer to employ the cathode ray tube as a measuring instrument. He was named head of the Electrical Engineering Department at Stanford in 1905 and remained there until his retirement in 1931. See the *NAS*, vol. 19, pp. 285–306 and the *DAB*, vol. 21, pp. 645–646. Also see Frederick Terman's paper on the history of electrical engineering education in the PROCEEDINGS OF THE IEEE of September 1976.

A Generation of the American Institute of Electrical Engineers---1884-1924

President's Address

BY HARRIS J. RYAN
Leland Stanford, Jr., University, California

THIS is the fortieth year of the Institute. A charter member who was twenty-five when it was organized is now sixty-five and his work has been substantially completed. While the American Institute of Electrical Engineers must endure as long as the Nation, no succeeding generation can encounter the unique experiences of its charter members. They were young with a wilderness of options for action before them. They had the mentality of early manhood, boundless energy and enthusiasm and a wonderful unity of purpose.

When the Institute began, the land telegraph had been in use forty years, the Europe-to-America submarine telegraph cable eighteen years, the direct-current generator twelve, the telephone eight, the arc light about five, and the incandescent lamp had been used three years. The direct-current motor had just arrived, and the alternator, transformer, synchronous and induction motors were soon to follow. New and wonderful electrical expediencies were arriving at a rate that was bewildering for the "man in the street." This acceleration of electrical progress took a strong hold on the imagination of our people. The boys who visited the Centennial Exposition at Philadelphia in 1876 reacted powerfully as boys always have and always will in regard to new things. They went home to play with batteries, magnets, telegraphs and telephones and a little later constructed direct-current dynamos and motors as the boys of today build radio equipment.

Some of the boys of 1880 quickly mastered an understanding of the critical speed of a direct-current shunt generator that had cost Edison and Hopkinson hours of intense mental effort. For the most part they were the men who, four years later, organized the Institute. Their wonderful work is going on today and always will. In the great turning movement that education is now making, the young men and women of 1884 are today helping their children to give the grandchildren, who are the boys and girls of today, a greater opportunity to see for themselves the things of highest interest and greatest worth-whileness for knowing and doing.

The "why not" spirit of the youth who founded the Institute is splendidly reflected in the frontispiece of Volume I of the TRANSACTIONS. It is lithographed in colors and labeled "American Institute of Electrical Engineers, September 1884. The Scientific Street as Applied to Broadway, New York."

It is recorded in the beginning of Volume I of the TRANSACTIONS that "The first steps toward the formation of the Society were taken in April, 1884," when a circular was prepared and published by Dr. Nathaniel S. Keith of New York City, inviting signatures to the proposal to organize "The American Institute of Electrical Engineers." The subscribers held a meeting in the rooms of the American Society of Civil Engineers in New York on May 13, 1884, adopted rules of procedure and elected officers for the ensuing year. Thus the American Institute of Electrical Engineers began forty years ago. The membership rapidly increased to 1000 and thereafter grew steadily to more than 15,000 today.

The first communication received and printed in the TRANSACTIONS was a letter from Mr. C. J. Kintner, Examiner, Class Electricity, U. S. Patent Office dated at Washington, D. C., May 12, 1884. He expressed his regret that he would not be able to attend the organization meeting of the Electrical Engineers and begged them to come to the rescue of the Patent Office which was in a deplorable state of neglect on the part of Congress. He invited helpful attention to the contents of an attached copy of an official report on the situation which revealed that his division was undermanned and housed in inadequate quarters poorly ventilated and lighted. Of his own office room he said there was no gas and that he was obliged to use kerosene oil which seriously vitiated the atmosphere. This was the state of things when inventors had paid for the maintenance of the patent office $2,500,000 in excess of operating expenses.

Thus, in the first hour of its existence, the American Institute of Electrical Engineers was forcefully made to take note of the fact that this is an ever practical world and that vital things are never what they should be if the men most concerned fail to make them so.

Without patents properly provided for, The American Institute of Electrical Engineers could not have been maintained as a forum for the reception of new ideas as to facts, principles and processes. The only alternative would have been procedure of individuals, or of small groups of individuals, bound to secrecy for the protection of their equities and rights.

The history of the ceramic arts is a specific illustration. Porcelain has been known for centuries, nevertheless the porcelain arts have advanced much more in the few years in which ceramists have abandoned secrecy and have depended upon the United States patent system for their protection, and upon their own cooperation through the American Ceramic Society.

The longing of the people of our Country to have an

Reprinted from *Trans. AIEE*, vol. XLIII, pp. 740–744, June 1924.

opportunity to come more intimately into contact with vast numbers of new things electrical in 1883, was clearly understood by the management of the Franklin Institute of Philadelphia. The home of the renowned Franklin had been in Philadelphia. In 1883 this city was also the home of two high school teachers who were becoming known throughout the world for their electrical achievements. In the circumstances, it was decided to hold the International Electrical Exposition of 1883 in Philadelphia under the auspices of the Franklin Institute.

The determination of the young men engaged in the electrical sciences and industries to organize a national society of electrical engineers was due in no small measure to this action of the Franklin Institute as may be learned by the following quotation from the circular that called the organization meeting: "An International Electrical Exhibition is to be held in Philadelphia next autumn, to which many of the foreign electrical savants, engineers and manufacturers will be visitors; and it would be a lasting disgrace to American electricians, if no American national electrical society was in existence to receive them with the honors due them from their co-laborers in the United States." Nothing was more natural, therefore, than to hold the first meeting of the Institute in Philadelphia for the presentation and discussion of papers in response to a cordial invitation from the Franklin Institute as recorded also in the early pages of Volume I of the TRANSACTIONS, as follows: "The Directors of the International Electrical Exhibition, Philadelphia, held under the auspices of the Franklin Institute, having kindly tendered to the Institute the free use of rooms in the exhibition building, so long as it should remain open, the offer was accepted by the Council, and the secretary was authorized to make the necessary preparations and to be present during the Exhibition as the representative of the Institute.

"The rooms were accordingly opened to members by Mr. Keith on September second, and remained open until October eleventh, during which time they were visited by several foreign electricians. On October seventh and eighth a meeting was held, in accordance with the rules, at the Continental Hotel and the Exhibition rooms, where the papers forming this volume were read and discussed. On the invitation of the Franklin Institute, the Exhibition was also inspected by the members in a body."

To Professor Edwin J. Houston belongs the never-to-be-forgotten honor of presenting to the Institute the first paper. It bore the modest title, "Some Notes on Incandescent Lamps" and dealt with the now universally known and highly valued "Edison effect." Referring to "the peculiar high vacuum phenomenon observed by Mr. Edison in some of his incandescent lamps, Prof. Houston said: "I wish to bring it before the Society for the purpose of having you puzzle over it." What a glorious result has come of "the puzzling." Little could those present have appreciated on the memorable morning of October seventh, 1884, in Philadelphia, the fact that some of them during this fortieth anniversary would witness through applications of the Edison effect the transmission of the human voice from airplane to airplane in flight and over land and ocean to the uttermost parts of the earth; that such Edison effect was due to a reaction at last of the long-sought-for electricity, that the effect was a characteristic reaction of matter within its own structural make-up and that it would be the means of producing an endless succession of most helpful facilities for research and progress.

Sir William Preece of England was present at the meeting and joined in the discussion of the Edison effect. He brought forward the fact that Dr. Oliver Lodge had recently demonstrated that fogs resulted from the formation of moisture around electrical nuclei. "He took a large glass globe, apparently chemically clean, exhausted the air, but with moisture in it. It remained perfectly transparent so long as no matter was admitted, but the moment a spark from an induction coil was passed between the two platinum electrodes in this apparently clean globe at once a cloud was formed throughout the whole globe."

It was nearly thirty years before important results were developed out of the Edison effect. The fortieth year is now passing since the fog forming nuclei were reported to the Institute and almost nothing has been done about it. What wonderful uses may yet be made of the fog effect in bringing water from the skies and in the control of fogs in places where they render traffic hazardous.

The overhead conductors to provide for the rapidly growing telephone and telegraph service in the business districts of our cities began to obscure the sky and to make city dwellers positively unhappy, besides being in fear of their lives because of the difficulties the conductors caused the fire department in fighting fires. An intense longing to get rid of overhead conductors possessed all city residents. It is natural, therefore, to find that the second paper presented to the Institute at the Philadelphia convention bore the title "Underground Wires," by W. M. Calender.

In 1884 many were living who had witnessed the almost desperate struggle, and final failure of Professor Morse, Alfred Vail and Ezra Cornell to install underground the first land telegraph between Baltimore and Washington. The enterprise was finally saved in 1884 by mounting the telegraph conductor in the air upon insulators supported on wood poles substantially as is the practise of today. These people had also witnessed the heart-breaking difficulties encountered by Cyrus W. Field and his co-workers in the Atlantic telegraph cable laying operations of 1858 and 1865, that finally culminated in complete success in the summer of 1866. As a result of such pioneer experiences there has existed almost to this day a mental momentum that has continued to place conductors overhead when they more properly belonged underground. The technical and economic difficulties have always been many. In telephonic as well as telegraphic communication on land the technical difficulties have now been overcome and the corresponding economic boundaries have been located with the result that at a comparatively early date all our great cities will be inter-connected with underground cables in the more densely populated areas of the United States and will no longer have their telephonic and telegraphic communications broken by severe winter storms.

Interconnection for power service is encountering much the same sort of progress in relation to underground conductors. Economic boundaries in the power service for the City of Cleveland and its environs have determined the use within and without the city of huge coal burning steam-power generating stations. The problem of a heavy power connection between such stations within the city limits has been solved by the use of an 8-mile section of 66,000 volt, three-phase underground cable.

Synchronism was the title of the third paper received by the Institute at its first convention. The value and remarkable possibilities of synchronism in communication was stressed and carefully defined. The expedients available in 1884 for maintaining synchronism at distant points in the electric circuit and of obtaining valuable results by means thereof were reported upon. It is of genuine interest today to note that after forty years, advances in physical chemistry and assiduous study of talented men well organized, have so improved and extended these synchronizing facilities in their application to telephotography that extraordinary success has been obtained. Because the telephotography of today when speeded up a few thousand times will result in "seeing at a distance" the usual person entertains the belief that "teleopto" will soon be an accomplished fact.

Other papers presented at the first convention carried the titles:

The Scientific Street,

Experimental Method for Testing a Dynamo Machine,

The Earth as an Electric Circuit Completer,

Telegraphy without Wires,

Chemistry of the Carbon Filament,

The Patent Office.

Of these, The Scientific Street and The Earth as an Electric Circuit Completer have surely made the least progress during the forty years that have past. A recent electrical engineer-graduate from one of our universities, was employed in municipal power service and assigned the job of planning a power feeder to run from the main receiving substation to a heavy demand district of a new and rapidly growing city. He encountered many trouble-making factors, among which were the heating limits of the underground cables that he would have to use and the heat liberation limitations of the ducts and adjacent earth in which such cables would have to be placed. He had never seen Volume I of the TRANSACTIONS, and had never heard of The Scientific Street nor of anything that had ever been done about it, nor had he ever heard of such a thing as organization red tape. What he knew was that years before the Chief of the municipal water supply had run a huge water main through the center of the street in which the power feeder would have to go. To the young man's mind the obvious thing to do was to place his power feeder just as close to the Chief's water main as possible, thermally connecting thereto in a 100 per cent fashion if possible, so that water in the main would carry away the heat from the power cable, a plan that would surely do away with the heating difficulty.

The young man had already discovered that most things in this world are "nailed down" and assumed, therefore, that he would have to get permission from somebody to discharge the heat from the power cable into the water main. It never occured to him that he would have to go to his immediate superior in the power bureau who would say "no" or in turn would pass the matter up to the chief of the power bureau who would in turn say "no," or broach the matter delicately to the chief of the water supply bureau who might say "no" or who would designate someone in his organization to look into the matter and report with recommendations. It occurred to the young man forthwith that he should be able to agree with the Chief of the water bureau upon the highly desirable plan in ten minutes and it should not be necessary to encounter any further "fuss and feathers." He called on the Chief of the water bureau, made a direct and immediate presentation of the proposition and thereby started a series of tremors in the two bureaus that have not altogether damped out to this day; and one must hope they never will, for when things tremble they slide more easily and that is often a helpful state on the road of progress. The young man demonstrated that the incoming generations can always be counted on to renew the "why not" spirit of the charter members of the Institute unless China's mistake is made by their successors of keeping the young people so occupied by the mental activities prescribed by their ancestors that no opportunity is permitted for activity on their own initiative.

Of the subjects brought to the attention of the members of the American Institute of Electrical Engineers at its first convention "The Earth as an Electric Circuit Completer" by Thomas D. Lockwood was the most important and has progressed the least through the intervening years. Pupin has recently taught us beautifully to have high regard for all terrestrial electrical phenomena.[1]

Modern radio and all related phenomena are interpreted in terms of the electromagnetic theory of light as formulated by Maxwell. Yet Maxwell never knew of the existence of electrons. Because it is now known

1. From Immigrant to Inventor, by Michael Pupin, 1923, pp. 301-302.

that every electric field is an aggregate of the elemental electric field fragments attached to their corresponding electrons, and that every electric field in motion is thereby also a magnetic field, it follows that there can be no electromagnetic activity without a corresponding activity of the electrons to which the electric field in action is attached. It is no wonder then that Pupin, the renowned physicist and radio pioneer, had to discover through Marconi's actual trial the importance of the ground connection in radio transmission and reception, and that he was not led thereto by Maxwell's theory.

In a classical experiment Michelson determined the velocity of light in the direction that the earth is moving around the sun and again in the opposite direction and found the two velocities to be identical. To account for this unexpected result some able persons have resorted to certain extraordinary assumptions. Pending the outcome of the far-flung attack of modern physics upon the problem of the fundamental character of light it should be far more helpful to remember the never failing presence of electrons in the earth's crust and in the atmosphere upon which may terminate an ample electric field required for the transmission of any light that originates on the earth, or that arrives from outer cosmic space. If such is the case, the earth must carry with it through space its own portion of the luminiferous electromagnetic medium in which the local transmission of light occurs much as a railway passenger coach carries with it a complement of air through which the passengers hear with no change in pitch due to the velocity of the train.

The fundamental lines along which our civilization has been advancing for a century will be broken utterly if the subterranean supplies of metals, coal, petroleum and chemicals can no longer be located in abundance as heretofore. It is a disturbing fact that the more these supplies are mined the more arduous the task of locating their new reserves. Aside from the slow and expensive prospector's drill, the only means available for "sounding" the contents of the superficial strata of the earth are those afforded by mechanical vibrations over a wide range of frequencies and by electric currents in every conceivable time relation, *viz.*, continuous current, audio and radio frequency currents and electromagnetic radio waves.

It is a problem for the engineer to define and attack and not for the scientist for the reason that it involves a close study of natural phenomena for the immediate purpose of promulgating practical results through the economic expenditure of capital.

Because the application of mechanical energy for subterranean reconnaissance can only be accomplished by means of vibrations, *i. e.*, by means of sound at audio or radio frequencies, and because the best facilities for the projection and echo-reception of such sounds are electrical, the solution of the problem will remain a job for the electrical engineer. Eventually, therefore, the electrical engineer will have to assume most of the burden of locating subterraneous treasures by mechanical or electrical means. The electrical engineers of the incoming generation are surely going to be called on to study the characteristic manner in which the earth transmits mechanical and electrical energy as affected by subterranean contents for the purpose of location of water, ore and fuel.

The first ten-year period of the Institute, 1884-1894, witnessed the initial arrival of the alternating-current facilities substantially as they are known today, the practical beginning of electric traction and a great extension of electric lighting.

This period occupied from the Philadelphia Electrical Exposition to the epoch making electrical exhibit of the World's Fair, Chicago, 1893. The second decade occupied the interval between the Chicago Exposition and the St. Louis Exposition of 1904. In this period the electric lighting industry was consolidated. Electric traction, communication and power transmission made large gains.

A thoroughgoing understanding of alternating-current phenomena as encountered in the transformer, electrical machinery and transmission lines was arrived at and a successful attack was begun upon transients. The new knowledge of alternating-current phenomena made possible the second cycle of evolution of the continuous-current generator, wherein the control of the troublesome field distortion and perfect commutation were finally accomplished. It was during this period that Mr. Andrew Carnegie made his generous offer to build a home for the American Engineering Societies and the United Engineering Society was organized to make the acceptance of Mr. Carnegie's offer practicable by undertaking to build and to maintain the Engineering Societies Building at 33 West 39th Street, New York City.

Toward the close of the period, students in electrical engineering in Universities and Colleges were accorded the benefits of the Institute as "enrolled students."

The third decade extended substantially from the St. Louis to the San Francisco Exposition in 1915, and witnessed spectacular advances in incandescent lighting and transmission of power, and the advent of the electron tube for conductor and radio communication. The period that extends substantially from the San Francisco Exposition to the present constitutes the fourth and last decade of the Institute's first generation. This decade has witnessed a magnificent consolidation of the activities of the electrical engineers within and without the Institute.

In a communication research laboratory where 3000 men were at work, the cathode ray oscillograph was being forced into its second round of evolution; it had encountered its first round about twenty years earlier. The problem developed of discovering better means for focussing the cathode ray, which henceforth should be known as the electron jet. It was carefully defined and submitted to all in the laboratory who were known to have an authoritative understanding of vacuum tube phenomena with requests for suggestions. Among the replies was the unique suggestion that a residual gas be used which would be ionized by collision of the electrons in the jet; the additional electrons thus liberated from the atoms of the gas would be dispersed and would be of no further consequence, while the heavy, slow moving positively charged atoms would remain in the path of the electrons drawing them centrally by their electric attraction to a closely restricted path or focus. This expedient, though never conceived before, was tried and worked beautifully. It would be difficult to find a better illustration of the huge benefits to be derived from the consolidation that has been going on during the recent decade.

Being the last decade of the first generation, it is necessarily the first decade of the Institute's mature membership, motivated by the mental characteristics of young and old, by the abundant working energy of the young under the cooperation and helpful guidance of the older men of profound knowledge and wide experience. It is promulgating those fine powerful drives for consolidation now engaged in by the older members with their splendid conception of what the Institute should be and with their invaluable contributions of time, effort and funds. They are mining the low grade ores and from them are refining the hard and tough metals with which they are reinforcing the working structure of the Institute, so that without failure it may carry its full load of duty for the maintenance of electrical progress for the benefit of the nation. To them the Institute exists for the advancement of electrical engineering in the highest sense. They know that in the end human character is everything, they know that it is the high privilege of the electrical engineer to understand mother nature and her ways; they know the work of the electrical engineers has required wide-spread cooperation at every turn; they have learned as no other group of men have learned the interrelation and interdependence of all things in nature, in the sciences, the arts and the social verities. They know that no group of men can achieve the amazing advances for the nation that the electrical engineers have without encountering an understanding of what should be done in the new order that their work has brought about, so that the gains shall be maximum and the losses minimum. They are aware of the higher duties thus revealed and that they may count upon the prestige and resources of the Institute to support them in their endeavors.

In a single generation and to an amazing degree the work of the electrical engineer is the cause of a vision of a new democracy and a new religion, the democracy of health and happiness and the religion that has an enhanced reverence for the present and a determination to make the future what it should be here on our earth where it has pleased God to put us and where he has asked us to be happy in the undertaking to make the best of life forever.

Dugald Jackson on Electrical Engineers and the Public

CONCERN over the social responsibility of engineers is not a new issue, as illustrated by Dugald Jackson's AIEE Presidential Address of 1911. Jackson argues that since engineers have introduced the technological basis of modern industrial and public service corporations, they cannot avoid "personal answerability to the public." He felt that it was the duty of knowledgeable engineers to defend public utilities against uninformed attacks and help overcome "the barrier of distrust." Jackson suggested that the engineer was especially qualified to solve "the mighty problems of a new age, for the reason that an efficient engineer must associate audacity and sobriety in his spirit."

Dugald C. Jackson (1865–1951) was born at Kennett Square, Pa. and graduated from Pennsylvania State College, where his father was a Professor of Mathematics. Jackson then studied electrical engineering at Cornell for two years before joining J. G. White and Harris J. Ryan in organizing the Western Engineering Company in Lincoln, Nebr., in 1887. He later worked for the Sprague Electric Company in New York before becoming head of the Department of Electrical Engineering at the University of Wisconsin in 1891. He developed an active consulting practice while at Wisconsin, and continued this after being selected to head the electrical engineering program at the Massachusetts Institute of Technology in 1907. He remained at M.I.T. until his retirement in 1935, and was responsible for several innovations such as a cooperative program and an honors program that stressed original research. He campaigned for adoption of a Code of Ethics for engineers while serving on the Ethics Committee of the Engineers Council for Professional Development. See the *NCAB*, vol. B, pp. 357–358. Also see "Professor Dugald C. Jackson," *Electrical World*, vol. 55, pp. 678–679, 1910. See also Frederick Terman's paper on the history of E.E. education in the PROCEEDINGS OF THE IEEE, September 1976.

An Address delivered at the 28th Annual Convention of the American Institute of Electrical Engineers, Chicago, Ill., June 27, 1911.

ELECTRICAL ENGINEERS AND THE PUBLIC

President's Address

BY DUGALD C. JACKSON

Members of the American Institute of Electrical Engineers are pleased to refer to electrical engineering as a Profession, and to the Institute itself as a Professional Society. When this occurs as a thoughtless repetition of fine sounding words, it has little meaning, since mere repetition of an alleged truth does not make it a real truth, and it can be established as a real truth only by tracing it to some adequate foundation. But when those statements arise from a ripe understanding that the word profession means more than a mere organized vocation for earning one's bread, it has a high and commendable meaning. The word profession "implies professed attainments in special knowledge, as distinguished from mere skill; a practical dealing with affairs, as distinguished from mere study or investigation; and an application of such knowledge to uses for others, as a vocation, as distinguished from its pursuit for one's own purposes." This sets the professional man in a position which demands from him an attitude of service and of leadership. He must have a masterly knowledge, in addition to skill in a vocation. He must deal practically in the affairs or needs of men. His duties must be performed with a touch of disinterested spirit in addition to the vocational spirit of earning his livelihood. Such men have a duty to the public; and in the performance of that duty they must exert their influence on that thought and practice of the day which affects the welfare and progress of the nation. We as electrical engineers cannot escape that duty in case we wish to maintain the professional character of our occupation.

It may be retorted that questions relating to the welfare and progress of the nation are matters of economics and sociology,

Reprinted from *Trans. AIEE*, vol. XXX, pp. 1135–1142, June 1911.

and not of engineering. The affirmation contained in this retort I will admit, but the negation I deny.

The theory of modern economics is built up under the influences produced by the introduction of steam power, with its potent agencies comprised in the steam railroad, ocean navigation, and the use of steam power in industrial operations. These agencies are the creatures of engineers. Watt, Stephenson, Fulton, Ericsson, Boulton, Arkwright, Nasmyth, Bessemer, Siemens, Corliss, Holley and the other fathers of our modern industrial economic conditions were engineers; and it would be folly to deny to the parents an interest in their offspring, and equally folly to assert that the further developments of economic theory are not largely dependent on those industrial changes which are continually produced by the inventive activities of the great body of engineers. When I speak of industrial operations or industrial conditions, it must be understood that I include among industrial affairs the great means for transportation and intercommunication which are comprised in railways, telegraphs and telephones, in addition to the manufacture and distribution of products which involve the application of mechanical power as distinguished from animal power, and the manufacture, accompanied by distribution by pipe or wire, of the media for providing illumination and power. The engineers have precipitated these affairs on the world by their inventions; these affairs are in a large measure the support of the engineering profession; and it is the duty of engineers to do their share in moulding their various economic creatures so that the creatures may reach the greatest practicable usefulness to society. In fact it would show a cowardly weakness to suggest that this duty should be avoided by men who are essentially responsible, as the engineers are, for the existing conditions. Theologians and physicians can practice their professions aloof from the ordinary affairs of the world, but the engineers associated with industrial events cannot. Moreover, such an avoidance of their duty by the engineers, even if avoidance of responsibility were possible, would be particularly unfortunate in view of the fact that the professed economists and sociologists apparently do not yet hold themselves subject to all the requirements of professional men, but still interpret their duties as being more confined to the field of study and investigation than to applying their knowledge to practical affairs.

It may again be retorted that the tenets which I am advocating will lead engineers out of a professional spirit and into

" commercialism ". It is worth while to pause here to reflect on that point. The word " commercialism " strictly means the characteristics of business or commercial life, but custom has made it applicable to any undue predominance of commercial ideas in a nation or community, and it has thereby come to infer a willingness to establish the strife for money in a position of precedence over reason and righteousness.

It has been alleged that learning loses of its dignity by becoming fashionable. It has also been alleged that learning loses of its dignity by becoming useful. Of the latter, at least, experience has proved the contrary,—happily for engineers who are proud of their profession, for engineering is necessarily an embodiment of the useful application of knowledge and learning. Engineering, relating, as it does, to the application of the powers of nature to useful purposes, must necessarily bring its followers into intimate contact with commercial affairs in an age when, as in ours, the industries dominate commerce, and the abatement of war has reduced the importance of military engineering. The tenets which I advocate do not tend to entangle the engineers in the depths of " commercialism " with which they may come in contact; but, on the contrary, those tenets propose that engineers should safeguard and nourish their professional spirit by assuming a part in public affairs in a spirit of disinterest, for the purpose of guiding the useful applications of natural forces to the greatest practicable service to society. A true engineer is a devoted follower after truth. He differs diametrically from the devotees of pure " commercialism ", who are strictly opportunists. He also differs from pure idealists, who are often notable for refusing to accept any advance unless it wholly meets their personal ideals. The spirit of the engineer rejoices in obtaining any move toward the truth, but is always seeking farther advance. This characteristic spirit has been manifested in men of great achievement in many walks of life. It is a part of the life of such men as Martin Luther, Gladstone and Lincoln.

Those who accept even in part the usual evolutionary doctrines which are summed up by Herbert Spencer in his view that progress occurs by successive differentiations and integrations producing development from the homogeneous to definite, coherent heterogeneity, will assent to the proposition that the modern giant corporation follows in the wake of the one-man business and the simple partnership in response to an inextinguishable natural law. This is a case of natural selection.

The progress of corporation development cannot be prevented. It is one of the manifestations accompanying improved means of speedy transportation and inter-communication. Of the influence of the latter agencies, a learned and distinguished historian says, " Of all inventions, the alphabet and the printing press alone excepted, those inventions which abridge distance have done most for the civilization of our species. Every improvement of the means of locomotion benefits mankind morally and intellectually as well as materially, * * *." The possibility of, and indeed a necessity for, great corporate organizations came in the train of leading improvements in the means of locomotion and other beneficial inventions which abridge distance and subjugate time. Men of this age do not desire to relinquish the benefits of the improvements. We must, therefore, adjust our mental attitude to dealing properly with the situation; and in making the adjustment we must return to the old and approved recognition that a misdeed is a personal thing, and remember that responsibility for it cannot be shifted from the personality of the man in responsibility to an impersonal aggregation entitled a corporation which he manages. In early days when English kings had great prerogatives in the government, and the doctrine of divine right, associated with the doctrine that the king can do no wrong, were still extant, the king was nevertheless limited to an administration of the affairs of the realm conducted, history tells us, in accordance with the laws, and, in case he broke those laws, his advisers and agents were held responsible, and they were made personally answerable to the courts. History also indicates that this personal answerability of the advisers and agents had a tremendous influence on the conduct of government and its relations to the public. In building up our industrial structure we must not overlook the plain guide board of history, and personal answerability must be established. But if we must establish personal answerability to the public, we must also establish fair and generous dealing by the public.

The building up of a great industrial nation in an honorable state of civilization is subject to many hazards,—an error may cause injury to the structure that takes years or even decades to eradicate. It is, therefore, desirable to go cautiously and utilize the mature reflection of straight-thinking men who will give their thought to the subject. The forward route is untested, and real progress can be made only by judiciously combining teachings from the records of yesterday with experience of to-day to make a working theory for to-morrow. It has been suggested that a theorist should be defined as a man who thinks he may learn to swim by sitting on the bank and watching a frog. Doubtless, there are many such men in the world, but they are not theorists. The definition is as inaccurate as defining a black object as an object without color. Such men are only inexperienced, superficial or foolish. Theory, as the word is used by engineers, means a working hypothesis founded on all known facts and experience, which may be used to guide progress beyond the margin of past experience. Every successful, progressive man is a constant user of theory in this proper sense of the term. Every progressive step is made according to a theory of the man responsible for the move. Theory is not antagonistic to practice but is founded on experience and is a guide to progress. Custom should be followed only when it has reason to support it. In the juncture now before us we must utilize the best theories of the corporation relations and the rights of persons and property, and cautiously extend our practices accordingly. No body of men are better equipped for this sound and scientific procedure than a body of professional engineers; and few others are so fully and adequately trained for such procedure as engineers, for the reason that this procedure is in accordance with the every-day steps of their business life. Moreover, the engineers of experience are well adapted to grapple with the mighty problems of a new age, for the reason that an efficient engineer must associate audacity and sobriety in his spirit.

If my premises are tenable, and I believe them to be incontestible, the engineers have a special duty, as professional men who are trained and experienced in straight thinking, to use their influence for the establishment and support of right and reason in the dealings between the public and the public service corporations. The problems surrounding the public service companies in American cities, and their relations to the citizens, should receive particular attention by members of our Institute, for those problems and those relations have been largely brought to their present importance and prominence through the activities of electrical engineers.

The public service corporations are the natural outcome of the demand of the civilized world for efficient and rapid transportation and intercommunication, and the concurrent need as communities become immersed in peaceful industrial pursuits for ample and conveniently provided supplies of water, gas and electric power. They compose a comparatively new and mighty force in the social organism and the organism must be adapted to efficiently utilize this force, but the force must be prevented from dominating or warping the organism. There is no danger of the public service corporations becoming despots as some people seem to fear, provided they are put under proper restraints, but society cannot afford to make restraints which of themselves are unnecessary or unfair. These corporations serve a beneficial end in our life, and their rights are as well founded and should be as well secured and held sacred as the rights of any citizens who are individually or collectively bent on any proper business pursuits.

Some people seem to believe that all public service corporation men are either wicked or are liars or thieves. This has as little foundation in fact as a belief that all men in Spain carry mandolins or that Spanish women always wear mantillas. If such unjust, superficial and improper opinions are to have influence in this nation, then only misfortune and woe can be the outcome. It is necessary for all men trained in straight thinking to combat such folly and to cry out for fair dealing, one with the other, as between the public service corporations and the public which they are established to serve. No engineer does his duty who does not stand with fidelity for equally square treatment FOR as BY these corporations. These corporations are not here as vampires on society, but are here to serve the needs of the people in a reasonable and business-like way; and their proper objects cannot be accomplished unless thay are treated with reason and established in confidence. They obtain their income from serving the public, and they cannot give generous service unless they are granted generous opportunities. When under reasonable restraints and supervision, as by properly constituted public commissions, they are more quickly responsive to public sentiment than could reasonably be expected of any publicly owned business organization of equal magnitude which could exist under our political conditions, and their usefulness is proved beyond contradiction. Perhaps no man is more likely to observe these things than one whose professional practice, like that which has come to me, makes him retained adviser in some instances to public service companies and in other instances to governments or municipalities, for he has to study fairness to each class of clients in all he does.

A barrier of distrust which exists between these servitors of the people and the people which they serve is presumably due, on the one hand, to a memory by the public of misdeeds which were perpetrated before recent demands for reform brought about the establishment of adequate public supervision in prominent centers, and to a fear of the repetition of misdeeds where supervision and publicity have not yet been prescribed; and, on the other hand, to a certain reluctance by corporation managers to exhibit full and convincing frankness for fear that such frankness may be made the opportunity by unscrupulous politicians or persons with interested motives to crowd them to the verge of insolvency. These particular conditions of distrust could be obviated by means of the public itself owning the public service properties and operating them in its own interest, but this is a drastic and undesirable alternative. Any fair-minded man of extended business experience who will study with unbiased intention the details of public ownership and public trading in the venerable and stable cities and states of continental Europe must be impressed with the reality that our inexpert and shifting governmental bodies are wholly unadapted to cope with such responsibilities, or to make an economic success equal on the average to that now accomplished by the privately managed service companies, whether the measure of success be taken on the basis of service provided for a unit of payment or on any other reasonable basis of comparison.

If the public could feel sure of the ingenuousness of corporation statements and statistics, and the corporations could be protected from unfair attacks made by ignorant, although, in many instances, educated, persons or persons with ulterior motives, the barrier of distrust to which I have referred would be dissipated as dampness is dissipated by the rays of the sun; but this cure requires a long step forward in the average line of progress, for it demands a supervision of the companies which imposes on them exact and ingenuous bookkeeping associated with the presentation to the public of accurate and luminous statements of their business, and it equally demands that the public shall be required to yield justice to the companies with the same ample fullness as individuals seek it for themselves. A progressive step of this nature is always accomplished slowly and hesitatingly. I have observed in Macaulay's writings a paragraph which is graphic in illustration of our present situation. " Everywhere," he says, " there is a class of men who cling with fondness to whatever is ancient, and who, even when convinced by overpowering reasons that innovation would be beneficial, consent to it with many misgivings and forebodings. We find also everywhere another class of men, sanguine in hope, bold in speculation, always pressing forward, quick to discern the imperfections of whatever exists, disposed to think lightly of the risks and inconveniences which attend improvements, and disposed to give every change credit for being an improvement. In the sentiments of both classes there is something to approve. But of both, the best specimens will be found not far from the common frontier. The extreme section of one class consists of bigoted dotards; the extreme section of the other consists of shallow and reckless empirics."

The public, misled or annoyed by the reluctance of some honest but overcautious managements to make frank public statements of financial results and present convincing statistics of operation, enraged by the acts of a few adventurers who from time to time have secured a speculative hold in the public service field, and enticed by the arguments of individuals with ulterior motives, are likely to follow the radical leadership of demagogues or of honest but false empirics. This is a danger which seriously exists in states where no public supervision of the service companies is provided, and also in a lesser degree in states where such supervision has been established. The danger must be rolled back by the exertions of fair-minded and right-thinking men. A serious menace to the welfare of the nation would be caused if unfair dealing toward the public service companies were established as a policy. A scrupulously frank and honest dealing with the public by the companies should be insisted on, but the public must be taught the importance of dealing on its part with an equally scrupulous fairness and a well-balanced generosity. It is here that I say lies a duty of electrical engineers to the public. It is to give of their time and brain to convincingly establish the facts (the *facts*, I repeat) which the public do not understand in regard to the business of the public service companies, to indicate the means for rightly treating these new influences which we and our fellow engineers have been creating by our works, and to aid in establishing measures which will favor and sustain mutual confidence and fair dealing between them and the public. This is an obscure and difficult problem on account of its touching the edge of men's ambitions and men's passions, and it seems at times to possess the opacity and insolubility of a mill-stone; but looking persistently and with care into what appears to be a mill-stone, not infrequently proves it to be composed of reasonably transparent material. The members of our Institute should take somewhat to themselves as professional men this obscure and difficult problem, and aid in its solution as a matter of their duty to the public.

AIEE Adopts Code of Ethics in 1912

TRADITIONALLY, a formal code of ethics that provides guidelines for relations with the public, clients, and professional peers has been part of the process of professionalization. The AIEE was the first of the four so-called founder engineering societies in the U.S. to adopt a formal code. As the brief historical appendix indicates, the impetus for a code came from two early AIEE Presidents, Schuyler S. Wheeler and Dugald C. Jackson. (See Paper 13.) According to Edwin T. Layton, author of a recent monograph on the history of social responsibility and the American engineering profession, the AIEE Code served mainly the ceremonial function of enhancing the status of electrical engineers and "no serious efforts were made to enforce it." Layton contends that the Code failed to take into account that most engineers were employees of large industrial corporations rather than independent consultants. He notes that the provisions of the Code that restricted criticism of fellow engineers or discussion of technical matters in the public media actually made it more difficult for an individual engineer "to take the side of the public in matters involving engineering." See Edwin T. Layton, Jr., *The Revolt of the Engineers: Social Responsibility and the American Engineering Profession* (Cleveland, Ohio, 1971), pp. 84–85 and passim. See also Ellis Rubinstein, "IEEE and the Founder Societies," *IEEE Spectrum*, pp. 76–84, May 1976.

CODE OF PRINCIPLES OF PROFESSIONAL CONDUCT

OF THE

AMERICAN INSTITUTE OF ELECTRICAL ENGINEERS

ADOPTED BY THE BOARD OF DIRECTORS, March 8, 1912.

A. General Principles.
B. The Engineer's Relations to Client or Employer.
C. Ownership of Engineering Records and Data.
D. The Engineer's Relations to the Public.
E. The Engineer's Relations to the Engineering Fraternity.
F. Amendments.

While the following principles express, generally, the engineer's relations to client, employer, the public, and the engineering fraternity, it is not presumed that they define all of the engineer's duties and obligations.

A. GENERAL PRINCIPLES

1. In all of his relations the engineer should be guided by the highest principles of honor.

2. It is the duty of the engineer to satisfy himself to the best of his ability that the enterprises with which he becomes identified are of legitimate character. If after becoming associated with an enterprise he finds it to be of questionable character, he should sever his connection with it as soon as practicable.

B. THE ENGINEER'S RELATIONS TO CLIENT OR EMPLOYER

3. The engineer should consider the protection of a client's or employer's interests his first professional obligation, and therefore should avoid every act contrary to this duty. If any other considerations, such as professional obligations or restrictions, interfere with his meeting the legitimate expectation of a client or employer, the engineer should inform him of the situation.

4. An engineer can not honorably accept compensation, financial or otherwise, from more than one interested party, without the consent of all parties. The engineer, whether consulting, designing installing or operating, must not accept commissions, directly or indirectly, from parties dealing with his client or employer.

5. An engineer called upon to decide on the use of inventions, apparatus, or anything in which he has a financial interest, should make his status in the matter clearly understood before engagement.

6. An engineer in independent practise may be employed by more than one party, when the interests of the several parties do not conflict; and it should be understood that he is not expected to devote his entire time to the work of one, but is free to carry out other engagements. A consulting

Reprinted from *Trans. AIEE*, vol. XXXI, part 2, pp. 2227–2230, 1912.

engineer permanently retained by a party, should notify others of this affiliation before entering into relations with them, if, in his opinion, the interests might conflict.

7. An engineer should consider it his duty to make every effort to remedy dangerous defects in apparatus or structures or dangerous conditions of operation, and should bring these to the attention of his client or employer.

C. Ownership Of Engineering Records And Data

8. It is desirable that an engineer undertaking for others work in connection with which he may make improvements, inventions, plans, designs, or other records, should enter into an agreement regarding their ownership.

9. If an engineer uses information which is not common knowledge or public property, but which he obtains from a client or employer, the results in the form of plans, designs, or other records, should not be regarded as his property, but the property of his client or employer.

10. If an engineer uses only his own knowledge, or information which by prior publication, or otherwise, is public property and obtains no engineering data from a client or employer, except performance specifications or routine information; then in the absence of an agreement to the contrary the results in the form of inventions, plans, designs, or other records, should be regarded as the property of the engineer, and the client or employer should be entitled to their use only in the case for which the engineer was retained.

11. All work and results accomplished by the engineer in the form of inventions, plans, designs, or other records, that are outside of the field of engineering for which a client or employer has retained him, should be regarded as the engineer's property unless there is an agreement to the contrary.

12. When an engineer or manufacturer builds apparatus from designs supplied to him by a customer, the designs remain the property of the customer and should not be duplicated by the engineer or manufacturer for others without express permission. When the engineer or manufacturer and a customer jointly work out designs and plans or develop inventions a clear understanding should be reached before the beginning of the work regarding the respective rights of ownership in any inventions, designs, or matters of similar character, that may result.

13. Any engineering data or information which an engineer obtains from his client or employer, or which he creates as a result of such information, must be considered confidential by the engineer; and while he is justified in using such data or information in his own practise as forming part of his professional experience, its publication without express permission is improper.

14. Designs, data, records and notes made by an employee and referring exclusively to his employer's work, should be regarded as his employer's property.

15. A customer, in buying apparatus, does not acquire any right in its design but only the use of the apparatus purchased. A client does not

acquire any right to the plans made by a consulting engineer except for the specific case for which they were made.

D. The Engineer's Relations To The Public

16. The engineer should endeavor to assist the public to a fair and correct general understanding of engineering matters, to extend the general knowledge of engineering, and to discourage the appearance of untrue, unfair or exaggerated statements on engineering subjects in the press or elsewhere, especially if these statements may lead to, or are made for the purpose of, inducing the public to participate in unworthy enterprises.

17. Technical discussions and criticisms of engineering subjects should not be conducted in the public press, but before engineering societies, or in the technical press.

18. It is desirable that first publication concerning inventions or other engineering advances should not be made through the public press, but before engineering societies or through technical publications.

19. It is unprofessional to give an opinion on a subject without being fully informed as to all the facts relating thereto and as to the purposes for which the information is asked. The opinion should contain a full statement of the conditions under which it applies.

E. The Engineer's Relations To The Engineering Fraternity

20. The engineer should take an interest in and assist his fellow engineers by exchange of general information and experience, by instruction and similar aid, through the engineering societies or by other means. He should endeavor to protect all reputable engineers from misrepresentation.

21. The engineer should take care that credit for engineering work is attributed to those who, so far as his knowledge of the matter goes, are the real authors of such work.

22. An engineer in responsible charge of work should not permit non-technical persons to overrule his engineering judgments on purely engineering grounds.

F. Amendments

Additions to, or modifications in, this Code may be made by the Board of Directors under the procedure applying to a by-law.

HISTORY OF THE CODE

At the Milwaukee Convention in May, 1906, Dr. Schuyler Skaats Wheeler delivered his presidential address on "Engineering Honor." It was the sense of the Convention that the ideas contained in this address should be embodied in a Code of Ethics for the electrical engineering profession, and to this end the following committee was appointed in October, 1906:

Schuyler Skaats Wheeler, *Chairman.*

H. W. Buck — Charles P. Steinmetz

In May, 1907, the committee reported a code to the President and Board of Directors for discussion at the June Convention at Niagara Falls. It was discussed and adopted by the Convention but later the adoption had to be set aside on account of the provisions of the Constitution prohibiting

Conventions from acting upon questions affecting the Institute's organization or policy.

It was taken up by the Board of Directors on August 30, 1907, revised, printed and submitted to the membership for suggestions to be sent to a new committee appointed by President Stott.

It lay dormant until June, 1911, when, in accordance with a resolution of the Board of Directors, President Jackson appointed a committee. The personnel of this committee, as reappointed by President Dunn in August, 1911, is as follows:

GEORGE F. SEVER, *Chairman.*

H. W. BUCK	CHARLES P. STEINMETZ
SAMUEL REBER	HENRY G. STOTT

SCHUYLER SKAATS WHEELER

This committee's work was presented in a report to the Board of Directors on February 9, 1912, when the code was tentatively adopted. After a month's careful analysis and consideration of numerous suggestions from the advisory members of the committee and others, the completed code was adopted at the meeting of the Board of Directors on March 8, 1912.

At the meeting of February 9, the title of the committee and of the code was changed from that of Code of Ethics to Code of Principles of Professional Conduct.

The committee was assisted by eighteen advisory members appointed by the President. Their names are appended.

WILLIAM S. BARSTOW	HENRY H. NORRIS
LOUIS BELL	RALPH W. POPE
JOHN J. CARTY	HARRIS J. RYAN
FRANCIS B. CROCKER	CHARLES F. SCOTT
DUGALD C. JACKSON	SAMUEL SHELDON
A. E. KENNELLY	WILLIAM STANLEY
JOHN W. LIEB, JR.	LEWIS B. STILLWELL
C. O. MAILLOUX	ELIHU THOMSON
RALPH D. MERSHON	W. D. WEAVER

Origins of the Institute of Radio Engineers

THE Institute of Radio Engineers was founded through a merger of two older societies of wireless enthusiasts in 1912 and enjoyed a spectacular success during its fifty-year existence. When the Institute merged with the older AIEE to form the IEEE in 1963, its membership of more than 100,000 was more than twice that of the AIEE. This paper includes a discussion of the prehistory of the IRE and some of the factors that led to its creation. As Whittemore points out, the list of early members "constitutes a veritable Who's Who in the early history of radio." A part of the appeal of the IRE was that its election process was more democratic than that of other American engineering societies and it was unique in its stress on scientific professionalism. For an informed discussion of the unusual emphasis of the IRE on science and professionalism, see Edwin Layton's paper in the special issue of the PROCEEDINGS OF THE IEEE for September 1976. Also see Layton's *Revolt of the Engineers: Social Responsibility and the American Engineering Profession* (Cleveland, Ohio, 1971), pp. 42–44 and passim.

Laurens E. Whittemore (1892–   ) was born in Topeka, Kans. and graduated from Washburn College in 1914. He received an M.A. degree in physics from the University of Kansas in 1915. After several years with the National Bureau of Standards, Whittemore served on the Radio Advisory Committee of the U.S. Department of Commerce before joining the Headquarters Staff of the American Telephone and Telegraph Company in 1925. He was active professionally in the IRE and was its Vice President in 1928. He retired from the Bell Company in 1957. See a biographical note in the PROCEEDINGS OF THE IRE, vol. 50, p. 1447, 1962.

*The Institute of Radio Engineers —Fifty Years of Service**

LAURENS E. WHITTEMORE†

FELLOW, IRE

A PREDICTION

(From an IRE information booklet, dated January 1, 1913, at which time the IRE had 109 members.)

The form of government of the Institute is thoroughly democratic and each member is given full opportunity to participate in all the advantages and privileges of the Institute. It is confidently expected that the already considerable membership will grow to the point where the greatest possible benefits will be extended to the greatest possible number of those interested in the development of radio-transmission.

INTRODUCTION

FOR THE INFORMATION of the members of the IRE, and especially those who are included in the one-third of the members who have joined during the past five years, there are here presented an outline of the beginnings of the Institute of Radio Engineers and some significant facts as to its organization, aims and recent accomplishments. Some statistics are included to show the growth of the Institute and the expansion of its activities over the 50 years of its existence. A perusal of these statistics shows how rapidly the Institute's activities are moving forward geographically, technically, professionally and in education.

This paper is in some respects a revision, and in some respects a repetition, of portions of the review of the first 45 years of IRE's service to its members, prepared under the sponsorship of the History Committee of the Institute and published in 1957.

Readers who are interested in more details of events, organization changes, etc., during the first 45 years of existence of the Institute, may refer to that paper which appeared in PROCEEDINGS OF THE IRE, May, 1957, pp. 597-635. As an appendix to both papers there is given a list of previous publications of significance relating to IRE history.

FORMATION OF THE INSTITUTE OF RADIO ENGINEERS

The Beginning

Radio people, apparently from the outset, have been characterized by a combination of two desires—1) to talk with one another about their accomplishments and their hopes, and 2) to write and read about the things which they and their

*Received by the IRE, November 17, 1961.

† Short Hills, New Jersey.

Reprinted from *Proc. IRE*, vol. 50, pp. 534–540 (extract of a longer paper), May 1962.

technical "brothers" have been doing. The desire to present papers about "wireless" and to hear and discuss them resulted in the establishment of two organizations, one in Boston in 1907, and one in New York in 1908, whose members became the nucleus of the Institute of Radio Engineers in 1912.

Society of Wireless Telegraph Engineers

The Society of Wireless Telegraph Engineers (SWTE) was formed in Boston, Mass., on February 25, 1907, by John Stone Stone as an outgrowth of seminars held by engineers on the staff of the Stone Wireless Telegraph Company. Membership was eventually opened to men from Fessenden's National Electric Signaling Company and some other organizations. Members of this society were familiarly known as "swatties." John Stone Stone was the first President of this society.

The Wireless Institute

Robert H. Marriott made what appears to have been the first specific attempt to form a radio engineering society composed of members from any and all companies. On May 14, 1908, he sent a circular letter to some two hundred persons interested in wireless asking their opinions regarding the formation of such a society. On January 23, 1909, a temporary organization was formed to draw up a constitution. The name of the new society was "The Wireless Institute."

Robert Marriott was elected first President of The Wireless Institute, and served in that capacity during the three years of its existence.

By 1911 the Stone Wireless Telegraph Company had gone out of existence and the National Electric Signaling Company had moved to Brooklyn so there was very little left of the SWTE. The Wireless Institute was also struggling to hold its membership and to keep out of debt.

Institute of Radio Engineers

It was early in 1912 that Robert H. Marriott and Alfred N. Goldsmith, representing The Wireless Institute, with John V. L. Hogan, who was very active in the Society of Wireless Telegraph Engineers, held an informal meeting to discuss the plights of both societies. Out of their discussions there developed a meeting on the night of May 13, 1912, at which members of TWI and SWTE gathered in Room 304 of Fayerweather Hall at Columbia University in New York City. A constitution was approved and an election of officers was held at which the following were chosen: Robert H. Marriott, President; Fritz Lowenstein, Vice-President; E. D. Forbes, Treasurer; E. J. Simon, Secretary; Alfred N. Goldsmith, Editor; and Lloyd Espenschied, Frank Fay, J. H. Hammond, Jr., John V. L. Hogan, and John Stone Stone, Managers.

As a name for the organization "The Institute of Radio Engineers" was chosen. The original membership roster of the IRE consisted of 46 members, 22 from SWTE and 25 from TWI. One member, Greenleaf W. Pickard, was the only charter member of IRE who had been affiliated with both of the preceding organizations.

Original Members of The Institute of Radio Engineers and their Affiliation with Parent Societies

Society of Wireless Telegraph Engineers

J. C. Armor	W. S. Hogg
Sewall Cabot	Guy Hill
W. E. Chadbourne	F. A. Knowlton
G. H. Clark	W. S. Kroger
T. E. Clark	Fritz Lowenstein
E. R. Cram	Walter W. Massie
G. S. Davis	G. W. Pickard
Lee deForest	Samuel Reber
E. D. Forbes	Oscar C. Roos
V. F. Greaves	J. S. Stone
J. V. L. Hogan, Jr.	*A. F. VanDyck

The Wireless Institute

William F. Bissing	Frank Hinners
A. B. Cole	James M. Hoffman
*P. B. Collison	Robert H. Marriott
James N. Dages	A. F. Parkhurst
*Lloyd Espenschied	G. W. Pickard
Philip Farnsworth	H. S. Price
Frank Fay	A. Rau
Edward G. Gage	Harry Shoemaker
*Alfred N. Goldsmith	*Emil J. Simon
Francis A. Hart	A. Kellogg Sloan
Robert L. Hatfield	C. H. Sphar
Arthur A. Herbert	Floyd Vanderpoel

R. A. Weagent

* Member of the IRE as of January, 1962.

After about a year, it was decided to incorporate the society. A meeting to decide details of this move was held at Sweet's Restaurant on Fulton Street in downtown New York on June 23, 1913. Those members with a more legal mind in the make-up of the Institute, drew up Articles of Incorporation, and on August 23, 1913, the organization was incorporated under the laws of the State of New York.

In brief, the expressed aims of the new association were:

"To advance the art and science of radio transmission, to publish works of literature, science and art for such purpose, to do all and every act necessary, suitable and proper for the accomplishment of any of the purposes or the attainment of any of the powers herein set forth, either alone or in association with other corporations, firms or individuals to do every act or acts, thing or things, incidental or appurtenant to or growing out of or connected with the aforesaid science or art, or power or any parts thereof, provided the same be not inconsistent with the laws under which this corporation is organized, or prohibited by the State of New York."

Growth of Membership of the Institute of Radio Engineers and its Predecessors, 1907–1914

	SWTE	TWI	IRE
February 25, 1907	11		
January 1, 1908	17		
January 1, 1909	27		
March 10, 1909		14	
January 1, 1910	36	81	
January 1, 1911	36	99	
January 1, 1912	43	27	
May 13, 1912	(22)	(25)	46
January 1, 1913			109
January 1, 1914			231

One of the most important functions of the Institute was to preserve its technical papers, and the remarks made regarding them, in published form. One of the early decisions of the Institute, therefore, was to publish a technical magazine which was named The Proceedings of the Institute of Radio Engineers. The first issue was dated January, 1913.

By the end of 1912, the Institute's membership had risen to 109 and during the succeeding year it more than doubled. The rapid increase in membership after consolidation, compared with the slow rate of growth of SWTE and TWI, bore out the wisdom of the founders who suggested the merger.

The original ledger book of the Institute, in which the names and dues payments of early members were recorded, constitutes a veritable *Who's Who* in the early history of radio. The names of many radio pioneers can be seen in the accompanying illustration showing the first few pages of the list of these members who joined the Institute during the first year.

Name of the Institute

In considering a name for the new organization the founders felt that something should be preserved from the names of both of the two component societies. The word "Institute" was borrowed from The Wireless Institute, and "Engineers" from the Society of Wireless Telegraph Engineers. Because the word "radio" was gradually supplanting "wireless," the title "The Institute of Radio Engineers" suggested itself. There was considerable temptation to add "American," particularly since TWI and IRE were modeled after the American Institute of Electrical Engineers in certain other respects. However, the temptation was resisted because it was expected that the IRE, as the only radio engineering society in existence, would be international in scope, an expectation that was promptly realized.

The Emblem or Symbol of the IRE

Neither of the emblems of the predecessor societies seemed readily adaptable to the new IRE. The SWTE emblem pictured a simple form of spark oscillator. The membership badge of TWI showed a spark gap functioning in the center of a dipole surrounded by a circular resonator provided with a micrometer gap for reception.

The founders of IRE decided not to use a representation of any specific form of equipment or physical structure but to devise a more general and perhaps perpetual symbol. It was realized that the Institute would always deal with electromagnetic energy, guided by conductors or passing through space, and that the distinguishing character of the transmission process was the existence of electrical forces and of their correlative magnetic forces. A representation of these forces

No.	Name
0001	Armor, J.C.
0002	Cabot, Sewall
0003	Chadbourne, W.E.
0004	Cram, Ernest R.
0005	Davis, Geo. S.
0006	DeForest, Lee, Ph.D.
0007	Forbes, E.D.
0008	Greares, V. Ford.
0009	Hill, Guy.
0010	Hogan, John L. Jr.
0011	Hogg, W.S. (Comm.)
0012	Knowlton, F.A.
0013	Kroger, F.H.
0014	Lowenstein, Fritz
0015	Massie, Walter W.
0016	Pickard, Greenleaf W.
0017	Reber, Samuel (Col)
0018	Roos, Oscar C.
0019	Stone, John Stone
0020	Sundberg E.W.
0021	Van Dyck, A.F.
0022	Bissing, Wm. F.
0023	Cole, A.B.
0024	Collison, P.B.
0025	Dages, Jas. N.
0026	Espenschied.
0027	Farnsworth, Philip.
0028	Fay, Frank.
0029	Gage, Edward G.
0030	Goldsmith, Alfred N.
0031	Hart, Francis A.
0032	Hatfield, Robert L.
0033	Hebert, Arthur A.
0034	Hinners, Frank.
0035	Hoffman, Jos. M.
0036	Marriott, Rob. H.
0037	Parkhurst, A.F.
0038	Price, H.S.
0039	Rau, Adolph.
0040	Shoemaker, Harry
0041	Simon, Emil J.
0042	Sloan, A Kellogg
0043	Sphar, Clark H.
0044	Vanderpoel, Floyd
0045	Weagant, Roy A.
0046	Brackett, Quincy A.
0047	Brill, O.C.
0048	Browne, A.P.
0049	Campbell, J.H.
0050	Clark, Geo H.
0051	Clark, Thos. E.
0052	Cohen, Louis.
0053	Terrill, W.D.
0054	Cowan, A.S. (Capt.)
0055	Irwin. Comm. N.E.
0056	Kolster. C.C.
0057	Kolster, F.A.
0058	Rawles, R.C.
0059	Thompson, Roy. E.
0060	Pegram, Geo B, Ph.D.
0061	Davis, F.C.
0062	Hallborg, H.E.
0063	Hammond, John H. Jr.
0064	Hodson, J.E.
0065	Langley, R.H.
0066	Lesh, Laurence.
0067	LeQuesne, Chas. A. Jr.
0068	Liebmann, M.N.
0069	Liebowitz, Benj.
0070	Meissner, Benj. F.
0071	Silverman J.A.
0072	Richards, Thos. S.
0073	Zeamans, Harold K.
0074	Benning, B.S.
0075	Bowen, Chas. F.
0076	Burnside, Don. G.
0077	Calvert, R. Neil,
0078	Campbell, J.E.
0079	Collins, Chas. H. Jr.
0080	Curtis, Austin M.
0081	Donle, Harold. P.
0082	Flenschneider, J.B.
0083	Engler, John.
0084	Gawler, H.C.
0085	Hale. W.H.
0086	Hanscom, W.W.
0087	Hensgen, W.O.
0088	Heatherington, W.H. Jr.
0089	Hoppough, C.I.
0090	Hobley, W.F.
0091	Jones, Jos. S.
0092	Kelly, C Merrill, Jr.
0093	Koehl, Jas. C.
0094	Leary, John J.
0095	Lewis, Geo H.
0096	Lindridge, C.D.
0097	Moore, H. Atherton.
0098	Pacent, L.G.
0099	Ryan, Fred. C.
0100	Shermerhorn, J.L.
0101	Secor, H.W.
0102	Seelig, Alfred E.
0103	Stevens, A.M.
0104	Stewart, Donald.
0105	Zwicker, Achilly C.
0106	Moore. E.B.
0107	Price, D.R.
0108	Ballou, H.Y.
0109	Kohn, Alfred. S.
0110	Israel, Lester.
0111	Waterman, Frank.
0112	Sarnoff. David.
0113	Kennelly, Arthur E, Ph.D.
0114	Page, Newell C.
0115	Austin, Louis W. Ph.D.
0116	Behr, F.J. (Capt.)
0117	Cadmus, Richard G.
0118	
0119	Duncan, R.D.
0120	Eastham, Melville.
0121	Pruden, Fred. H.
0122	Spangenberg, Lester
0123	Walton, Capt. John. Q.
0124	Wood, A.A.
0125	Barth, Julian.
0126	Ford, Ried G.
0127	Laurent, J.D.
0128	McDowell, C.S. (Lieut.)
0129	Trapnell, Thos T.
0130	Priess, Wm H.
0131	Proctor, J.A.
0132	Weinberger, Julius.
0133	Woodworth, F.B. (Lieut.)
0134	Alexanderson, E.F.W.
0135	Kahant, Chas. G.
0136	Marshall, Cloyd.
0137	Montcalm, S.R.
0138	Packman, M.E.
0139	Wright, Geo. B.
0140	Apgar, Chas. E.

Pages from the first IRE Record book listing the earliest members.

IRE OFFICERS, 1912–1962

Year	President	Vice President	Secretary	Treasurer	Editor	Hdqs. Manager
1912	R. H. Marriott	Fritz Lowenstein	E. J. Simon	E. D. Forbes	A. N. Goldsmith	
1913	G. W. Pickard	R. H. Marriott	"	J. H. Hammond, Jr.	"	
1914	L. W. Austin	J. S. Stone	"	"	"	
1915	J. S. Stone	G. W. Pierce	David Sarnoff	W. F. Hubley	"	
1916	A. E. Kennelly	J. V. L. Hogan	"	"	"	
1917	M. I. Pupin	"	"	L. R. Krumm	"	
1918	G. W. Pierce	"	A. N. Goldsmith	Warren F. Hubley	"	
1919	"	"	"	"	"	
1920	J. V. L. Hogan	E. F. W. Alexanderson	"	"	"	
1921	E. F. W. Alexanderson	Fulton Cutting	"	"	"	
1922	Fulton Cutting	E. L. Chaffee	"	"	"	
1923	Irving Langmuir	J. H. Morecroft	"	"	"	
1924	J. H. Morecroft	J. H. Dellinger	"	"	"	
1925	J. H. Dellinger	Donald McNicol	"	"	"	
1926	Donald McNicol	Ralph Bown	"	"	"	
1927	Ralph Bown	Frank Conrad	"	"	"	J. M. Clayton
1928	A. N. Goldsmith	L. E. Whittemore	J. M. Clayton	Melville Eastham	"	"
1929	A. H. Taylor	Alexander Meissner	"	"	W. G. Cady	"
1930	Lee de Forest	A. G. Lee	H. P. Westman	"	A. N. Goldsmith	H. P. Westman
1931	R. H. Manson	C. P. Edwards	"	"	"	"
1932	W. G. Cady	E. V. Appleton	"	"	"	"
1933	L. M. Hull	Jonathan Zenneck	"	"	"	"
1934	C. M. Jansky, Jr.	B. van der Pol, Jr.	"	"	"	"
1935	Stuart Ballantine	G. H. Barkhausen	"	"	"	"
1936	L. A. Hazeltine	Valdemar Poulsen	"	"	"	"
1937	H. H. Beverage	P. P. Eckersley	"	"	"	"
1938	Haraden Pratt	E. T. Fisk	"	"	"	"
1939	R. A. Heising	P. O. Pederson	"	"	"	"
1940	L. C. F. Horle	F. E. Terman	"	"	"	"
1941	F. E. Terman	A. T. Cosentino	"	Haraden Pratt	"	" (Jan.–Oct.) J. D. Crawford (Nov.–Dec.)
1942	A. F. Van Dyck	W. A. Rush	"	"	"	J. D. Crawford (Jan.–Mar.) L. B. Keim (Apr.–May) W. B. Cowilich (Oct.–Dec.)
1943	L. P. Wheeler	F. S. Barton	Haraden Pratt	R. A. Heising	"	W. B. Cowilich
1944	H. M. Turner	R. A. Hackbusch	"	"	"	"
1945	W. L. Everitt	H. F. van der Bijl	"	"	"	G. W. Bailey
1946	F. B. Llewellyn	E. M. Deloraine	"	W. C. White	"	"
1947	W. R. G. Baker	Noel Ashbridge	"	R. F. Guy	"	"
1948	B. E. Shackelford	R. L. Smith-Rose	"	S. L. Bailey	"	"
1949	S. L. Bailey	A. S. McDonald	"	D. B. Sinclair	"	"
1950	R. F. Guy	R. A. Watson-Watt	"	"	"	"
1951	I. S. Coggeshall	Jorgen Rybner	"	W. R. G. Baker	"	"
1952	D. B. Sinclair	H. L. Kirke	"	"	"	"
1953	J. W. McRae	S. R. Kantebet	"	"	"	"
1954	W. R. Hewlett	M. J. H. Ponte	"	"	J. R. Pierce	"
1955	J. D. Ryder	Franz Tank	"	"	"	"
1956	A. V. Loughren	Herre Rinia	"	"	D. G. Fink	"
1957	J. T. Henderson	Yasujiro Niwa	"	"	"	"
1958	D. G. Fink	C. E. Granquist	"	"	J. D. Ryder	"
1959	Ernst Weber	D. B. Sinclair	"	"	"	"
1960	R. L. McFarlan	J. N. Dyer	"	"	F. Hamburger, Jr.	"
1961	L. V. Berkner	J. A. Ratcliffe J. F. Byrne F. Ollendorff	"	S. L. Bailey	"	"
1962	P. E. Haggerty	A. M. Angot T. A. Hunter Ernest Weber	"	"	T. F. Jones, Jr.	"

was adopted as part of the symbol; the electrical force being represented by a vertical arrow and the magnetic force by a circular arrow surrounding the electrical line and in the conventional relationship to it. The shape of the resulting drawing lent itself to a triangular placement of the letters, I, R, and E. This, in turn, led to the selection of a triangular emblem. Incidentally, the letters I, R, and E also symbolize the fundamental quantities, current, resistance and electromotive force, as well as the name, Institute of Radio Engineers.

Officers

The officers of the IRE from the beginning have been President, one or more Vice-Presidents, Secretary, Treasurer, Editor, and Directors with the infrequent addition of an Assistant Secretary or an Assistant Treasurer. The President and Vice-President(s) have always been elected by the IRE membership as have part of or all the members of the Board of Directors. Beginning in 1915, the elected members of the Board were authorized by the Constitution to choose several additional persons to complete the Board membership. The Secretary and Treasurer have for many years been elected by the Board.

An accompanying table shows, for each year since the formation of the IRE, the names of the Officers who served during that year.

Beginning in 1930, it became the custom for the Vice-President of the IRE to be a member who resides in a country other than the United States. In 1957 neither the President nor the Vice-President was a resident of the United States. In 1960, the IRE established two Vice-Presidents. Initially, one resided in North America and the other resided outside of North America. In 1962, the former was replaced by a Vice-President, elected by the voting members, whose function is to assist the President. Also a new Vice-President, elected by the Annual Assembly, was created to serve as a coordinator between the Professional Groups and the Executive Committee. The office of the Vice-President residing outside North America will terminate at the end of 1962.

As the IRE membership increased in numbers and geographical distribution, there developed an appreciation by the members of the Board that some specific measures should be adopted, possibly of an organizational nature, to make more certain that the contacts between the Board and the IRE membership would always be close and continuous and that the Board would comprise a truly democratic representation of the IRE membership. After several years of consideration of this problem, the Board of Directors recommended an amendment to the Constitution which was adopted by the Institute membership in 1947, providing for Regional Directors, selected specifically to represent designated regions of the United States and Canada from which they came and whose memberships had elected them.

Membership

Growth in Numbers

It is fitting to consider as the "Charter Members" of the IRE those members of the two parent societies who became the first members of the IRE when it was organized on May 13, 1912. The formal charter of the Institute of Radio Engineers, however, was granted on August 23, 1913.

The 1914 Year Book gives an analysis of the geographical distribution of the 271 Members and Associate Members of the IRE as of March 1, 1914, showing that there were members in eight countries other than the United States. The 1916 Year Book shows that the membership of the Institute immediately began to increase and by January 1, 1916, was only slightly under 1000.

The Constitution, as adopted in May, 1912, provided that the names of applicants for membership in the IRE should be sent out to each member of the Institute, prior to their acceptance as members. The members who were elected were required to subscribe personally to the Constitution of the IRE.

From the beginning, membership was open not only to radio engineers (Member grade) but also to those who had a real interest in radio engineering even though they were not professionally engaged in this field (Associate Member grade).

In 1914 the grade of Junior Member was established for persons under 21 years of age.

One of the outstanding aspects of IRE membership is the substantial number and generally increasing proportion of its membership living in countries outside of the United States. The IRE has taken special steps from time to time to recognize outstanding members living in other countries and to stimulate membership in such areas. At the end of 1961 the IRE had members in 90 countries outside of the United States.

Significance of Membership Grades

In the IRE, as is customary in professional societies, the several grades of membership are intended as a basis for giving recognition to the experience and achievements of the members in radio engineering and the related technical fields.

From the beginning, the Fellow grade has been intended to represent high attainment.

Beginning with the year 1940, all entries to the Fellow grade have been by invitation rather than upon application. The custom has been established of presenting the Fellow awards at meetings of the local Sections of which the recipients are members, and 78 persons were being honored in this way in the early part of 1962. About one per cent of the IRE's total membership of over 96,000 are of the Fellow grade.

The Senior Member grade was established in 1943 as a means of providing a higher grade than the Member grade into which members of the IRE might advance on the basis of their experience and training. This enabled the Institute to keep the Fellow grade as a special recognition.

Recognizing the desirability of encouraging engineering students to become affiliated with the Institute of Radio Engineers, a Student grade of membership in the IRE was established by constitutional amendment effective in 1932.

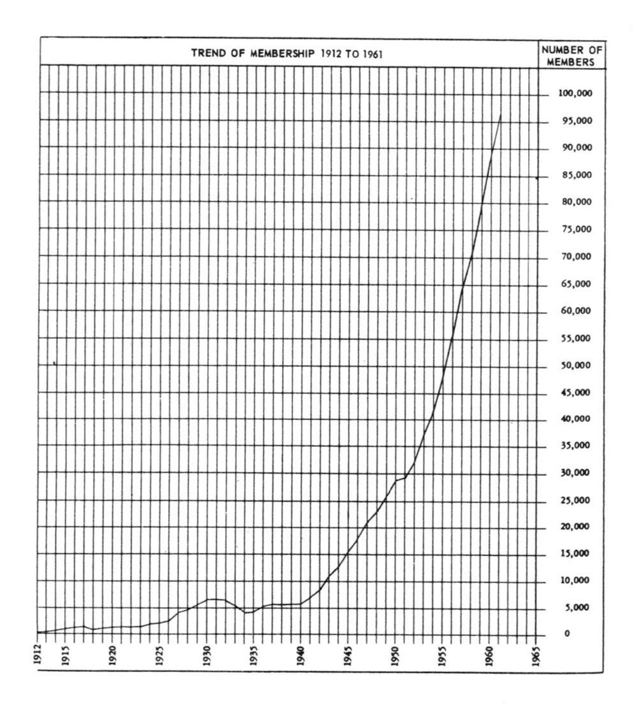

MEMBERSHIP GROWTH IN U.S.A., CANADA AND ABROAD

Year	U.S. and Possessions	Canada	Abroad	Total
May 13, 1912	46			46
Dec. 31, 1927	3550	184	476	4210
" 1936	3975	178	1042	5195
" 1946	15,898	978	1278	18,154
" 1956	51,551	2085	1858	55,494
" 1961	88,956	3758	3837	96,551

MEMBERSHIP GROWTH BY GRADES

Date	Junior	Student	Associate Non-Voting	Associate Voting	Member	Senior Member	Fellow	Total
May 13, 1912				46				46
Dec. 31, 1936	34	299		4092	637		133	5195
Dec. 31, 1946		2252	9890	1701	2330	1763	218	18,154
Dec. 31, 1956		10,384	18,491	388	19,110	6486	635	55,494
Dec. 31, 1961		19,167	13,566		52,284	10,570	964	96,551

◆◆◆ ◆◆◆ ◆◆◆

Section II-B
Electric Light and Power Systems

Designs For The Two-Phase Generators at Niagara.

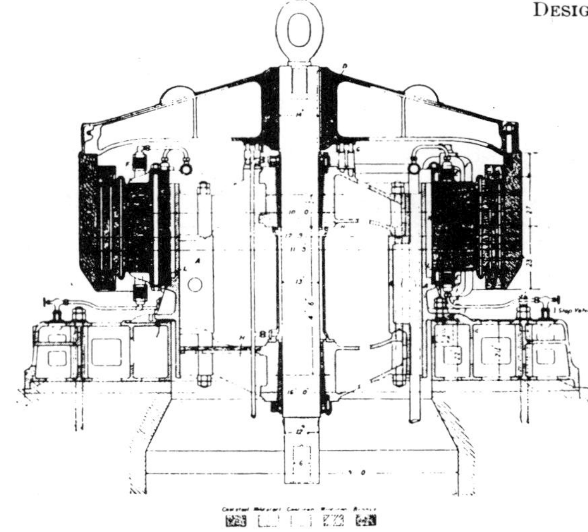

Forbes Design.

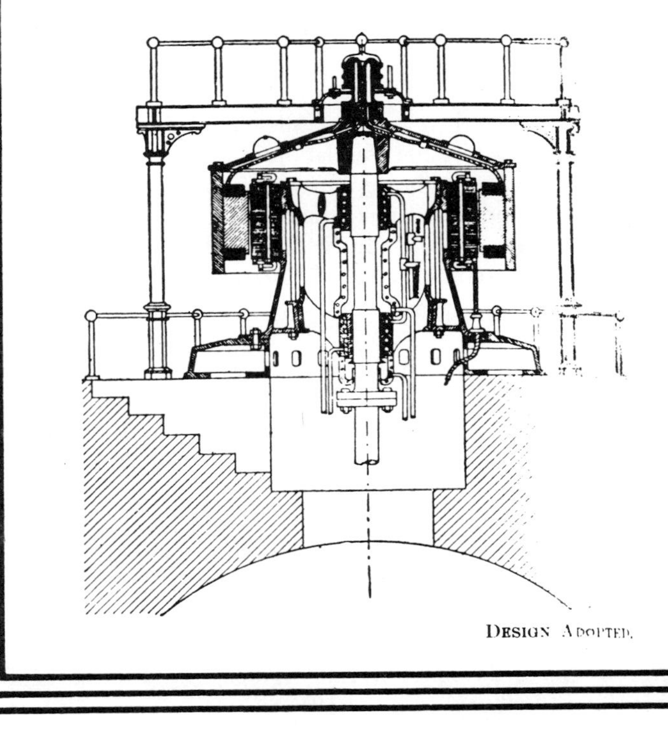

Design Adopted.

Chesney's Assessment of Six E. E. Pioneers

C. C. CHESNEY contributed this reflective essay to a special issue of *Electrical Engineering* that commemorated the 50th anniversary of the founding of the AIEE. (The issue is a useful source document in electrical history. It contains biographical information on many engineers, as well as reminiscences.) Chesney selected six men who had made outstanding contributions to alternating-current technology during the late 19th century. As a pioneer innovator in the same field, he was well qualified to evaluate the work of his distinguished peers. For biographical notes on Thomson, Tesla, and Steinmetz, see Papers 17, 19, and 20, respectively. For a perceptive analysis of the "battle of the systems" (ac versus dc), see a paper by T. S. Reynolds and T. Bernstein in the special issue of the PROCEEDINGS OF THE IEEE of September 1976.

Cummings C. Chesney (1863–1947) was born in Selinsgrove, Pa., and received a degree in chemistry from Pennsylvania State College in 1885. After three years of teaching, he was hired by William Stanley and worked on the design of several of the first alternating-current plants in the U.S. The S.K.C. system for changing phase and frequency of alternating current was named for Stanley, John. F. Kelly, and Chesney. A biographical note on Chesney published in 1901 characterized him as "always a man of action rather than of words, but when the latter prove necessary they are forcible and cogent." When the Stanley Company was sold to General Electric in 1902, Chesney decided to join the G.E. engineering staff and remained until his retirement in 1930. See the *NCAB*, vol. 38, pp. 238–239. Also see "Electrical Engineers of the Day: Cummings C. Chesney," *Electrical World and Engineer*, vol. 38, p. 880, 1901.

William Stanley (1858–1916) was born in Brooklyn, N.Y. and studied at Yale College, but did not graduate. He worked with Hiram Maxim and Edward Weston before establishing his own laboratory in Englewood, N.J. in 1883. During a brief period with the Westinghouse Electric Company, he designed the pioneering alternating-current generating plant at Great Barrington, Mass. before resigning to organize the Stanley Electric Company. His company was later acquired by the General Electric Company in 1905. See the *DAB*, vol. 17, pp. 514–515. Also see L. A. Hawkins, "William E. Stanley, A Gentleman and Genius," *General Electric Review*, vol. 39, pp. 169–170, 1936.

Oliver B. Shallenberger (1860–1898) was born in Rochester, Pa., and attended the Naval Academy where he graduated in 1880. He resigned from the Navy in 1884 to join the Westinghouse Union Switch and Signal Company. He remained with Westinghouse until 1895. Shallenberger organized the Colorado Electric Power Company in 1897 shortly before his untimely death. See Charles A. Terry, "Oliver Blackburn Shallenberger: A Memorial," *Transactions of the AIEE*, vol. 15, pp. 744–748, 1898.

Benjamin G. Lamme (1864–1924) was born on a farm in Ohio and graduated from Ohio State University in 1888. His first job was with the Philadelphia Natural Gas Company owned by George Westinghouse, but he soon transferred to the Westinghouse Electric Company where he spent the rest of his career. He was Chief Engineer at Westinghouse for many years, and played a major role in the design of the first power plant at Niagara Falls. See the *DAB*, vol. 10, pp. 561–562. Also see *Benjamin Garver Lamme: An Autobiography* (New York, 1926) and Benjamin G. Lamme, *Electrical Engineering Papers* (East Pittsburgh, Pa., 1919).

Some Contributions to the Electrical Industry

By C. C. Chesney, President A.I.E.E. 1926–27

Some outstanding individual contributions of the past 50 years to the electrical industry are reviewed here by a past-president who himself has made no small contribution; he says: . . . "as we of the electrical fraternity hope for continued progress, we must remember that our hopes can be fully realized only by remaining true to the greatest of our traditions, 'to produce and to serve.'"

AS WE LOOK BACK into the history of the electrical industry and visualize the past 50 years, we can hope, yea, expect that future accomplishments in the electrical world will be fully as eventful as the unmatched events of the past. Promises that come from home and abroad are filled with predictions of continuous progress

This optimistic sentiment, emanating not from one but from all of the many responsible sources throughout the world, applies not only to the business side, but also to the scientific side of the industry—to the central station business for furnishing light and power, the core of the industry with its investment values already reaching the $10,000,000,000 mark, a value greater than the combined value of the industries of England of the Gladstone period when Michael Faraday made his fundamental discovery of magnetic induction in 1832. It applies also to the possible future accomplishments of the research laboratories, forecasts of which are to be found in the accomplishments already given to the world by these institutions.

These forecasts are full of hope, so far as it is given to fallible man to read the future, and they may well bring pride to the heart of the electrical engineering fraternity as well as to the whole world. Coupled with that pride is a spirit of gratitude on the part of the present generation of engineers toward those who have given their lives and their leisure in establishing the fundamentals on which electrical science and industry are built.

In that spirit I am prompted at the outset to dwell upon the versatile achievements of Thomas A. Edison. However, as my association has been entirely with that part of the art which had to do with the manufacture of generating machinery for the transmission of power by the use of alternating currents, I propose to review the early history of the electrical profession for outstanding individual contributions peculiar to the development of the science and art of transmitting power by the use of alternating currents.

The salient feature of the art of generating and distributing power at the present time is the superpower system, that is, an interconnection of existing and prospective generating and distributing systems. The broad idea of the superpower system must continue to grow more and more, because it is economically sound. It brings about an improvement in the load factor of the generating system; it allows the metropolitan markets for power to be connected in a continuous system with remote power reserves, and makes the exchange of energy from one part of such a system to another a practical, reliable, and everyday occurrence. Many are the engineering problems involved in the safe operation and satisfactory service of an interconnected system. However, electrical engineers already have solved these problems or are well advanced in their solution. For instance, the spreading of the troubles of one system to the next (bugbear of the past) is prevented by proper relaying and sectionalizing.

The holding of the proper voltage at different points, and the prevention of the flow of wattless current, are accomplished by adjusting automatically, if necessary, the ratio of the transformers so that the voltages at the point of connection may be of the same value and have the same phase relations. The interconnection of electric systems constitutes also an important progress in civilization, because it aims to allow electric energy, like sunlight, to become available everywhere.

It is well known that a discovery in the sciences is not an isolated event. The laws of nature have ordained that progress or change is never by leaps or revolutions. This is true, of course, of electrical engineering and the branch of it that deals with long distance transmission of power by means of alternating current. It has grown as does a snowball, by the process of almost infinitesimal additions. Practically every experiment or new development in the generation, transmission, and conversion of electric power is a modification of an experiment that has gone before. Almost every new theory is built through the contributions of many workers, of many different elements, one adding a little here and another a little there; thus to the observer in retrospect, progress seems to be continuous and uniform.

I wish, however, to emphasize the fact that the changes introduced into the art during the early '90's of the last century by the engineers of that period have placed the whole structure of electrical art of today as applied to light and power, firmly on the use of alternating currents. These changes have made economically possible the generation of large amounts of power in suitably located central stations, and its conversion and transmission to those points where it can be used most advantageously by

Reprinted from *Elec. Eng.*, vol. 53, pp. 726–729, May 1934.

industry to operate and to increase the capacity and the economy of our mills and factories; to provide electrical transmission to the small town and country; to extend and improve the processes of metallurgy; and now to place in the homes of the great agricultural classes, through the use of electric power, the comforts and conveniences of the city, and to place in the hands of the farmer the opportunity to extend the economy of the farm to a point where it may compare favorably in efficiency and effectiveness with the factory and the workshop. Thus will the nation be prepared, through the aid of the superpower systems, for a complete decentralization of industry, which is needed ultimately to relieve the economic stress of both farm and city.

Nevertheless, to me the outstanding accomplishments of this period which made for the greatest progress were: the broad generalization of electrical phenomena, and the mathematical formula for the design of alternating current machinery by Charles P. Steinmetz; the invention and development of the modern transformer by William Stanley; the invention of the induction motor by Nikola Tesla;

The original transformer built by William Stanley in 1885

the induction meter by Oliver Shallenberger; the dynamo-electric machine by Benjamin G. Lamme; and the numerous contributions to all branches of electrical machinery by Elihu Thomson.

In 1886, William Stanley, in the first alternating current plant in America, which was engineered and built by him at Great Barrington, Mass., demonstrated how electric power could be generated at a low voltage, transformed into a higher voltage, transmitted at the higher voltage, retransformed to a lower voltage, and used at this voltage as might be required. This feature of adapting the voltage to varying requirements, and of maintaining it substantially constant, irrespective of the load, rendered possible the enormous development and progress in the distribution and transmission of electric energy that have taken place since.

This capability of voltage transformation lies in the transformer itself, insignificant though it may appear. Stanley always spoke of the transformer as the "heart of the alternating current system." Naturally the great development of the art has been accompanied by a similar development of the transformer. Very early Stanley had properly visualized the fundamentals of transformer design, and correctly solved many of its problems in the Great Barrington installation. This revealed a thorough understanding on his part of electromagnetic induction, rather surprising for 50 years ago. The same ability in handling these laws as applied to transformers was shown by Stanley in the construction of the inductor alternator, which had no windings on the rotor, a feature considered of much value at the time. The inductor alternator, as well as the Stanley induction motor and the Stanley induction meter, did not survive; but the transformer did, and is substantially the same as the one originally built by Stanley.

The possibilities of the alternating current system early appealed strongly to the imagination of electrical engineers, both at home and abroad, but they appealed to none more strongly than to William Stanley. At 30 years of age he had a full conception of the alternating current station idea of manufacturing power, that is, the manufacture of power in

An early Stanley commercial transformer with front plate removed showing arrangement of fuses

large volume in some suitable location, transmitting and distributing it to points of consumption by the use of alternating current. With this idea firmly fixed in his mind, and fully determined to find out at

Pittsfield, Mass., whether there were any limits in sight barring the use of line potentials higher than the 2,000 volts then generally employed, he instructed me to design and build, in 1892, transformers and a line for 15,000-volt operation. To this end we erected a pole line, built a transformer house, and set up the transformers. These increased the potential of the town circuit from 1,000 volts to 15,000 volts. We connected the line to this high potential supply, sent the current around a farm and back to the same transformer house, then retransformed the line potential to 1,000 volts, and operated the distribution transformers of the local company. This little plant was operated during a New England winter with entire success, and the engineering data obtained were the reason for subsequent recommendations by the Stanley Electric Company for the use of potentials much higher than 15,000 volts. I recall these facts only to emphasize the undeveloped state of the art of that early period of which I speak, and how limited and provincial was its outlook compared with our present-day accomplishments.

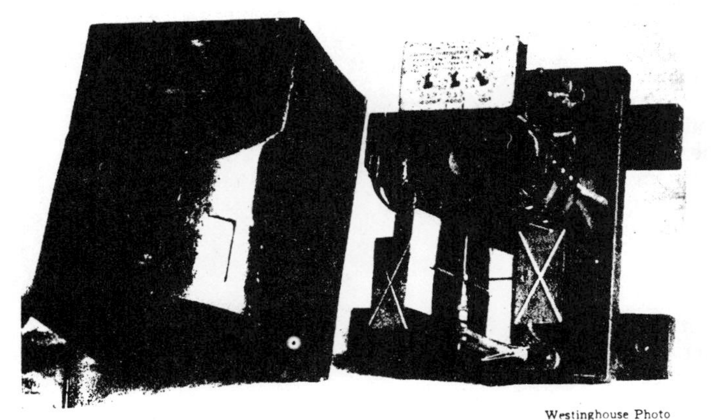

Westinghouse Photo

Induction type ampere-hour meter invented by Shallenberger in 1888. This was the first a-c integrating meter, and it is the parent of all a-c watthour meters now in use

Nikola Tesla invented the induction motor in 1888. This invention was a great step forward, and it has been stated frequently that the invention of this motor was one of the greatest advances made in the industrial application of electricity. This statement without doubt is true, but the development of the motor was long and costly, and as late as 1895 it was still in the experimental stage. Vital reasons for its slow progress, development, and application were: First, it was primarily a polyphase motor, and the alternating current plants of the day were single phase; second, these plants operated at a frequency of 133 cycles per second, and subsequent studies revealed that this frequency was not well suited for that type of motor. By 1895, however, its development through the aid of many other electrical engineers, was far enough advanced so that a good commercial motor became available.

The invention of the induction meter by Oliver Shallenberger was vital and important in the growth of the electrical industry. Until the invention of this interesting and much needed device, there was no instrument to measure the quantity of alternating current supplied to the consumer. While the meter operated on the same fundamental principles as the Tesla motor, Shallenberger invented the meter entirely independently of Tesla. While Shallenberger was observing the movement of a spring in an alternating current arc lamp, under the influence of a shifting magnetic field, the idea of the induction meter came to him. Within 2 weeks after he had conceived the idea, he designed and built a most successful alternating current meter of the induction type. This meter was accepted immediately as a success by the electrical industry and the public—a long time before the induction motor was accepted as such. While Shallenberger's particular meter has long been discarded, its influence on the development of the struggling industry was great indeed. He died in 1898, before he had an opportunity fully to appreciate or to enjoy the success of his labors.

The engineering talents of the late Benjamin G. Lamme, an engineer and inventor endowed with unusual ingenuity, resourcefulness, and good judgment, presented to the art the synchronous converter, the rotary condenser, and also the electrical design of the 5,000-hp generator—a far bigger generator than had ever been built up to that time—

Westinghouse Photo

Early laboratory model of Tesla motor with wound rotor and slip rings

which inaugurated the hydroelectric power development at Niagara Falls in 1895. This type of generator has persisted to the present day. The single-phase railroad motor and the introduction of the squirrel-cage induction motor with high starting torque, were the individual works of Mr. Lamme.

Of the contributions of Charles P. Steinmetz there is little to tell electrical engineers; they all know, and they knew him and recognized him among the leaders of modern science. His personal contributions to the science and the art of long distance power transmission by alternating current were many and valuable. To me, however, it has always seemed that his greatest contributions to the electrical art of our day were his writings, embracing the results of theoretical and experimental scientific investigations. In these is laid an invaluable mathematical foundation for the design of electrical machinery. His work in this respect has no equal in our day.

Elihu Thomson's contributions to the electrical industry have been so many and are so generic in character that it is almost impossible to select any one contribution from his work of the last half century which overshadows in importance and value any of his others. His remarkable depth and range of scientific knowledge have influenced in a major way the development in every field of electrical endeavor. A master of industrial research, he invented many early types of lightning arresters, magnetic blowout switches, the induction regulator, and the single-phase repulsion motor. From a power transmission and distribution standpoint, one of his most important contributions was his proposition to use oil for insulating and cooling transformers, a practice now universal the world over and upon which the success of the progressive and ever increasing power transmission potentials has depended.

Steinmetz observing artificial lightning flash in his laboratory, February 1922

I have selected these men as the most outstanding among all the electrical engineers and inventors of that pioneer period, the closing decade of the last century; their accomplishments more than those of any other group, made possible the high state of the art of transmitting and distributing electric energy as we find it and as we enjoy it today. It was through their insight into, and their solution of, the technical problems that beset them, and by their foresight in reading the future promises of their time, that they were able to blaze the trail not alone for their contemporaries, but for future generations as well. It was their glory to be able to catch a glimpse of what was before us while the rest of the world wondered. On the traditions of the past, a great future for electricity and for the transmission of electrical energy is predicated. However, as we of the electrical fraternity hope for continued progress, we must remember that our hopes can be fully realized only by remaining true to the greatest of those traditions, "to produce and to serve," and by cherishing the ideals and emulating the ceaseless activities of such pioneers as those mentioned.

Thompson induction regulator of the late '90's

All 6 of these men possessed the scientific spirit. They were truly men of research, with patience and vision, always seeking earnestly and hopefully for new knowledge, more fully to understand Nature's laws as they are, and more effectively to use those laws for the benefit of humanity.

Thomson and Houston Assess Electrical Power Transmission

Two well-known pioneers considered the economics of long-distance electric power transmission in this paper from the *Journal of the Franklin Institute* of 1879. They concluded that the extreme pessimism of some was unjustified and that "an enormous quantity of power may be transferred to considerable distances." Their assumption of an unavoidable loss of about 50 percent was a commonly held misconception at the time that was soon to be challenged by Thomas Edison and his assistant, Francis Upton.

Elihu Thomson (1853–1937) was born in Manchester, England, but his family moved to the U.S. prior to the Civil War and settled in Philadelphia. Thomson graduated from the Central High School (a technically-oriented school with laboratories and an academic department) in 1870. Houston was one of his teachers. Thomson was hired by the school and soon began to attract attention through his research on Edison's "etheric force" and a thorough study of dynamos carried out in 1878. His patents on an arc-lighting system became the basis for creation of the Thomson-Houston Electric Company that became one of the most successful in the field of arc lighting and railway motors during the 1880's. The Company merged with the Edison General Electric Company to form the General Electric Company in 1892. Thomson was a prolific inventor and was issued approximately 700 patents during his career. See Chesney's discussion of Thomson in Paper 16. Also see the *NAS*, vol. 21, pp. 143–179 and the *DSB*, vol. 13, pp. 361–362. See also *Selections from the Scientific Correspondence of Elihu Thomson*, edited by Harold J. Abrahams and Marion B. Savin (Cambridge, Mass., 1971).

Edwin J. Houston (1847–1914) was born in Alexandria, Va., and graduated from the Central High School in Philadelphia in 1864. After teaching at Girard College and spending a brief time in Germany, Houston accepted a teaching position at Central High School in 1867. In addition to collaborating with Elihu Thomson in various researches, Houston served as Editor of the *Journal of the Franklin Institute* and was author or coauthor of a large number of textbooks in electricity. He shared a consulting engineering practice with A. E. Kennelly from 1894 until his death. He presented the first technical paper at the organization meeting of the AIEE. (See Ryan's discussion in Paper 12 and Paper 50.) See the *DAB*, vol. 9, p. 261.

ON THE TRANSMISSION OF POWER BY MEANS OF ELECTRICITY.

By Profs. Elihu Thomson and Edwin J. Houston.

The statements recently made as to the size and cost of the cable that would be needed to convey the power of Niagara Falls to a distance of several hundred miles by electricity, have induced the authors to write the present paper, in the hope that it may throw light upon this interesting subject.

As an example of some of the statements alluded to, we may cite the following, viz.: That made by a certain electrician, who asserts that the thickness of the cable required to convey the current that could be produced by the power of Niagara, would require more copper than exists in the enormous deposits in the region of Lake Superior. Another statement estimates the cost of the cable at about $60 per lineal foot.

Fig. 1.

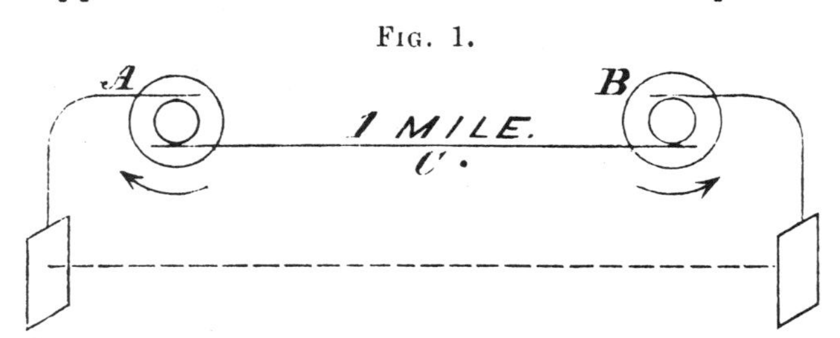

As a matter of fact, however, the thickness of the cable required to convey such power is of no particular moment. Indeed, it is possible, should it be deemed desirable, to convey the total power of Niagara, *a distance of* 500 *miles or more, by a copper cable not exceeding one-half of an inch in thickness.* This, however, is an extreme case, and the exigencies of practical working would not require such restrictions as to size.

The following considerations will elucidate this matter. Suppose two machines connected by a cable, of say 1 mile in length. One of these machines, as, for example, *A*, Fig. 1, is producing current by

Reprinted from *J. Franklin Inst.*, vol. LXXVII, 3rd ser., pp. 36–39, Jan. 1879.

the expenditure of power; the other machine, B, used as an electrical motor, is producing power, by the current transmitted to it from A, by the cable C. The other terminals, x and y, are either put to earth, or connected by a separate conductor.

Let us suppose that the electromotive force of the current which flows is unity. Since by the revolution of B, a counter-electromotive force is produced to that of A, the electromotive force of the current that flows is manifestly the difference of the two. Let the resistance of A and B together, be equal to unity, and that of the mile of cable and connections between them, the ·01 of this unit. Then the current which flows will be $C=\frac{E}{R}=\frac{1}{1{\cdot}01}$. If now an additional machine, A', Fig. 2, and an additional motor, B', and an additional mile of cable, be introduced into the above circuit, the electromotive

FIG. 2.

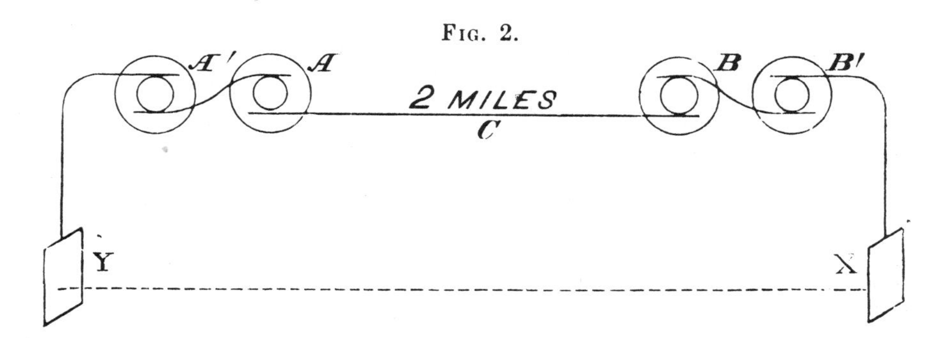

force will be doubled, and the resistances will be doubled, the current strength remaining the same as $C=\frac{E}{R}=\frac{1+1}{1{\cdot}01+1{\cdot}01}=\frac{2}{2{\cdot}02}$. Here it will be seen that the introduction of the two additional machines, A' B', has permitted the length of the cable c to be doubled, without increasing the strength of the current which flows, and yet allowing the expenditure of double the power at A A', and a double recovery at B B' of power, *or, in other words, a double transmission of power, without increase of current.* Increase, now, the number of machines at A to say one thousand, and of those at B in like proportion, and the distance between them, or the length of the cable, one thousand, or in the case we have supposed, make it one thousand miles, its diameter remaining the same. Then although the same current will flow, yet *we have a thousand times the expenditure of power at one end of the cable, and a thousand-fold recovery at the other end, without increase of current.* And the same would be true for any other proportion.

Since the electromotive force is increased in proportion to the increase of power transmission, the insulation of the cable and machines would require to be proportionally increased.

As an example, it may be mentioned that a dynamo-electric machine used for the purpose of A in the figure, may have a resistance of say 40 ohms, and produce an electromotive force of say 400 volts. Such a machine might require from three to five horse-power when used in connection with a suitable motor B, for recovery of the power transmitted.

If the resistance of the motor B, be say 60 ohms, and the cable transmitting the currents a distance of one mile, be one ohm, then the current $C = \frac{400}{60 + 40 + 1} = \frac{400}{101}$. If now, one thousand machines and one thousand motors, and a thousand miles of cable, each of the same relative resistances be used, the current $C = \frac{1000 \times 400}{1000 \times 101}$, which has manifestly the same value as before. If our supposition of the power used to drive one machine be correct, then from three to five thousand horse-power would be expended in driving the machines, and possibly about fifty per cent of this amount recovered. Then we have from 1500 to 2000 horse-power conveyed a distance of 1000 miles. What diameter of copper cable will be required for such transmission? Since this cable is supposed to have the resistance of one ohm to the mile, calculation would place the requisite thickness at about $\frac{1}{4}$ inch. If, however, the distance be only 500 miles, then the resistance per mile may be doubled, or the section of the cable be decreased one-half, or its diameter will be less than the $\frac{1}{5}$ inch.

For the consumption of 1,000,000 horse-power, a cable of about 3 inches in diameter would suffice under the same conditions. However, by producing a much higher electromotive force, the section of the cable could be proportionally reduced, until the theoretical estimates, which we have given in the first part of this paper, might be fulfilled. The enormous electromotive force required in the above calculation, would, however, necessitate such perfect insulation of the cable, that the practical limits might soon be reached. The amount of power required to be conveyed in any one direction, would, of course, be dependent upon the uses that could be found for it; and it is hardly conceivable that any one locality could advantageously use the enormous supposed power we have referred to.

Stripped of its theoretical considerations, the important fact still remains, that with a cable of very limited size, an enormous quantity of power may be transferred to considerable distances. The burning of coal in the mines, and the conveyance of the power generated by the flow of rivers, may therefore be regarded as practicable, always, however, remembering that a loss of about 50 per cent. will be almost unavoidable.

It may be mentioned that Dr. C. W. Siemens, and Sir William Thomson, have recently made statements that are in general accordance with the views here expressed.

Edison's Jumbo Dynamo

THIS paper was presented at a meeting of the American Society of Mechanical Engineers in 1882, the year that Edison's historic Pearl Street electric power plant began operation. The Jumbo dynamo described was a "2nd generation" machine capable of supplying about 1200 lamps as compared to the 50 lamp capacity of the earlier bipolar dynamo known as the "long-waisted Mary Ann." The Jumbo had been something of a sensation during the Paris Electrical Exposition of 1881. As indicated in this paper, its design followed the tradition of the American System of Manufacturing by using duplicate parts and special machine tools calculated to reduce costs and facilitate assembly and repair. The paper is unusual in that Edison rarely published the results of his work in engineering journals. (He was one of only four life members of the ASME listed in 1881.) The paper was delivered by Porter, and Edison may not even have attended the meeting since he did not participate in the discussion of the paper. It may also be noted that Edison was listed as a Ph.D. He had been awarded an honorary doctorate by Union College in 1878, probably in recognition of his invention of the phonograph. The Porter–Allen steam engine that had been designed to drive the Jumbo dynamo was also notable for its high speed and smooth performance over a wide range of speeds. An earlier model had been a sensation at the Paris Exposition of 1867. Unfortunately, the Porter–Allen engines did not work well when two or more dynamos were parallel connected, but hunted so badly that one observer recalled it as a "terrifying experience." The problem was solved by substituting Armington–Sims engines for the Porter–Allen type.

Thomas A. Edison (1847–1931) was born in Milan, Ohio and had minimal formal education, although he later became an avid reader of Faraday and other scientific authors. Edison worked as an itinerant telegrapher and electrical inventor during the 1860's. He was a partner in possibly the first electrical consulting firm in the U.S., Pope, Edison and Co., established in 1869. The partnership soon was dissolved and Edison achieved both a reputation and substantial financial rewards for such inventions as his quadruplex telegraph system and an improved telephone transmitter. One of his most important contributions was toward the institutionalization of industrial research by the creation of a laboratory at Menlo Park, N.J. in 1876. It was there that Edison and a team of gifted associates created the elements of a successful incandescent lighting and power distribution system. Edison later moved to a new facility at West Orange, N.J. in 1887. During his inventive career, Edison received nearly 1100 U.S. patents. See the *NAS*, vol. 15, pp. 287–304. The best book-length biography on Edison is Mathew Josephson, *Edison* (New York, 1959).

Charles T. Porter (1826–1910) was born in New York and graduated from Hamilton College. He practiced law until 1854 when he made what Otto Mayr has termed "an astonishing decision" by deciding to become an engineer–inventor. He began his new career by inventing and patenting an improved governor for steam engines. This, when combined with a valve invented by John Allen whom he met in late 1860, held the key to a high-speed reciprocating steam engine that was perfected in time for the Paris Exposition of 1867. After several years in England spent developing and promoting the new engines, Porter returned to the U.S. in 1868. He was a founder of the ASME and highly regarded in mechanical engineering circles. See the *NCAB*, vol. 20, p. 494. Also see Otto Mayr, "Yankee Practice and Engineering Theory: Charles T. Porter and the Dynamics of the High-Speed Steam Engine," *Technology and Culture*, vol. 16, pp. 570–602, 1975.

DESCRIPTION OF THE EDISON STEAM DYNAMO.

BY

BY T. A. EDISON, PH.D., NEW YORK CITY,

AND

CHARLES T. PORTER, PHILADELPHIA, PA.

THE central Edison station of the first district in New York city will, when fully equipped, be supplied with twelve dynamos, each of which is nominally rated as a 1200-light machine, at 16 candle-power incandescence, but is capable of supplying 1400 lights of this power continuously, and with high economy, without heating the armature, or burning or injuring the commutator or brushes. This increased capacity is due to improvements in the lamp itself.

The armature of each dynamo is driven by a Porter-Allen engine, of $11\frac{3}{16}''$ diameter of cylinder by 16″ stroke, directly connected, and making 350 revolutions per minute, giving a piston travel of 933 feet per minute.

The steam is supplied by eight Babcock & Wilcox boilers of 2000 aggregate horse-power, and which will work under a pressure of about 120 pounds. These occupy the basement of the building. Over them, the first and second floors being removed, an iron superstructure is erected entirely separated from the walls of the building, and on this the combined dynamos and engines are placed.

One-half of this equipment is now nearly ready for service, and the remainder is expected to be completed during the coming season.

The armature of the dynamo is of the form commonly known as the Siemens armature, but in its construction and "connecting up" it differs radically from all others.

The foundation of the armature, or the iron core which is built upon the shaft, is made up of sheet-iron disks, separated from each other by sheets of tissue-paper, and bolted together. This has all the advantages of a solid iron core in strengthening the magnetic field, while it completely prevents the great loss of power by local currents, which would circulate in the iron if it were solid. In the place of insulated wires, the cylindrical face of the armature is made up of heavy copper bars, trapezoidal in section, each bar being insulated, and also separated from its neighbors and from the iron core underneath by an air-space.

The connection between the bars on opposite sides of the armature, to form the electrical circuit, is made by copper disks, of the same diameter as the core. At each end of the core are one-half as many of these copper disks as there are bars, each disk being insulated from its neighbors, and the whole being bolted together in such

Reprinted from *Trans. ASME*, vol. III, pp. 218–225 (extract of a longer paper), 1882.

a manner as to form, with the disks of sheet-iron constituting the core, one solid mass. Each disk is formed with projecting lugs on its opposite sides to which the two bars are connected.

The connections between the opposite surfaces of an armature are of no benefit in generating an electric current, but are a necessary evil, introducing useless resistance into the circuit. By using for this connection copper disks in the manner described, a great weight of copper is disposed in a limited space; and so this useless resistance, and consequent loss of energy. is reduced to a minimum.

This method, moreover, reduces the work to a simple machine construction, in which all the parts are duplicates, and the operations can be much cheapened and facilitated by the use of special tools.

The spaces between the armature bars admit of a free circulation of air, thereby preventing the accumulation of heat, and increasing to an enormous degree the capacity of the machine. The armature is at intervals wound with piano wire over the bars to resist the centrifugal force developed by their revolution.

The commutator and brushes of an electrical machine are the parts subject to the greatest depreciation. In this machine all parts of the end of the armature are so constructed as to be easy of access, and they can be quickly and cheaply repaired, or removed and replaced by new parts, when necessary. Any accident would require but a short stoppage for repairs.

Provision is made for keeping a continuous and rapid circulation of air over the entire face of the armature.

This armature is 27.8″ in diameter by 61″ long. The commutator adds 18″ to this length, and is itself 12¾″ in diameter. The armature shaft is of steel, 7¾″ in diameter, having a total length of 10′ 3″. The journals are 6½″ in diameter by 15″ long, and run in Babbitt-metal bearings, in pillow-blocks of the box form, giving the greatest stiffness with minimum of weight.

Provision is made for continuous water circulation underneath the boxes, and for continuous lubrication, with traps to prevent the creeping of the oil along the shaft and reaching the commutator, and drains to receive it as it runs through the bearings and convey it to a drip pan.

The magnet is made up of two immense cast-iron "pole pieces," between the semi-cylindrical faces of which the armature revolves, twelve cylindrical soft iron cores attached to these pole pieces, and made magnetic by an electrical current circulating in the wire around them, and four soft iron keepers connecting the back ends of these cores. Eight of the cores are attached to the upper pole piece, and four to the lower one.

The width of these "poles" is 49″, and their height 61½″. The length of the twelve soft iron cores is 57″, the diameter of the four upper ones is 8″, and of the eight lower ones 9″.

The four soft iron keepers are each 11″ wide by 9″ in thickness, and the total length of the magnet is 94″.

The magnet is insulated by cast zinc bases 3″ in thickness.

The weight of the dynamo is as follows:

Armature and shaft, ...	9,800 lbs.
Two pillow-blocks, ...	1,340 "
Magnet, complete, ...	33,000 "
Zinc bases, ...	680 "
Total, ...	44,820 "

The copper is distributed as follows:

In the armature bars, ...	590 lbs.
" " disks, ...	1350 "
In the magnet wire, ...	1500 "
Total, ...	3440 "

Mr. Edison was early impressed with the conviction that to give steady and reliable motion to these armatures it would be necessary to connect an engine to each one of them directly. This combination has been termed by him the Steam Dynamo.

In adapting the Porter-Allen engine to this service a special construction in some respects was found to be called for. These special features will be briefly described.

It seemed important to avoid a rigid connection between the engine and the armature shafts, which would require the entire series of bearings to be maintained absolutely in line. In place of this, therefore, a self-adjusting coupling (see Fig. 5) has been introduced, which will permit of considerable errors of alignment without any abnormal friction being produced in the bearings.

The point of difficulty was the backlash, the engine having no fly-wheel, except the heavy armature itself, which was to be driven through the coupling. Provision was made for taking this up by steel-keys of a somewhat peculiar form, between which the tongues of the coupling move freely, while they themselves are immovable. These keys are held between set-screws threaded in wrought-iron rings covering the flanges on the ends of the shaft. All the faces liable to move upon each other are oiled from a central reservoir. This coupling is a very compact affair, without a projection anywhere above its surface, and gives every promise of completely answering its purpose.

The engine is made with a forked bed and two shaft bearings and a double crank, and so is completely self-contained. It is shown in plan and elevation, Figs. 1 and 2.

The shaft having no support beyond these bearings on either side, unusual stiffness was required in the crank-pin to prevent deflection under the great strains to which it is subjected.

A novel form of pin (see Fig. 3) was proposed by Mr. Richards, which is found to possess all the rigidity required. It is provided with flanges, which are let into each crank, and held each by four screws, as shown, while the shanks of the pin are also forced firmly into the cranks.

Special appliances enabled the work of putting the cranks together in this manner to be done with extreme and uniform accuracy.

The engine is so arranged as to have the valve gear on the side furthest from the dynamo. The engineer has not to go between the engine and the dynamo, when running, for any purpose.

The connecting-rod (Fig. 4) is of steel, and the crank-pin boxes are formed directly in the end of it.

This end is finished from a solid forging, and chambered out for Babbitt metal. The bolts are then fitted, after which it is parted and holes are drilled for holding the Babbitt securely.

In the connecting-rods for single crank engines of this type permanent length of rod is secured by forming the crank-pin end solid, and taking up the wear by a wedge closing up the inside box. In these double crank engines this construction is impracticable, but the same object is attained by forming the cross-head end in the manner shown, in which the strap is made permanent, and the inside box is closed up by a key bearing against a steel plate.

The weight of the reciprocating parts of this engine is as follows:

Piston, with rod, ...	83 lbs.
Cross-head, ...	42 "
Connecting rod, ...	109 "
Total, ...	234 "

The initial acceleration of this mass, or the force required, on the dead centres, to give it the motion necessary to relieve the crank from strain is as follows:

$$350^2 \times .66 \times .000341 = 27.57,$$

or 27.57 times the weight of the mass, which gives

$$234 \times 27.57 = 6451 \text{ lbs.}$$

The formula is $R^2\, l\, c$, when

R = the revolutions per minute;

l = the length of the crank in decimals of a foot; and

c = the coefficient of centrifugal force.

The connecting-rod is 48″, or six cranks, in length. This affects the initial acceleration, making this to be on the dead centre farthest from the crank 7526 lbs., and on the dead centre nearest to the crank 5376 lbs., a difference of 40 per cent.

The area of the cylinder is 98.2 square inches.

The area of the piston-rod, $1\frac{3}{4}$ inches diameter, is 2.4 square inches, leaving area of cylinder at crank end 95.8 square inches.

The initial accelerating forces are therefore as follows, viz.: At the end of the cylinder farthest from the crank 77 lbs., and at the end of the cylinder nearest to the crank 56 lbs., on the square inch of piston area.

The counterweight was after some trials fixed at 135 lbs. This leaves 99 lbs. of the reciprocating parts running unbalanced. It is found that this is not sufficient to disturb the stability of the engine, while on the other hand the counterweight is not so great as to exert an objectionable strain in the vertical direction.

The total weight of the engine is 6445 lbs.

The engine and dynamo are mounted on a cast-iron base plate, made for convenience in two parts, and bolted together.

The dimensions of this base plate are as follows: Length, 14 feet; width, 8 feet 9 inches; and its weight is 10,300 lbs. The entire weight is therefore as follows:

Base plate, ...	10,300 lbs.
Dynamo, ...	44,800 "
Engine, ...	6,450 "
Total, ...	61,550 "

The large engraving is a perspective view of the Dynamo and Engine combined. (Not included).

The last and most careful test of one of these dynamos gives the following results, as shown by the indicator diagrams, which are here reproduced full size; scale, 80 lbs. to the inch.

The lamps used in all the trials were of the older construction, of which $8\frac{1}{2}$ lamps, at 16 candle-power incandescence, require one horse-power of electrical energy.

Since these were placed for experimental uses, improvements in the lamp have increased their economy, so that one horse-power is sufficient to maintain fully 10 of the present lamps at 16 candle power incandescence.

Diagram No. 1 shows the friction of engine and dynamo at 350 revolutions per minute, requiring 13.63 H. P.

Diagram No. 2 shows the resistance with the magnet circuit on = 19.17 H. P.

Field 5.78 ohms, 103 volts.

The increased resistance due to the magnets was 5.54 H. P.

Of this, the calculated energy developed in the magnets was

$$\frac{103^2 \times 44.3}{5.78 \times 33,000} = \quad . \quad . \quad . \quad . \quad . \quad 2.46 \text{ H. P.}$$

Leaving energy to be accounted for by local currents in iron core of armature, and in armature bars, 3.08 H. P.

Diagram No. 3 shows the work done in maintaining 300 lamps.

These, in the ratio of $8\frac{1}{2}$ to 10, were equal to 353 lamps of the present construction. The pressure was maintained also at 102 volts, representing 25 candle power, in place of 98 volts, representing 16 candle power incandescence, which requires the number of lamps to be increased in the ratio of 102^2 to 98^2, or to 382 lamps.

The pressure of the armature was 104 volts, showing a loss in the conductor of 2 volts, which would increase the number of lamps as 104 : 102.*

The total correction is therefore as follows:

$$300 \times \frac{10}{8.5} \times \frac{102^2}{98^2} \times \frac{104}{102} = 389 \text{ lamps.}$$

The power exerted was 60.6 H. P., which gives to the indicated horse power

$$389 \div 60.6 = 6.42 \text{ lamps.}$$

The magnet circuit had now a resistance of 5.28 ohms with 104 volts pressure, representing

$$\frac{104^2 \times 44.3}{5.28 \times 33{,}000} = \quad . \quad . \quad . \quad . \quad . \quad 2.75 \text{ H. P.}$$

Substituting this in place of 2.46 H. P. in the first trial, we have 19.46 H. P., which, deducted from 60.6 H. P., leaves net 41.14 H. P.

This gives $389 \div 41.14 = 9.45$ lamps per H. P.

Diagram No. 4 shows the work done in maintaining 700 lamps.

The pressure at the lamps was maintained, as in the preceding trial, at 102 volts, which required at the armature a pressure of 105 volts.

The total correction in this case is therefore

$$700 \times \frac{10}{8.5} \times \frac{102^2}{98^2} \times \frac{105}{102} = 919 \text{ lamps.}$$

The power exerted was 115.83 H. P., giving to the indicated horse power $919 \div 115.83 = 7.93$ lamps.

The resistance of the magnet circuit was now 4.78 ohms, with 105 volts pressure, representing,

$$\frac{105^2 \times 44.3}{4.78 \times 33{,}000} = \quad . \quad . \quad . \quad . \quad . \quad . \quad 3.1 \text{ H. P.}$$

---

* The conductors were insufficient, occasioning a loss that increased with the increase in the number of lamps.

Substituting this in place of 2.46 H. P. in the first trial, we have 19.81, which, deducted from 115.83 H. P., leaves net 96.02 H. P.

This gives $919 \div 96.02 = 9.57$ lamps per H.P.

Diagram No. 5 shows the work done in maintaining 1050 lamps.

The pressure at the lamps was maintained in this trial at only 99 volts, but this required at the armature a pressure of 108 volts, showing a loss of 9 volts in conduction.

The total correction in this case is thus

$$1050 \times \frac{10}{8.5} \times \frac{99^2}{98^2} \times \frac{108}{99} = 1375 \text{ lamps.}$$

The power was 168.4 H.P.

Giving to the indicated horse power

$$1375 \div 168.4 = 8.16 \text{ lamps.}$$

The resistance of the magnetic circuit was now 3.28 ohms, with 108 volts pressure, representing

$$\frac{108^2 \times 44.3}{3.28 \times 33{,}000} = \quad . \quad . \quad . \quad . \quad . \quad 4\,77 \text{ H.P.}$$

Substituting this in place of 2.45 H. P. in the first trial, we have 21.48 H. P., which, deducted from 168.4 H. P., leaves net 146.92 H.P.

This gives $1375 \div 146.92 = 9.36$ lamps per H.P.

It will be seen that the losses of efficiency due to undiscovered resistances are only

In the first case, $10 - 9.45 = .55$ H.P. per lamp,

In the second case, $10 - 9.57 = .43$ H.P. per lamp, and

In the third place, $10 - 9.36 = .64$ H.P. per lamp,

Averaging 5.4 per cent.

The friction in the journals of the armature, when driven in this manner, does not increase with the resistance, and, on account of the action of the reciprocating parts of the engine, that in its bearings is also nearly a constant quantity, whatever the load may be.

The above figures show this very clearly, the subtraction of the friction diagram in each case exhibiting substantially the same net power per lamp.

••• ••• •••

Nikola Tesla's Alternating-Current Motors

THE introduction of alternating-current power distribution systems that overcame the distance limitation of the Edison system was a major concern for electrical engineers during the 1890's. A key innovation of the new systems was the alternating-current motor that is the subject of this paper by Albert Schmid. He sought to correct the impression that the Frankfort–Lauffen project of 1891 had been the first to demonstrate polyphase transmission and motors. Schmid, a Westinghouse engineer, credited Tesla with inventing the polyphase motor and reported that the Westinghouse Company had already developed such motors in sizes of up to 1000 hp.

Albert Schmid (1858–1919) was born in Zurich, Switzerland. After spending approximately five years working in a machine shop, he graduated from the Polytechnic School at Zurich in mechanical engineering in 1879. He had worked as an engineer in Belgium, England, and France before being persuaded by George Westinghouse to take charge of designing and directing his shops at the Pittsburgh works. He played a major role in the mechanical design of the generators manufactured for the first Niagara Falls Power Plant by Westinghouse. Schmid's designs were known for their "symmetry of form and beauty of line." His explanation was that "all you have to do is place the metal where it belongs and the machine becomes beautiful of itself." See "Electrical Engineers. Albert Schmid," *The Electrical Engineer* vol. 11, p. 623, 1891. Also see Charles A. Terry, "A Tribute to Albert Schmid," *The Electric Journal*, vol. 17, pp. 40–41, 1920.

Nikola Tesla (1856–1943) was born in what is now Yugoslavia and studied science and engineering at the Polytechnic College in Gratz. While in school he became familiar with the Gramme dynamo and decided that its commutators and brushes were a serious defect. After working for a time as an engineer on the government telegraph system, Tesla went to Paris in 1881 and joined the Continental Edison Company. He came to the U.S. in 1884 and soon left Edison to organize the Tesla Electric Company to develop his electrical inventions. His proposed new system was announced in an important AIEE paper in 1888, entitled "A New System of Alternate Current Motors and Transformers." His several patents were purchased by Westinghouse in July 1888 and became the basis for the strong position of the Westinghouse Company in the ac field. After a brief period with Westinghouse, Tesla resigned to resume his research at the frontiers of electrical knowledge. His later work included some extraordinary investigations of high-frequency and high-voltage phenomena using the Tesla coil, as well as unsuccessful efforts to achieve wireless transmission of power by means of an induced distortion of the earth's magnetic field. See the *DAB*, vol. 23, pp. 767–770 and "Electrical World Portraits. Nikola Tesla," *The Electrical World*, vol. 15, p. 106, 1890. Also see Gordon D. Friedlander, "Tesla: Eccentric Genius," *IEEE Spectrum*, pp. 26–29, June 1972. For Chesney's comments on Tesla as one of six pioneers, see Paper 16. For a paper on the history of the induction motor by one of Tesla's successors, see Philip Alger's paper in the PROCEEDINGS OF THE IEEE for September 1976.

THE

Electrical Engineer.

VOL. XIII. MARCH 9, 1892. No. 201.

THE TESLA MULTIPHASE CURRENT MOTORS.

BY

Albert Schmid

HE prominence which has been given to power transmission by multiphase currents in the experiments between Frankfort and Lauffen, seems to have led to the impression, even among those otherwise well informed, that this was the first example of this method of transmission. It may therefore be well to put on record the fact that in this country multiphase motors manufactured by the Westinghouse Co. have been in successful commercial operation for several years, and a three phase power plant has been in daily service in the shops of the Westinghouse Co. at Pittsburgh for some time past, driving the shafting and machinery of the winding department.

As is well known, Mr. Nikola Tesla is the inventor of the multiphase motor, and his patents in this country are controlled by the Westinghouse Electric & Manufacturing Company, who have been quietly standardizing the different sizes and have, within the last year or two, placed a large number of them in service.

The generators for these motors are of the Westinghouse multipolar type with toothed, drum armatures wound to give currents of sixty or ninety degrees difference of phase. These machines, one type of which is shown in Fig. 1, are built either entirely self-exciting and self-regulating or separately-exciting and self-regulating. The machine shown is designed for 60 kilowatt capacity.

The motors are constructed with internal poles and grooved armatures. The larger motors have armatures with three collecting rings, and the fields are excited with direct current. The smaller ones have armatures wound with coils closed upon themselves, doing away with all collecting rings and commutators. The engraving Fig. 2, shows a 10 h. p. Tesla motor and Fig. 3 exhibits the armature removed from the enclosing field magnets. These multiphase motors are built in sizes of from 1 to 1,000 h. p.

FIG. 4.—THREE-PHASE 1,000 H. P. TESLA ALTERNATING MOTOR.

A 1,000 h. p. three-phase generator or motor, for the same machine can be made either as a motor or generator, is shown in the engraving Fig. 4 and when run as a generator will give 150 amperes at 5,000 volts. In mechanical design this machine is similar to the multipolar direct current railway generators built by the Westinghouse Company. The armature is of the drum type with slots for the wire and is without bands.

With the Tesla multiphase system the power can be subdivided to any extent for motor or lighting purposes, and high pressures may be employed which can be reduced to whatever voltage is desired. The multiphase motors

Reprinted from *The Elec. Eng.*, vol. XIII, pp. 243–244, Mar. 9, 1892.

start under load, and run at constant speed. The smaller ones are made entirely alternating, and the larger ones are self-exciting after reaching synchronism.

As the starting torque with the Tesla multiphase motors is comparable with shunt wound direct current machines, they may be applied in every case where direct current shunt wound motors can be used. At starting, the direct current motors require dead resistance in series, which means considerable loss, but the alternating current motor has inductive resistance, and therefore the loss at starting is much less. Multiphase motors can be used for operating mines, factories, and street railways, and the generating station may be located wherever most convenient for obtaining the best results, and if the pressure is not over 5,000 volts, the motors may be supplied direct from the mains. The Tesla motor has a great advantage over other types of motors in that it requires no commutator, and consequently is sparkless. The motor may be boxed up, and kept free from dust and grit, and high pressures can be employed successfully. No starting resistance is required and the motor is small and compact, and requires no attention. Tests made at the Westinghouse Co.'s works demonstrate that the Tesla motors are fully as efficient as the best direct current motors.

FIG. 1.—GENERATOR FOR OPERATING TESLA ALTERNATING MOTOR.

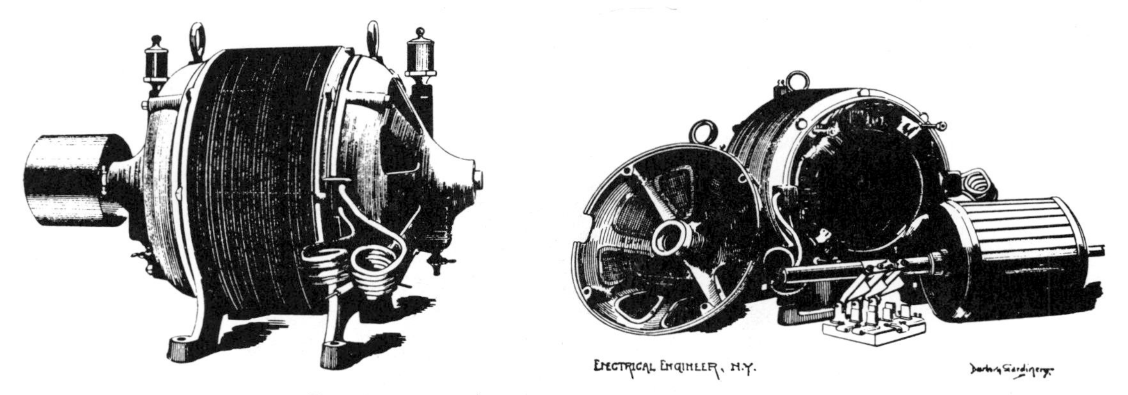

FIGS. 2 AND 3.—10 H. P. TESLA ALTERNATING MOTOR.

Steinmetz Proposes a Law of Magnetic Hysteresis

BY the time this paper appeared in 1890, it was becoming apparent that the phenomenon of hysteresis was a major factor limiting the efficiency of power transformers and other alternating-current apparatus. Charles Steinmetz was just beginning his meteoric rise to prominence in electrical engineering when he used the method of least squares and the empirical data of the British scientist, J. A. Ewing, to arrive at his famous 1.6 power law of hysteresis. Two years later Steinmetz published the results of an exhaustive theoretical and experimental investigation of hysteresis in two papers in the *Transactions of the AIEE*. As the first member of the electrical engineering profession in America to undertake investigations that were comparable in quality and sophistication to the best in Europe, Steinmetz became both a leader and a legend.

Charles Proteus Steinmetz (1865–1923) was born in Breslau, Germany as Karl August Rudolf Steinmetz (later changed when he came to America). He enrolled at the University in Breslau to prepare for a career in mathematics in 1882. He had completed a doctoral thesis in theoretical mathematics by 1888, but was forced to flee to Switzerland to avoid arrest because of his participation in a student socialist movement. He studied mechanical engineering at the Polytechnic School in Zurich for a year before seeking his fortune in the U.S. in 1889. His first employer was another German immigrant, Rudolf Eickemeyer, who owned a manufacturing plant in Yonkers, N.Y. With the encouragement of Eickemeyer, Steinmetz completed his classic hysteresis experiments at the factory. When Eickemeyer sold out to General Electric early in 1893, Steinmetz became a member of G.E.'s calculating department and remained with G.E. for the rest of his life. He played a major role in the introduction of complex quantities into ac analysis during the 1890's. His seminal paper on this subject, entitled "Complex Quantities and their Use in Electrical Engineering," was delivered at the International Electrical Congress held in Chicago in 1893. He received 195 patents and was responsible for the creation of an electrical engineering program at Union College in Schenectady. See the *DSB*, vol. 13, pp. 24–25 and my essay on Steinmetz in the *Encyclopedia of American Biography*, pp. 1035–1036, 1974. Also see P. L. Alger and C. D. Wagoner, "Charles Proteus Steinmetz," *IEEE Spectrum*, pp. 82–95, April 1965. Also see Paper 27 and my paper on Steinmetz and Alexanderson in the PROCEEDINGS OF THE IEEE for September 1976. A number of nontechnical papers by Steinmetz are included in *Steinmetz the Philosopher*, compiled by Philip L. Alger and Ernest Caldecott (Schenectady, N.Y., 1965).

NOTE ON THE LAW OF HYSTERESIS.

BY CHAS. STEINMETZ.

THE magnetism of a magnetic circuit will vary periodically, if subjected to a periodically varying magnetomotive force. The variations of the magnetism, however, will not be simultaneous with the variations of the magnetomotive force, but show a certain lag, so that the curve of magnetism, as a function of the magnetomotive force, forms a kind of loop, the well known curve of hysteresis.

This phenomenon proves, that in the production of the magnetic circuit by the conversion of electric energy into magnetic energy, and in the destruction of the magnetic flow by its reversion into electric energy, a certain amount of energy has been lost, that is, converted into heat.

The amount of energy converted into heat by hysteresis in a full magnetic cycle depends on the maximum magnetization. It increases with increasing magnetization, but faster than the magnetization, so that, when for a magnetization of $\mathbf{B}=3{,}000$ (3,000 lines of magnetic force per square centimetre) the loss by hysteresis amounts to 736 absolute units or ergs per cubic centimetre (10^7 ergs $=1$ watt-second); for four times as high a magnetization, or $\mathbf{B}=12{,}000$, the loss is 6,720, that is, more than nine times as high. On the other hand, the loss increases more slowly than the square of the magnetization, because the square law would require a loss of 11,776 for $\mathbf{B}=12{,}000$.

A great number of experimental researches on the loss of energy due to hysteresis, with different magnetizations, have been made by Ewing; but that law of nature is still unknown, which gives the dependence of the hysteresis upon the magnetization.

In trying to find at least a clew to this law, I subjected a very complete set of Ewing's observations on the hysteretic energy, made on a soft iron wire, and consisting of ten tests from a magnetization of 1,974 lines of magnetic force per square centimetre, up to 15,560 lines per square centimetre, to an analytical treatment by the method of least squares, to ascertain whether the losses due to hysteresis are proportional at all to any power of the magnetization, and which power this is.

The results of this calculation seem to me interesting enough to publish, in so far as all those observations fit very closely the calculated curve, within the errors of observation, and the exponent of the power was so very nearly 1.6, that I can substitute 1.6 for it, and combine those observations of Ewing in the formula

$$\mathbf{H}=.002\,\mathbf{B}^{1.6},$$

where $\mathbf{H}$ is the loss due to hysteresis, in ergs per cubic centimetre ($=10^{-7}$ watt-second) per cycle, and $\mathbf{B}$, the maximum magnetization (number of lines of magnetic force per square centimetre).

In Table I., in the first column, are given the values of the magnetomotive force $\mathbf{F}$, in absolute units; in the second column is given the maximum magnetization or induction, $\mathbf{B}$, in lines per square centimetre; in the third column the magnetic conductivity $\mu=\frac{\mathbf{B}}{\mathbf{F}}$; in the fourth column the hysteretic loss $\mathbf{E}$, in ergs per cubic centimetre, as observed by Ewing, but in the fifth column the hysteretic loss calculated by the formula:

$$\mathbf{H}=.002\,\mathbf{B}^{1.6}.$$

The sixth column gives the differences of the observed and the calculated values, $\mathbf{E}-\mathbf{H}$; the seventh column gives these differences in per cents. of $\mathbf{E}$.

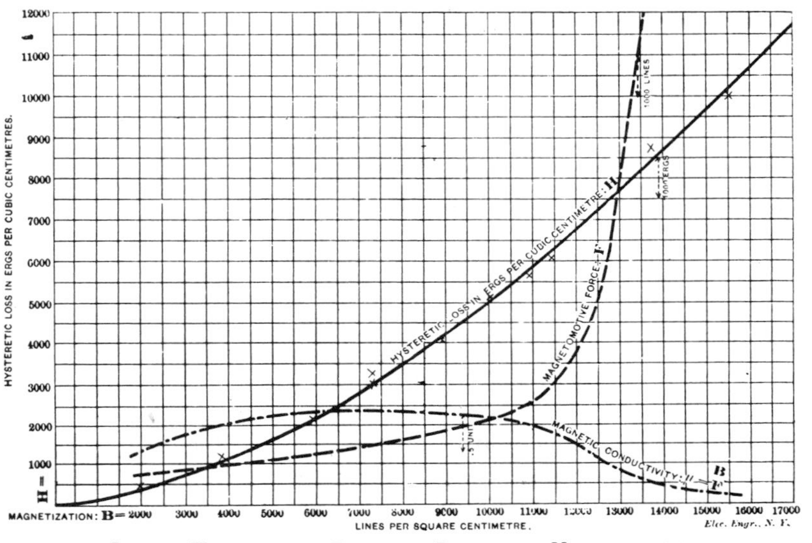

LOSS BY HYSTERESIS AT DIFFERENT DEGREES OF MAGNETIZATION.

In the diagram these calculated values, $\mathbf{H}$, of hysteretic loss are shown in the curve; the black crosses show the values of hysteretic loss $\mathbf{E}$ observed by Ewing.

For comparison there are shown, in dotted lines, the curves of magnetomotive force $\mathbf{F}$ and of magnetic conductivity, $\mu=\frac{\mathbf{B}}{\mathbf{F}}$, as functions of the magnetization $\mathbf{B}$.

It will be seen that the observed values of hysteretic loss are very near the calculated curve through the whole range of observation, and do not show any tendency to deviation, which justifies my considering this coincidence as something more than a mere accident, and, indeed, as an indication of a general law, although certainly this law might be more complicated than the formula.

In Table II. are given the values of hysteretic loss, calculated by the formula:

$$\mathbf{H}=.002\,\mathbf{B}^{1.6}.$$

To one interesting fact I wish to draw attention: The hysteretic loss seems to be independent of the magnetomotive force $\mathbf{F}$, and only dependent upon the magnetiza-

Reprinted from *The Elec. Eng.*, vol. X, pp. 677–678, Dec. 17, 1890.

tion **B**; it therefore shows no special singularity at the point of the beginning of magnetic saturation, but increases in the last two observations in Table I., which, for an increase of **B** by 3,500, require an increase of **F** by 68, showing high saturation, according to the same rule as in the first eight observations, where **B**=12,000 corresponds to **F**=7. Therefore the "knee" of the magnetic curve or "characteristic,"

$$\mathbf{B}=f\,(\mathbf{F}),$$

is no singular point of the curve of hysteresis $\mathbf{H}=.002\ \mathbf{B}^{1.6}$, as the diagram shows.

From this formula we get the loss due to hysteresis per cubic inch of soft iron and for the maximum magnetization of M lines of magnetic force per square inch, when n = the number of complete periods of the exciting alternate current:

$$\mathbf{H} = \tfrac{5}{3} \times 10^{-10}\, n\, M^{1.6} \text{ watts.}$$

TABLE I.

Comparison of Ewing's observed values of **E**, the energy consumed by hysteresis in soft iron, with the values calculated by the equation:

$$\mathbf{H}=.002\ \mathbf{B}^{1.6}.$$

F	B	μ	E	H	E–H	Diff. in %.
1.50.	1,974.	1,330	410.	375.	35.	8.5%.
1.95.	3,830.	1,964	1,160.	1,082.	58.	5.0%.
2 56.	5,950.	2,324	2,190.	2,190.	0.	.0%.
3.01.	7,180.	2,385	2,940.	2,956.	–16.	–.5%.
3.76.	8,790.	2,340	3,990.	4,080.	–90.	–2.3%.
4.96.	10.590.	2,133	5,560.	5,510.	50.	.9%.
6.62.	11,480.	1,734	6,160.	6,260.	–100.	–1.7%.
7.04.	11,960.	1,700	6,590.	6,690.	–100.	–1.5%.
26.5.	13,700.	517	8,690.	8,310.	380.	4.4%.
75.2.	15,560.	207	10.040.	10,190.	–150.	–1.5%.

TABLE II.

Energy consumed by hysteresis, in absolute units (=10⁻ watt-seconds) per cycle and per cubic centimetre soft iron, for the induction of **B** lines of magnetic force per square-centimetre, calculated by the equation:

$$\mathbf{H}=.002\ \mathbf{B}^{1.6}.$$

B	H	B	H
1,000	126	11,000	5,850
2,000	382	12,000	6,720
3,000	736	13,000	7,640
4,000	1,160	14,000	8,600
5,000	1,658	15,000	9,610
6.000	2.220	16,000	10,660
7,000	2.840	18,000	12,870
8,000	3,520	20,000	15,230
9,000	4,240	25,000	21,760
10,000	5.020	30,000	29,280

Electrification of Niagara Falls

In this progress report on the first electric power plant at Niagara Falls, George Forbes, who had designed the plant, described it as "one of the greatest engineering works in the world." This power project that held the record for its power capacity for many years captured the interest of electrical engineers all over the world. It demonstrated conclusively the feasibility of transmitting large amounts of energy for long distances through the use of high-voltage and multiphase techniques. The plant was designed to employ ten 5000-hp alternators of a novel design. They were manufactured by the Westinghouse Electric Company using the design of Forbes, a British consultant to the Cataract Construction Company that had been organized in 1890 to undertake what was regarded at the time as a high-risk project. In his comments, Forbes refers to the "typically American" boldness of the capitalists who backed it and the manufacturers who began buiding factories to use the power before the plant was completed. See a special issue of *Cassier's Magazine* on the Niagara Falls project, vol. 8, 1895. Also see Edward Dean Adams, *Niagara Power: History of the Niagara Falls Power Company, 1886–1918*, 2 vols. (Niagara Falls, N.Y., 1927). Also see Robert Belfield's paper on the Niagara System in the PROCEEDINGS OF THE IEEE for September 1976.

George Forbes (1849–1936) was born in Scotland and educated at St. Andrews University and Cambridge. He became a Professor of Science at Anderson's College in Glasgow in 1872. Forbes began his electrical engineering career in 1881 with the British Electric Light Company. In 1890 he submitted a plan to use a low-frequency two-phase system at Niagara Falls. His plan was adopted in 1892 when he was hired as consulting engineer on the project. See a biographical note in the *Journal of the Institution of Electrical Engineers*, vol. 79, p. 693, 1936.

THE
Electrical Engineer.

Vol. XIX. JANUARY 16, 1895. No. 350.

NIAGARA TO-DAY.

THE accompanying engravings, reproduced from photographs, show the actual conditions of the work at Niagara of which we have already described many features. Fig. 1 shows the exterior of the power house with the inlet canal and Fig. 3 gives a view of the interior, exhibiting the base of two of the great 5,000 H. P. alternators in position.

FIG. 1.—THE POWER HOUSE.

The first of these machines as erected at the shops of the Westinghouse Company in Pittsburg is shown in Fig. 4 and it may be of interest to recall the details of its construction given in our issue of Sept. 26, 1894. The sectional view of the machine shown in Fig. 2 will serve to make the description clear.

To a circular foundation is bolted a vertical cast iron cylinder, provided with a flange on which the stationary armature rests. The inner part of the cylinder is bored to the shape of an inverted cone and serves as a bearing for another conical piece of cast iron, supporting the shaft-bearings. The armature core is made of thin, oxidized iron plates, held together by 8 nickel-steel bolts. In the outer edge of the plates are 187 rectangular holes to receive the armature winding.

The outer rotating field magnet consists of a wrought steel ring to which are bolted the 12, inwardly projecting, massive cast iron polepieces. The ring constituting the field magnet is supported by a six-armed cast-steel spider keyed to a vertical axis. The field-magnets act also as a flywheel. The shaft rests on two bearings supported by four arms projecting from the inner adjustable cast-iron cylinder. The bushings of the bearings are made of bronze provided with zig-zag grooves in which oil constantly circulates. On the outer side of the bushing there are also grooves into which cold water may be pumped, if required.

The armature-conductors are rectangular copper bars 32 x 8 millimetres and each of the 187 holes of the armature contains two of these bars, surrounded with mica. The upper and under sides of the armature are connected by means of V-shaped copper bars, riveted to the ends of the bars which project out behind the ends of the armature. The connections are made so as to give two independent circuits, a pair of cables connecting each circuit with the switchboard. The magnet-winding is also composed of bent copper bars, air-insulated, inclosed in brass boxes, two of which are fastened to each pole-piece. Continuous current for exciting the field magnets is obtained from a rotary transformer.

The current is conducted to the field coils by means of a pair of brushes and two copper rings fixed to the top of the shaft of the generator. At a speed of 250 revolutions per minute the machine produces two alternating currents, differing in phase 90 degs. from each other, each of 775 amperes and 2,250 volts pressure. The alternations are 50 per second. The height from base of bed plate to top of machine is nearly 13½ feet.

An interesting letter summing up the situation, from Prof. George Forbes, has recently appeared in the London *Times*. We reprint it below:—

"Nearly three years ago you published a letter from this place in which I gave some account of how the dreams of the engineer were in the act of being realized, and without injury to the natural beauties of the spot. Three years have passed, my work is ended, and it seems natural to continue the narrative and tell what these three years have brought forth. I am perched on the top of a small Eiffel Tower, lately erected, and, casting my eyes up the river, over the housetops and beyond the town, I see a new world created. There is a wide canal leading water into that gigantic power house where three turbines are set up to drive three dynamos of 5,000 H.P. each. There is the bridge to carry cables across to the transformer house. Inside the power house the water is carried down pipes 7½ ft. diameter into the turbines, and then it passes through a 7,000 ft. tunnel under the town, emerging below the Falls, and capable of developing 100,000 H.P. Far as the eye can reach extend the Company's lands, with here and there a huge factory either now using the water-power, or waiting for the electric supply. One of them uses 3,300 H.P., another 300, a third one 1,500, and that unfinished mill requires 1,000. You can see, far away, the model village for working men, and improved sewage works with drainage, pumps for water supply, electric light, and well-paved streets. There, again, is the dock where ships from all parts of the great lakes can unload, and there a huge expanse of reclaimed land; while the whole is swept by the Company's railway, seven miles long, connecting every factory with the great trunk lines. The power is transmitted by electricity, and the

The Tunnel Outlet.

Reprinted from *The Elec. Eng.*, vol. XIX, pp. 43–44, Jan. 16, 1895.

first work is to produce aluminum with 1,500 H.P. New types of machines have been devised for this work, as also for every other purpose. All criticism as to cost of electric works has been swept away by the results achieved, and the efficiency of each type of machine is greater than has been attained before. All the machinery for the first

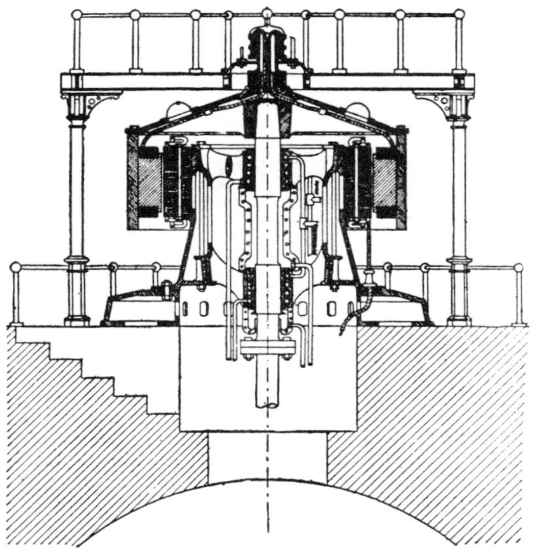

FIG. 2.—NIAGARA 5,000 H. P. TWO-PHASE ALTERNATOR.—SECTION.

working has been made and tested in the shops, and the last parts are now being set up. The plans for carrying the power to Buffalo, 18 miles distant, are complete. In a month or two factories will be in full operation; in a year Buffalo will be supplied ; in two years the same company will be working the Canadian side of the Falls, and in ten

FIG. 3.—VIEW OF INTERIOR OF THE NIAGARA POWER HOUSE.

years (shall we say ?) the whole of the 100,000 H.P. which can be supplied by the existing hydraulic works will be giving power to smokeless manufacturing towns. The period of planning the transmission scheme, of designing the greatest dynamos in the world, and of construction of the first plant now closes. The financial period commences with the new year. The earning of dividends and the ordering of duplicate machinery is the future work of the company. In conclusion, it is difficult for me to say who were the boldest, the capitalists who embarked on the scheme before any plans were matured, or the manufacturers who moved their factories to this field before a single result had been achieved. The action of both was typically American, but their confidence was not misplaced. Their success is now assured."

The turbines built by the I. P. Morris Co., are in position and all the galleries, the ladders, the elevators and the electric lights have been placed in the wheelpit.

Men are now at work putting in the turbine oiling apparatus, with the tanks, filters, pipes and pumps. A novel way of operating the force pumps which take the oil up through the pipes to the upper tank has been developed. All the water which leaks into the wheelpit is collected by a ring around the pit into a reservoir about a third of the way down the pit and is used to run the

FIG. 4.—NIAGARA 5,000 H. P. TWO-PHASE ALTERNATOR.

pumps. A pressure of 90 pounds to the inch is secured by this surplus water.

The main switch-board room in the power house is also well under way. It is to be constructed of white enameled brick and when finished will be 58 feet long, 13 feet wide and 8 feet high. The front will be composed of 10 heavy plate-glass windows, while the top of the inclosure will be an observation platform, inclosed in a brass railing, for visitors. This room will contain some of the heaviest switches thus far constructed and will be one of the most interesting features of the power plant.

The transformer building which stands over the mouth of the electric subway is also well up to the roof. It is of the same style of architecture as the main power building, massively constructed.

It will thus be seen that within probably one or two months the dynamos at Niagara will be running and what must be considered to be one of the greatest engineering works in the world will have become an accomplished fact.

The Niagara Dynamo Controversy

THIS paper from *The Electrical Engineer* of 1895 refers to a controversy that had apparently originated during a heated discussion of a paper by George Forbes at the British Institution of Electrical Engineers in 1893. Sebastian Ziani de Ferranti, himself a pioneer in alternating-current power, alleged that Forbes' Niagara Falls alternator embodied features of a design that C.E.L. Brown had submitted to the Cataract Construction Company. Ferranti also accused the Cataract Company of having never intended to have the machines manufactured outside the U.S. despite its solicitation of designs from European manufacturers. Brown was not present at the meeting, but did contribute to the discussion by letter. At the time he did not lend his support to Ferranti, but was critical of Forbes' design for being too low in frequency and for adopting an external revolving field that gave poor ventilation and made the armature too inaccessible. The controversy was rekindled in 1895, as indicated by this paper. Actually, as later pointed out by B. G. Lamme of Westinghouse, "the design finally adopted was a compromise between those of the Westinghouse Company and of Professor Forbes, embodying what appeared to be the most satisfactory features of both." See B. G. Lamme, "Early Work on First 5000 Horse-Power Alternators," in Edward Dean Adams, *Niagara Power*, vol. 2 (Niagara Falls, N.Y., 1927), pp. 409–421.

C. E. L. Brown (1863–1924) was born in Switzerland and attended a technical school in Winterthur. After spending a year in a machine shop, he joined the Oerlikon Electric Company in 1884 and became head of its Electrical Department two years later. He designed much of the apparatus used in the Frankfort–Lauffen polyphase power transmission experiment of 1891. He was invited by E. D. Adams, President of the Cataract Construction Company, to establish a manufacturing plant at Niagara Falls, but declined and instead helped organize the Brown-Boveri Company in Switzerland. His innovative designs of large slow-speed alternators were adopted widely for use in large power plants during the late 19th and early 20th centuries. In 1912, Brown became one of a small number of foreign engineers to be elected an Honorary Member of the AIEE. See "Electrical World Portraits. C. E. L. Brown," *Electrical World*, vol. 18, p. 284, 1891. Also see B. A. Behrend, "The Debt of Electrical Engineers to C. E. L. Brown," *Electrical World*, vol. 53, p. 812, 1934. See also James E. Brittain, "The International Diffusion of Electrical Power Technology, 1870–1920," *Journal of Economic History*, pp. 108–121, March 1974.

THE NIAGARA DYNAMO CONTROVERSY.

THE controversy as to the authorship of the design adopted for the 5,000 H. P. alternators now being installed at Niagara has broken out afresh. In our issue of Sept. 26, 1894 we illustrated the machine actually constructed and that originally specified by Prof. Forbes for this work. In the London *Electrical Review* of March 15, Mr. C. E. L. Brown, of Baden, Switzerland, publishes the design submitted by him to the Niagara Construction Corporation on Sept. 20, 1892 which we reproduce below, in connection with the design of Prof. Forbes and that actually adopted, for the purpose of comparison. Mr. Brown bases his claims for priority on the following:—

"1. A design in which the employment of a vertical generator shaft, which is a prolongation of the turbine shaft, is rendered practicable.

"2. The 'umbrella shape' form of the field magnet system. This design obviates the necessity for a bearing above the field magnets, by bringing down the poles with their exciting coils into the same horizontal plane, with a bearing placed just above the floor line.

"3. The mode of winding the field coils with flat copper strip edge-on, so that the insulation cannot be crushed or otherwise damaged by the excessive centrifugal forces which, in this instance, will be exerted on the winding.

"4. The method of armature winding. In the scheme submitted I proposed that rectangular copper bars should be buried in slots in the armature, two in each slot; that the insulation should be mica; that the whole winding, including the end projections and connections, should lie on a cylindric surface. This latter arrangement facilitates removal in case of repair, reduces the number of bends, and likewise the number of joints to be soldered or otherwise connected, thus rendering more perfect insulation possible.

"In addition to the above, there are several minor points which it is not necessary to discuss here.

"It is quite true that Prof. Forbes has seen no machine of mine with an exterior rotating field magnet system, but I may in passing mention that, while developing the scheme I submitted, I had this arrangement in view (although I, at that time, in no way considered it a novel feature of design; for example, the same device will be found described and illustrated as employed by Patin in *L'Industrie Electrique* for May 25th, 1892), but its adoption for the case in question seemed to me unwise, for reasons already pointed out in the discussion of Forbes's paper, read before the Institution of Electrical Engineers.

"A comparison of the illustrations in THE ELECTRICAL ENGINEER of New York, January 16th, 1895, with the drawing I publish here, will establish the fact that certainly the first two points for which I claim priority have been introduced into the design of the Niagara alternators. Regarding the third point, the description is not sufficiently minute to say definitely, and as for the fourth, that too is indefinite. Allow me to point out that Figs. 2 and 4 do not agree with each other. In Fig. 2, V-shaped connections are shown, while Fig. 4 leads one to suppose that my method of end connections has been adopted."

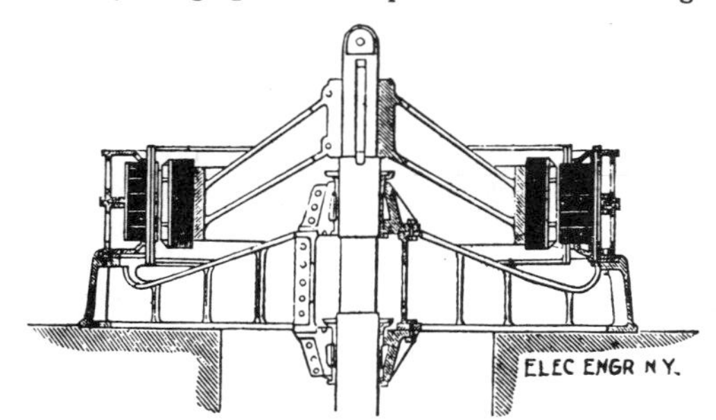

BROWN'S DESIGN FOR NIAGARA 5,000 H. P. ALTERNATOR.

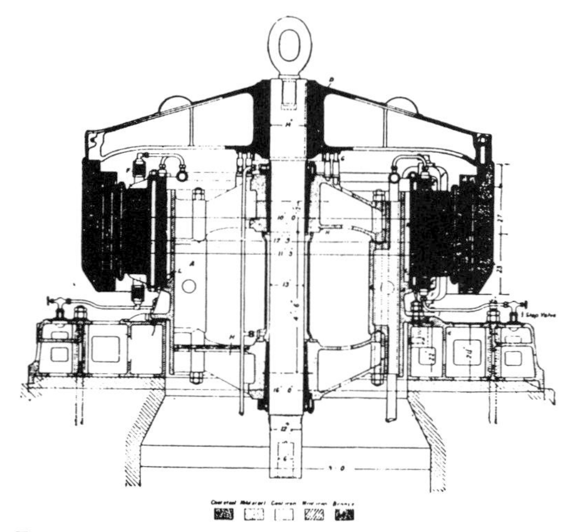

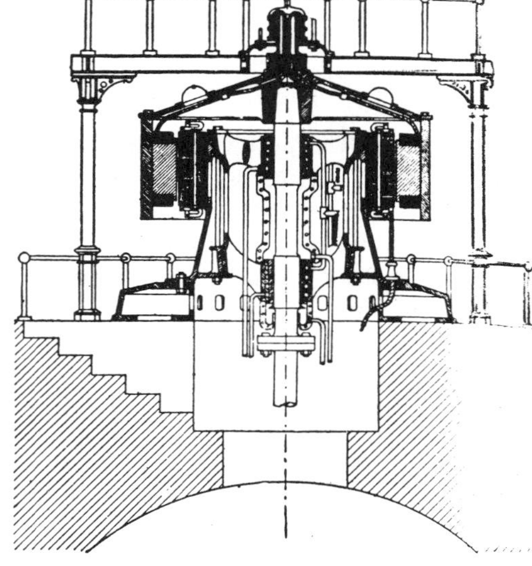

FORBES DESIGN. DESIGNS FOR THE TWO-PHASE GENERATORS AT NIAGARA. DESIGN ADOPTED.

Reprinted from *The Elec. Eng.*, vol. XIX, p. 308, 1895.

The Scott Connection

IN this retrospective paper from *The Electric Journal* of 1919, one of the leading pioneers in alternating-current engineering recalled the circumstances that had led to his invention of a way to effect a transformation from two-phase to three-phase power systems. Scott quotes at length from his 1894 paper that had served to introduce the method to the profession. He also notes its significance for the Niagara Falls to Buffalo transmission line.

Charles F. Scott (1864–1944) was born in Ohio and was the son of a Professor of Greek at Ohio University. After two years at Ohio University, Scott transferred to Ohio State University where he received a B.A. degree in 1885. He then did graduate work at the Johns Hopkins University before joining the Westinghouse Electric Company in 1888. Scott worked with Nikola Tesla on development of the Tesla motor (see Paper 19) and was involved in the design of a pioneering ac installation at Telluride, Colo. that he described in an AIEE paper in 1892. He was also one of the Westinghouse engineers assigned to the Niagara Falls project (see Paper 21). Scott was unusual in that the second half of his professional career was devoted to engineering education. He headed the electrical engineering program of the Sheffield Scientific School at Yale from 1911 through 1933. See the *NCAB*, vol. 13, pp. 207–208. Also see "Charles Felton Scott: Engineer, Educator, Co-ordinator," *Electrical Engineering*, vol. 59, pp. 349–353, 1940.

The Engineering Evolution of Electrical Apparatus - XXXVI

The Development of the Two-Phase, Three-Phase Transformation

CHAS. F. SCOTT

ALTERNATING-CURRENT installations made during the latter eighties were single phase of about 120 to 133 cycles, for lighting circuits only. There were no motors, as the recently invented Tesla rotating or shifting field motor was not suitable for these circuits. Attempts were made to use this principle for the operation of split-phase motors from a single-phase circuit and also to operate polyphase motors at high frequency, but neither of these attempts was successful. Later, when the frequency was reduced and polyphase generators were used, selection had to be made between two-phase and three-phase. Both were produced and there were certain conditions where one was better, while for other conditions the other was preferable.

At that time, when polyphase generators were installed the motor service was usually incidental. The lighting service continued by single-phase circuits and it was simpler to supply such circuits in two groups from a two-phase generator than it was to divide them into the three groups necessitated by the three-phase generator. The interaction between the phases was less and it was simpler to obtain the desired voltage regulation from the generator or from voltage regulators.

Some industrial installations were made for both light and power and in such cases it was usually found that the two-phase system was to be preferred. The long distance transmission circuit with its lesser cost for a three-phase transmission line was scarcely a factor in the central station work, which consisted primarily of single-phase lighting circuits.

The Niagara Falls Power Company, in the first large and pre-eminent alternating-current power plant, employed two-phase generators as it was expected that a large part of the power would be used locally as single-phase for the operation of electric furnaces and electrochemical processes.

Some of the early forms of generator windings and the original form of the polyphase motor designed by Mr. Tesla lent themselves much more readily to a construction for two-phase than for three-phase. The greater simplicity in measuring instruments and in the use of two instead of three transformers for reducing the voltage were also contributing influences in the preference for the two-phase system. In those early years when lighting circuits were being changed from 133 to 60 cycles and the single-phase system was being supplanted by the polyphase system to accommodate induction motors, the inherent regulation of small generators and the kind of regulating and control devices available were very different from present standards. In short, the features of the apparatus and its use for lighting and for incidental power work from central stations and in factory plants gave the preference to two-phase.

One afternoon Mr. L. B. Stillwell came into the Westinghouse Laboratory and said that in a certain negotiation then pending the Company was at a decided commercial disadvantage in proposing a ten mile transmission by a four-wire, two-phase circuit, as a competitor offered three-phase apparatus with a saving in the cost of transmission line of some $10 000. He was considering whether a proposal should be made for supplying the three-phase apparatus. I pointed out some of the manufacturing objections to supplying three-phase equipment when our ordinary products were two-phase, and also recounted the advantages in having a two-phase distributing system. He reiterated, however, that a $10 000 saving in the transmission line was a handicap which would be hard to overcome. It occurred to me that the ideal arrangement would be a three-phase transmission line for supplying two-phase distributing circuits; to which he agreed.

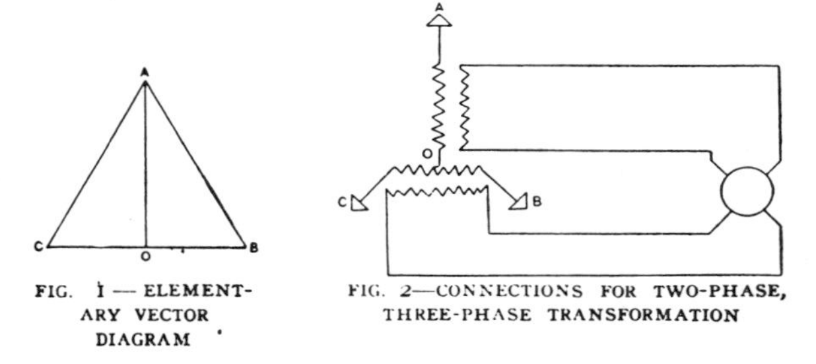

FIG. 1 — ELEMENTARY VECTOR DIAGRAM

FIG. 2—CONNECTIONS FOR TWO-PHASE, THREE-PHASE TRANSFORMATION

I went to my desk, drew an equilateral triangle, then in a half musing, half mechanical way, without having formulated a definite purpose, drew a line from the upper apex to the middle of the lower side of the triangle. This line made a right angle with the side which it intersected and in an instant my imagination made these two lines represent a two-phase system combined with the triangle representing the three-phase system. I rushed back into the adjoining room, exclaiming "I've got it." "What?" "The way to get from three-phase to two-phase." I don't think I had been away more than a minute.

Thus when the commercial necessity was presented, the engineering problem was formulated and a simple vector diagram was the key to the solution.

I was at that time preparing a paper* for the National Electric Light Association in which was described the effect of self-induction and its bearing upon

*See *N.E.L.A. Proceedings* for 1894; also THE ELECTRIC JOURNAL, Volume II, p. 713.

Reprinted from *The Elec. J.*, vol. XVI, pp. 28–30, 1919.

voltage drop in transmission circuits—a matter which was not then generally understood. The method of phase transformation was added to the paper, which was presented at Washington, D. C. on March 1st, 1894. The description then given follows:

"In considering the marked advantages of the two-phase system for distribution and of the three-phase system for transmission, it occurred to me that a combination of the two systems might secure the advantages of both, and I have worked out a simple and effective method of accomplishing this result. If two e.m.fs. differing in phase be connected in series, the resulting e.m.f. will in general, differ in value and in phase from either of its components. If two e.m.fs. differing in phase 90 degrees be connected in series the resultant e.m.f. is represented in direction and magnitude by the hypotenuse of a right angle triangle, of which the two sides are the two component e.m.fs. Thus in Fig. 1 if AO and OB are two e.m.fs. at right angles and these e. m. fs. be connected in series the resultant is the line AB, of different phase from either of the components.

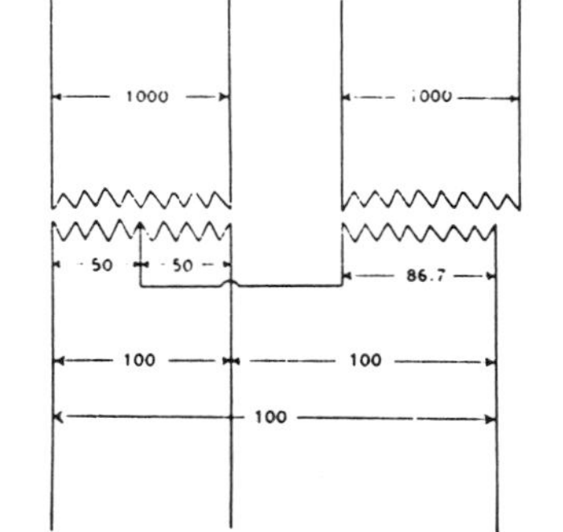

FIG. 3—APPLICATION OF TWO-PHASE, THREE-PHASE PRINCIPLE TO TRANSFORMERS

It is a simple matter to so proportion the components that OB is equal to one-half of AB, as shown in the diagram. In a similar manner it is readily seen that the same e.m.f. OA may be combined with OC, which differs from it by 90 degrees (but is equal and opposite to OB) in such a way as to give AC equal to AB, but differing in direction. BO and OC added together give BC. The e.m.f., BC, may therefore be combined with the e.m.f. AO. at right angles to it, in such a way as to give additional e.m.fs., AB and CA, which in connection with BC, give three equal e.m.fs. 120 degrees apart. This is the relation of e.m.fs. in the three-phase system.

"The application of this arrangement to transformers is illustrated in the accompanying Figs. 2 and 3. The primaries of two transformers are connected to a generator giving two-phase current. The secondary e.m.f's., therefore, differ 90 degrees. One secondary is made equal to 100 turns and a loop is brought out at its middle point giving 50 turns at each side. The second secondary has 87 turns, which is approximately equal to 50 multiplied by $\sqrt{3}$. One end of the secondary circuit is connected with the middle point of the secondary of the first transformer, as shown, and the three free terminals will then deliver e.m.f's. differing in phase 120 degrees. If the e.m.f. on each primary be 1000 volts and on one secondary 100 volts and on the other 87 volts, then the e.m.f. measured between any two secondary terminals will be 100 volts. This three-phase circuit is adapted for operating three-phase motors. In a system of transmission two-phase currents at the generator may be converted into three-phase currents, as shown in Fig. 4 and the windings may be such that the e.m.f. is raised for transmission. The currents are then transmitted by three phases, effecting economy in copper. At the other end of the line a similar arrangement of transformers may be used for converting from the three-phase to the two-phase system. The two-phase currents may then be used for the operation of two-phase motors or the circuits may be independently loaded with lamps or otherwise. If lamps be placed on the transformer which supplied current directly from its two terminals, the transmission is directly from the generator without affecting in any way the other circuit, as the generator terminals are connected directly to the first primaries, and the secondaries of the raising transformer are connected directly to the

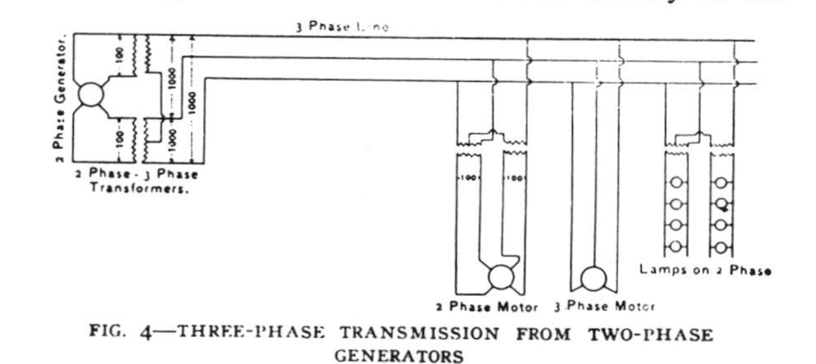

FIG. 4—THREE-PHASE TRANSMISSION FROM TWO-PHASE GENERATORS

primaries of the lowering transformer, and the current from this is taken directly to the load. On the other hand, if the other circuit be loaded, the action here will also be on its own generator circuit without affecting the first. The current from the raising transformer in this circuit passes to the middle of the secondary of the other transformer, where it divides and flows in parallel through the two parts of the coil and two of the lines. As one-half of the current flows through each part of the secondary coil in opposite directions, the self-induction is completely neutralized and the transformers in this circuit are independent of the operation of the other circuit. It is to be noted that under this condition the line e.m.f. delivered by this transformer is only 87 percent of that delivered by the first transformer. The lower e.m.f. however, is compensated for by the fact that the current on one side is passed through two of the lines in parallel, thus reducing the resistance of the circuit and compensating for the slightly lower e.m.f. The effect upon the regulation of the generator when two-phase circuits at the end of a three-phase transmission line are independently loaded is found both by theory and test, to be the same that prevails when the load is placed directly upon the corresponding circuit of the generator.

"A modification of this system is found in the arrangement where the three-phase current is produced in the generator and transmitted over three wires to the reducing transformers. These transformers may be arranged as described for producing two phases (Fig. 5). Loads may be placed upon either of the two-phase circuits, and practically the same regulation in the generator will result that would have resulted if the generator itself had been wound for two phases and one of these circuits loaded. In this way it is possible to place a lighting load upon a three-phase generator in two instead of three units, and to avoid the bad regulation in the generator due to unequal loading.

"A similar arrangement of two transformers may be used for converting three phases of one potential into three phases of another potential, as shown in Fig. 6.

"The efficiency of two transformers arranged for converting from two-phase to three-phase is reduced below that when working independently on ordinary loads by an insignificant amount. If the efficiency in ordinary working is, say, 97.5 percent, it would be reduced to 97.4 percent in converting from one number of phases to the other.

"The Tesla polyphase system, adapts itself with marvelous facility not only to all branches of electrical industry, but also by the transformation of its phases to

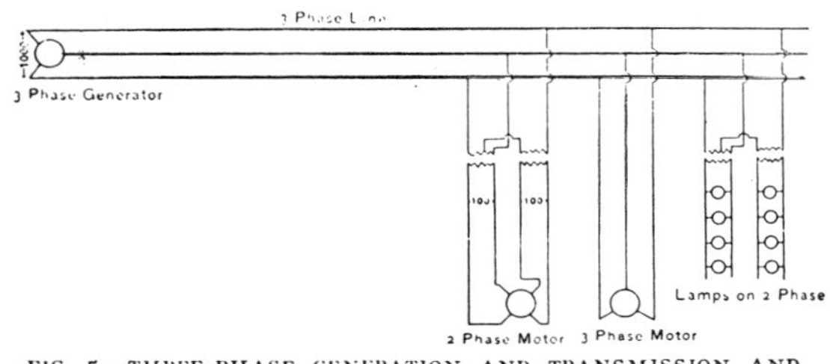

FIG. 5—THREE-PHASE GENERATION AND TRANSMISSION AND TWO-PHASE DISTRIBUTION

the utilization of three phases for gaining the highest economy in transmission and of two phases for securing the maximum advantages in distribution."

One of the early applications of this method of transformation was made in connection with the first transmission lines from Niagara Falls to Buffalo. These circuits were at the time, in point of amount of power transmitted, the most important transmission of that period, although there were smaller amounts of power transmitted in some cases at a higher voltage, or over a greater distance. Transformation from the two-phase generators was made for transmitting the power to Buffalo over three-phase circuits.

In connection with the development of the transformation from two-phase to three-phase, I asked B. G. Lamme if he knew of any method of accomplishing this result. Mr. Lamme at that time was working on the construction of induction motors on the basis of distributed primary windings instead of the polar type. As was the practice in those days in alternating-current generators, a closed coil, two-circuit type of winding was tried on the earliest distributed field induction motors and there were four taps on the winding for quarter phase. He suggested in an offhand manner that the only way he could think of was to put three-phase taps on the primary winding of one of the two-phase induction motors, as with such an arrangement, if two-phase current was supplied to the primary three-phase could be taken off. However, he did not consider this a very practical scheme, and it was not deemed worth patenting. The interesting feature of the suggestion is that this additional method of phase transformation is the only one, aside from the use of transformer connections, which has since been used to any extent.

I had previously noted that when a two-phase induction motor is running idle with one circuit open, so that it is being supplied with single phase, there is a voltage on the terminals of the idle winding of the motor which is approximately equal in value, but is 90 degrees from the impressed voltage on the other circuit. The idle motor was therefore a phase converter and the voltage of its idle winding, combined with the voltage of the supply circuit, furnished two voltages differing 90 degrees which constitute a two-phase circuit and could have been employed for operating small two-phase motors. At the time it did not occur to me that

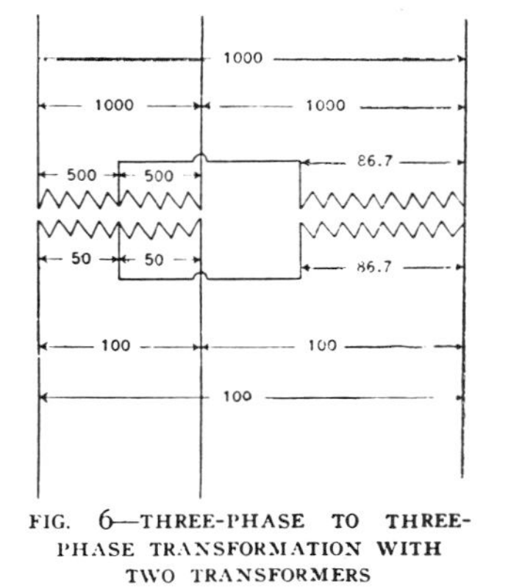

FIG. 6—THREE-PHASE TO THREE-PHASE TRANSFORMATION WITH TWO TRANSFORMERS

this principle was one which might be useful. It is at present, however, being employed on the electric locomotives of the Norfolk and Western Railroad which are the most powerful electric locomotives in service. On these locomotives, the single-phase trolley supplies current to a rotative phase converter which is in fact a two-phase induction motor operated from one phase only and producing in its second winding a ninety degree electromotive force which, in conjunction with that of the trolley circuit, constitutes a two-phase supply to a pair of transformers for transforming from two-phase to three-phase. The three-phase circuit then operates the three-phase propulsion motors.

Possibly if a commercial need for this sort of transformation had been definitely presented, the importance of this transformation from single-phase to polyphase by an idle induction motor might have been recognized years earlier. Under the circumstances, however, a patent would never have amounted to much, so far as the locomotives are concerned, as they were not built until after the patent would have expired.

Sprague on the Electrification of Urban Transportation

THIS paper is the transcript of a talk by Sprague in 1891 shortly after completion of his pioneering electric streetcar system in Richmond, Va. Sprague was the leading innovator in a revolution in urban transportation that led to the replacement of thousands of horses and mules by electric motors. In this talk, he reviewed the circumstances that had led him to become an electrical enthusiast and some of the enormous technical difficulties encountered during the successful effort to introduce electric propulsion for the Richmond system. Sprague's amusing play on the word "providence" is a reflection of the fact that the site of his talk was Providence, R.I. See Carl Condit's paper on railroad electrification in the U.S. in the PROCEEDINGS OF THE IEEE for September 1976.

Frank J. Sprague (1857–1934) was born in Milford, Conn. and graduated from the Naval Academy in 1878. He served a two-year tour aboard the "Richmond" (a striking coincidence?). In 1882 Sprague attended an electrical exhibition in London where he became acquainted with E. H. Johnson, an employee of Edison. Johnson persuaded Sprague to resign from the Navy and join the Edison Company. Pursuing his interest in electric motors, Sprague soon resigned from Edison to organize the Sprague Electric Railway and Motor Company in partnership with Johnson. The Company enjoyed considerable success with industrial motors while Sprague began experimenting with electric propulsion. An opportunity to demonstrate the advantages of electric streetcars on a large system came in 1887 when Sprague signed a contract to install a system in Richmond. It was completed the following year and became the pattern for future installations that proliferated during the next decade. The Sprague Company was absorbed by Edison General Electric in 1890. Sprague continued his role as an innovator by developing vertical urban transportation in the form of electric elevators. His experiments with elevator control led to his invention of multiple-unit control for elevated electric transit systems, an innovation that he introduced on the South Side Elevated Railway in Chicago during 1898. See "Electrical World Portraits. Frank J. Sprague," *The Electrical World*, vol. 14, p. 163, 1889. Also see the *DAB*, vol. 21, pp. 669–670 and Dugald C. Jackson, "Frank Julian Sprague," *Scientific Monthly*, vol. 57, pp. 431–441, 1943. See also Harold C. Passer, "Frank Julian Sprague: Father of Electric Traction," in *Men in Business*, edited by William Miller (Cambridge, Mass., 1952), pp. 212–237.

THE PRESIDENT: First, almost, in the field of transmission and utilization of electrical energy for power purposes, and, I believe, the first to make a practical application and demonstration for the propelling of our street cars, is a gentleman whom we have with us to-day, so well known to you that I shall simply introduce him by calling his name—Frank Sprague.

SPRAGUE ELECTRIC RAILROAD.

MR. SPRAGUE: Mr. President, ladies and gentlemen: I feel somewhat embarrassed by being called upon to speak here to-day. The request that I should do so has been somewhat sudden. While the hum of the motor may be called the song of emancipation, like the hum of many another song, it is best when quietest. Since my early musical education was neglected, I am not able to compose music to go with it. I once made the remark that I owed much to Providence energy, and the industries of Providence, for some of the earliest success in electric railway work. A gentleman looked at me and said in a somewhat sarcastic manner, "I think you do owe much to Providence." Our thoughts to-day seem to run in a reminiscent past, and I will indulge in a little bit in that direction myself, simply because the commercial history of the electric railway is well known, while, perhaps, some of the inside history of it you do not know, and the story that I shall tell, in a very brief way, is nothing more or less than the story which is common to every earlier enthusiast in the commercial development of electric railway or electric light. I remember in 1879 I was on duty on board of a naval

Reprinted from *Proc. Nat. Elec. Light Assoc.*, vol. IX, pp. 150–158 (extract of a longer paper), 1891.

vessel of the United States on the coast of Japan. I had before that time been at the naval academy, graduating there in 1878. Just at that time Edison, Bell, and other electricians were prominent in the field, and there grew up in my mind a love for electrical matters, and the desire to be associated, in some way, with electrical developments. The United States Navy Department at that time was not very encouraging to any such ambition. I went the way of all other midshipmen, and drifted out to Asia, but the stories of the discoveries being made in electric lighting by Thomson, Brush and Bell drifted out there, and I became very nervous under it. At that time I was trying to invent an electric motor. I meditated very seriously sending a request to the Navy Department to be allowed to come home on a sailing vessel, as I could not well afford to come home in any other way.

Well, by some good intervention, I was ordered back to the United States, and I went to Ansonia, Conn., and saw an experiment in the transmission of power which greatly impressed me. Mr. Wallace, of Wallace & Sons, of Ansonia, is a man whose position in the history of this enterprise has never been fully appreciated. I went there, knowing him personally, and was taken through his factory, and he said, "Sprague, let me show you something." He had one of those old four-pole armature machines. He had connected it with a dynamo. He turned on the current, and pretty soon this thing was running two or three thousand revolutions a minute. He said, "It is no good." I said, "Why?" He said, "I will show you." And he turned to one of the men—a large, strong man—and said, "Get hold of the armature." This armature, I suppose, was about ten inches in diameter. He got hold, with the brake on the arm of it, and finally began to bring it down, lowering its speed, and finally brought it to a dead rest. In that simple experiment he outlined and he overlooked the most important fact that he had to deal with—that the energy which was being developed was not the sole question to be considered, but there was also to be considered its torsional effort multiplied by the speed with which it was traveling. Had Mr. Wallace seized on that single point, or had I, or had anybody else seized upon it at that time, the transmission of power would then have been very much advanced.

Mr. Wallace at that time, if I remember rightly, was building a machine invented by Mr. Farmer; certainly one of the earliest and most esteemed pioneers of electric science in this country. Mr. Farmer was then a crippled man, confined to his chair. He was attached to the torpedo station at Newport, and he had invented a machine for use in torpedo work on board ship. One of his machines was one of the first machines that I ever had the pleasure of dealing with. With a lack, perhaps, of some little formality, which is necessary in official life, I applied for the whole second floor of a machine shop in the Brooklyn Navy Yard. I think that was in 1880 or 1881. I got one lathe, and in that lathe I set up a Wallace-Farmer machine, and tried the first experiment I ever tried in the building of a dynamo or a motor. My work from that time fell off. My duties called me out of the United States, and it was not until 1882, at the Electrical Exposition in London, that I was brought to a consideration of the tremendous advance that electricity was making.

Few of you are familiar, perhaps, with the inside difficulties with which we had to deal in the history of the electric railway enterprise, and the amount which it was necessary to conceal from the general public. The contract for the road in Richmond was taken in 1887.

I had never seen the road. I remember to this day the impression that came over me. I had been talking with John Stephenson, the pioneer car builder of the United States, and he said, "I do not believe that any self-propelling car can operate under street car conditions. I am acquainted with 14 of them in the city of Paris." I said, "Mr. Stephenson, there has been one difficulty with those; not one of them utilizes the weight of the car for purpose of traction." The first machine that was built, was built for storage battery experiments on the West End in Boston. Subsequently, when I went to Richmond, I saw a grade of 10 per cent. My heart fell within me, and I said, "It is utterly impossible for any car to climb that hill." I came back to New York, and had some consultation in a retired room in New York that settled this question: Is it possible for any car to mount a 10 per cent. grade? Everybody said it is not, and I confess that I myself thought it was not. Shall we use a cable on that 10 per cent. grade and trust to electricity for the balance, or shall we change the gearing of the machine? We had a single geared machine. Mr. Johnson, who, I think, is a sort of John the Baptist in the electrical profession—a sort of forerunner of good things—said we had better find if we could get up that grade at all. Back to Richmond I posted. I started out one night when there had been an election, and when the streets were filled with a crowd of drunken men. We went down in great shape, and then just before us was this grade. I said to the Superintendent, "We cannot mount it." "Well," said he, "It will go up, and I will bet you five dollars I will take you up." We started, and we mounted the hill. When we got to the top, it had settled the question of traction on a 10 per cent. grade. It had settled, likewise, that the motors we had were altogether too small and too light, because they were hotter, perhaps, than the furnace in an electric welding apparatus. On getting to the top, we thought the best thing we could do was to stand still a little while. I remember one of my assistants, Mr. Green, was with me. I said to him, "Green, I think we had better send for some instruments, I think a little accident has happened to one of the machines." So we lay down on the bottom of the car until he brought those instruments. They were four strong mules, for it was necessary to get that car back into the car shed that night. As I say, we are indebted very much to Providence skill and energy for the success of that road. That fact brought me face to face with the fact that it was necessary to make a complete change on 60 machines in Richmond, in the matter of gearing. I was thoroughly at a loss what to do. There was no factory to which I could go to get that built, excepting one, and that was the shop of Brown & Sharpe in this city. I said to the foreman, "We have met the worst obstacle I have ever seen, in Richmond. We have got 60 cars we are under contract to run. If we fail there it will delay the electrical development in this country; we, likewise, personally, will go up. Now," I said, "I have got everything at stake. My associates have got every dollar at stake. The electric railway in this country is liable to be at stake. The road has got to go, and go it must. Will you do for me what I ask, and put as many men and as much money and material at my command for 24 hours a day as will be needed until I recover the position we have lost?" He said, "I will." They went to work and made probably the most difficult piece of gearing they were ever called on to make. In the course of four or five weeks, we started again in Richmond on the 2d of February, 1888. The gearing had all stuck for lack of oil. Then the next thing we found was that the armatures were too large, and so we had to resort to electrocution ; we had to cut off their heads. We were using a 450 ground circuit, a machine that was covered with water and mud. Every field magnet had been wound and rewound three or four times. Our next experience was, that our commutators were going to pieces. We had tried no less than 65 different kinds of trolleys, and had not got one that would work reasonably satisfactorily. We were attempting to run motors in both directions, and with a brush that was difficult to handle, and, after trying a great variety of brushes, we finally settled upon square bars of brass about three-eighths of an inch square, and every time a car would go through the street you would see a shower of brass dust, so that we had the whole road marked from one end of the town to the other with a sort of golden path. The copper we were using was costing us nine dollars a day, and it was raw material. Subsequently we adopted the use of carbon.

This inside history of some of the work at Richmond has, of course, not ordinarily been told. I think the time has come, now, when there is no harm in its being told. I think that people would have looked on electric railways as out of the question, if they had known of the straits we were going through. In the last three years you have become perfectly familiar with the first geared machine ; that is, a machine that is concentrated on the driving axle and geared to the car axle. It was built by the Brown & Sharpe Company. They had one of the elevated railroad trucks in their shops some five or six weeks, and I think they are satisfied it will be the last they will have.

The Armington & Sims engine was the first engine we ever had, and the engines put in in Richmond have never to my knowledge been changed or caused the slightest trouble from that day to this.

The wire we first used was made by the American Electric Company in this city. The forgings we used were from Bridgeport, near by. So, almost all the material and the work done at Richmond either emanated from Providence or from places within a short distance of it.

There is a profession which owes much to the efforts of the electric railway people ; that is the legal fraternity. I see a representative here of an enterprise and industry with which we had some conflicts—the Bell Telephone Company. I have heard since I have been here a criticism that I should have been an advocate of an overhead wire—one only, remember—and had taken upon myself to get out an injunction against the Metropolitan Bell Telephone Company of New York. About two or three weeks ago we had a very providential storm, and in what I considered one of the most beautiful portions of New York there was a magnificent line of poles, with 170 beautiful wires in all directions, and one to cap them, making 171 wires. A flaky fall of snow occurred, and we found this army of poles lying flat in the street. We thought that the telephone company had so well built its line that Providence never would aid us. But we immediately had a meeting of property holders and by the next morning we had an injunction against the erection or maintenance of any poles or wires on the street. I have been somewhat criticised, but I think it nothing more than turning the tables on that company which, for the past two years, has been so successful in sitting down on the infringers of its patents. I think the public has some reason to be satisfied, because the grounded circuit telephone is very largely disappearing from use. A more perfect telephone service is being adopted. Sixty per cent., if I remember rightly, of the lines in New York City are now underground, and we hope to see the balance of them there. As to the future of the electric railway, I heard with some interest what Mr. Monks stated. His objections to the double equipment have been based somewhat upon the cost. I think we all know the experience that the electric lighting apparatus has gone through, and none of us can doubt from that experience, that it is going to be made more simple, lighter, stronger, more reliable and at much less expense; so that the objection which he raises, I think, is not entirely a sound one. It is based upon past results, not upon future promises. The gearing of the present motor will disappear from the electric motor in the future, and it is a very near future. It will also be an electric motor driving direct without any reduction whatever. Its revolutions will be coincident with the revolutions of the wheel. The day of the gearing is fast approaching, and those who have had experience of gears with electric motors can surely feel encouraged by that fact.

Of late, I have been in exactly the position of Mr. Thomson, where the infant swallowed up the larger specimen, and I have also been very much in the position where the infant got very thoroughly swallowed up. But the electric railway problem in New York City—the question of rapid transit—is coming to the front. That we shall have there electric railway transit I think no man who knows the future of this industry will doubt. What form it will take no man will tell. While I am an advocate of one wire, if necessary overhead, I am an advocate of everything being put underground as far as possible. I do not believe in the theory of God's free air in New York City, where the air is monopolized by three things—buildings from 10 to 14 stories high, the elevated railroads and the Bell Telephone Company. But I believe in God's free earth, where we can go independent of grades and independent of weather conditions. We have a most remarkable example of the perfection of that operation in London. I had the pleasure of visiting the Suburban and South London Railway under the guidance of Mr. Hopkinson, the engineer, and I told him there could be no question of the success of that enterprise. I think information since received of the progress has fully justified that prediction. I thank you, gentlemen, for your kindness. (Applause.)

♦♦♦ ♦♦♦ ♦♦♦

Behrend Describes a Revolution in Electrical Generators

IN this paper from the *Electrical World* of 1913, a leading designer of large alternators of both the old and new types gave his impressions on the revolutionary changes in machine design during the past decade. These had been associated with the rapid replacement of giant slow-speed alternators of the type introduced by C. E. L. Brown (see paper 22) by high-speed alternators driven by steam turbines. Behrend seemed slightly nostalgic for the older style generators, and suggested that a quasi-Darwinian "principle of natural selection" was "leading to the evolution of types no better than necessary to hold their own." His comments on the dangers of hidden flaws in large forgings reflects an episode from his own engineering experience. The rotor of a large generator that he designed for a Niagara Falls power plant had exploded during a runaway test in 1908. As a result, he had introduced the plate rotor to enable use of rolled plates instead of large forgings, an innovation that he suggests "marks what one might almost be tempted to call an epoch."

Bernard A. Behrend (1875–1932) was of German parentage, although he was born in Switzerland where his parents were temporarily residing. He graduated from the Engineering College in Charlottenburg in 1894 and worked for a year as assistant to the well-known consulting engineer, Gisbert Kapp. Behrend then took a position as an electrical engineer with the Oerlikon Company in Zurich, Switzerland where he became familiar with the designs of C. E. L. Brown (see Paper 22). He came to the U.S. in 1898 and was hired by the Bullock Electric Company in Cincinnati as a designer of ac machinery. He designed a large slow-speed alternator fondly known as "Big Reliable" to furnish electrical power for the St. Louis Exposition of 1904. Behrend also designed an experimental radial-slot turbogenerator that won him a gold medal at the same Exposition. When the Bullock Company was acquired by Allis-Chalmers in 1904, Behrend became chief electrical engineer of the Allis-Chalmers Company and remained in that position until late in 1908. In 1909 he was hired by the Westinghouse Company which almost immediately adopted his plate rotor design for its large turbogenerators. Two years later he opened a consulting engineering practice in Boston with Westinghouse as a principal client. See "Electrical Engineers of the Day. B. A. Behrend," *Electrical World and Engineer*, vol. 38, p. 4, 1901. Also see the *DAB*, vol. 21, pp. 65–66 and the *NCAB*, vol. 35, pp. 275–276.

Developments in Electrical Machinery

By B. A. Behrend

As it is easier to summarize the past than it is to forecast the future, so it is easier to study evanescent types than to make predictions as to those which will replace them. With the sounding of the death knell of the reciprocating steam engine in large units, the virtual disappearance of large engine-type generators has commenced. It seems pathetic that the labors of the last thirty years, culminating in the development of almost perfect types of slow-speed electric generators, have not been of more permanent value. Their effect has been chiefly instructive, as far as the present types of electric machinery are concerned. The gas engine has not taken the important place which its friends predicted for it half a dozen years ago.

Although statistical figures are not available, it is not unreasonable to assume that the aggregate horse-power of gas engines produced per annum is only a small fraction of the horse-power of steam turbines produced in a corresponding period. The numerous attempts at compounding these generators which used to be made have been altogether given up and, if there is any notable tendency in regard to the regulation of these generators, it is in favor of coarser regulation, rather than finer regulation, leaving the task of keeping the voltage constant to the switchboard operator.

Low-Cost Tendencies

There is a general tendency, noticeable here as elsewhere, aiming at a reduction of cost and, while this tendency is of course natural, it is probably accentuated by the knowledge of the relatively short time during which any particular type of machine has been able to hold its own. Quick succession of types directly tends to a desire to install short-lived machinery of low cost, rather than long-lived machinery of high cost, as new ideas and improvements may lead to obtaining economies and advantages which could not be obtained with the older types. This is really to-day a characteristic of engineering development in all fields. Rapid progress has unsettled our opinions and played havoc with our standards. Quality has to yield to cheapness as permanence is called in question, and the guiding principle of natural selection, leading to the evolution of types no better than necessary to hold their own, finds application here as elsewhere.

We may now turn from this field to another. The energies of the electrical designer have been concentrated upon the development of the high-speed generator for direct connection to steam turbines. The progress achieved in this direction is brought out clearly by the accompanying table, giving the maximum rating obtained for certain speeds during the last eight years.

DATA ON HIGH-SPEED GENERATORS DIRECTLY CONNECTED TO STEAM TURBINES

Year	R.P.M.	Maximum Rating, Kw
1904	3600	400
1909	3600	2,500
1912	3600	5,000
1904	1800	1,500
1909	1800	10,000
1912	1800	20,000

The tendency to push the limit of speed farther and farther and to reduce the cost of the unit and increase its economy of operation seems to have reached a goal. Not only have almost all known special alloys been called into use and taxed to the limit of their strength, but also it seems as if a further increase in speed would increase the losses and thus lead to lower efficiency, while the need of higher grades of material and greater care of workmanship and inspection would enhance the cost of production.

Special Materials

Thanks to the stimulation of the automobile, alloy steels have been produced of an elastic limit of 100,000 lb. per square inch, and of an ultimate strength of 125,000 lb. per square inch, showing a ductility better than that ordinarily obtained in the regular carbon steels, and this even in comparatively large forgings. Although such large forgings, which have to be used extensively for turbo-generators, have been improved from year to year, and although their manufacture is now less of a speciality than it used to be, yet the difficulty of heat treatment and the danger of concealed flaws is such as to make the careful designer prefer the use of rolled materials wherever possible to large steel

Reprinted from *Elec. World*, vol. 61, pp. 9–10, Jan. 4, 1913.

forgings. A rolled plate receives a more thorough work through its texture than a forging, insuring greater homogeneity, which is of prime importance. Thus, in the building of turbo-generators, the use of steel plates marks what one might almost be tempted to call an epoch.

Ventilation

It is evident that the great development of heat, due to the concentration of comparatively large losses in single units, represents a formidable problem for the engineer. In a 20,000-kw unit the losses dissipated in heat will reach approximately 400 kw, and this heat is developed in so small a compass that the old methods of carrying it away from the generator are no longer efficacious. It has been found necessary to subdivide the iron and the copper, not only through the regular ventilating ducts, but also through longitudinal ducts through the cores of both the rotor and stator. In fact, the tendency toward the utilization of such axial ducts for carrying off the heat is clearly in evidence. We are passing in these machines from radial ventilation to axial ventilation and, while heretofore powerful fans, mounted on the ends of the generator shafts, were sufficient to pass air currents through the generators to carry off the heat, in the latest units it has become advisable to install separate blowers, which can be operated at speeds more economical than the generator speeds, leading to a greater efficiency on the part of the ventilating mechanism, besides allowing a more scientific design unhampered by the constructive features of the turbo-generator. Future power plants will have large blower plants carrying away the hot air developed by the electric generators, similar to the present installations used for air-blast transformers.

Friction Losses

With the high rotative speeds of turbo-generators there has come as a grave difficulty the great loss due to aerial friction. This loss, usually styled windage loss, has become so great that it constitutes a serious argument against further increase in speed. The gain in efficiency due to the better utilization of steam in the steam turbine must, at a certain point of speed of the generator, be counterbalanced in the increased windage losses of the generator proper. This is a matter of calculation for each size and speed where the efficiency of the steam turbine is accurately known beforehand. There is no doubt that ideas at present seemingly abortive, as, for instance, operating the generators in a vacuum, irresistibly suggest themselves to the designer, and he must remember that here as elsewhere new ideas are the heresies of to-day and the superstitions of to-morrow.

Direct-Current Turbo-Generators

The direct-current turbo-generator has been a will-o'-the-wisp during the past few years. Attempts to adapt the unipolar type to it have not been altogether unsuccessful, and this seems really the best that can be said. More and more we begin to realize that large direct-current units are not necessary, and that they are certainly not desirable. The necessity of distribution of electric energy leads logically to the generation of alternating currents which are transformed into direct current by means of synchronous converters and motor-generator sets. For the time being, we must see the solution of the problem of generating direct current at high rotative speeds in this method of transformation, without considering it at all as final.

Electrical Devices for Automobiles

To sum up, the progress in the development of electrical machinery during the past year has been largely in the refinement of details, and even there it cannot be said that the year 1912 has been remarkable for notable improvements. The attention of the electrical engineer has been directed toward the development of small electrical devices for the automobile industry, as the electric lighting of automobiles, the starting of the gasoline motor and the combination of these two electric devices with the ignition of the gasoline motor. Marked and ingenious work has been done in this field, and, in view of the great commercial and industrial importance which the automobile has assumed, the manufacture of these auxiliary devices for the automobile is the most important event in the electrical industry during the year 1912.

Coolidge on the Discovery of the Ductile Tungsten Process

IN this paper a pioneer industrial researcher reported the discovery of a process that revolutionized the design of incandescent lamps. The paper also well illustrates the growing importance of organized industrial research in the electrical industry during the early 20th century. It will be noted that Coolidge credited the innovation to a cooperative effort involving some 20 research chemists and numerous assistants. The importance of information transfer between the G.E. Research Lab and the lamp factory is also mentioned. Coolidge received a patent on the process in 1913. An interesting sequel occurred in 1927 when it was announced that Coolidge had been awarded the Edison Medal by the AIEE for the discovery of ductile tungsten and improvements in X-ray tubes. Coolidge declined to accept on the grounds that a recent court decision had declared his ductile tungsten patent invalid as not constituting an invention in the sense required by patent laws. Nevertheless, Coolidge received the award the following year "for his contributions to incandescent electric lighting and the X-ray arts."

William D. Coolidge (1873–1975) was born in Hudson, Mass. and received a B.S. degree in electrical engineering from M.I.T. in 1896. He received a scholarship to take graduate work in Germany and received a Ph.D. degree from the University of Leipzig in 1899. Coolidge then taught at M.I.T. until 1905 when he joined the staff of the G.E. Research Laboratory at Schenectady. In addition to his work on tungsten, Coolidge made significant contributions to the design of electronic amplifiers and X-ray tubes. He became Assistant Director of the GERL in 1908 and was Director from 1932–1944. Coolidge's published papers and patents are listed in John Anderson Miller, *William David Coolidge: Yankee Scientist* (Schenectady, N.Y., 1963). Also see the *NCAB*, vol. G, pp. 124–125 and his obituary in *IEEE Spectrum*, pp. 108–109, April 1975.

A paper presented at the 249th meeting of the American Institute of Electrical Engineers, New York, May 17, 1910.

DUCTILE TUNGSTEN

BY W. D. COOLIDGE

When work was first started on the problem of producing a ductile form of tungsten, the metal looked very uncompromising. It was so hard that it could not be filed without detriment to the file, and was, at ordinary temperatures, very brittle.

It was of course known from the start that, at the operating temperature of a tungsten lamp, the metal was soft; but this fact seemed unavailing, for there was no tool that could be used for working the metal at such temperatures, and materials from which such tools could be made were lacking.

To a man ignorant of our success, the problem would certainly look more hopeless to-day than it did then. For since that time millions of tungsten filaments have been produced from all available tungsten ores, by widely differing methods, and by different groups of men. And each manufacturer has been fully alive to the fact that he must strive for the highest attainable purity. Yet all of the filaments made have been brittle. They are elastic and flexible as spun glass, but, like the latter, are incapable of taking the slightest permanent set.

Not only was there nothing in the past history of tungsten to encourage us, but, in the natural periodic system of the elements, the metal belonged to a family no member of which had been brought into a ductile state. The other members of the family are chromium, molybdenum, and uranium, elements which had always been characterized by hardness and brittleness. A study of the periodic system shows that, in a general way, elements of the same family do resemble one another in point of ductility, as well as in their other physical and chemical properties. For example, copper, silver, and gold are all in one family and are all very ductile.

Reprinted from *Trans. AIEE*, vol. XXIX, part II, pp. 961–965, May 17, 1910.

Little encouragement could be drawn from the achievement of Dr. Von Bolton with tantalum, because of the fact that this element is in a different family. And the two families differ markedly in both physical and chemical characteristics.

The only arguments on which we could base the hope that tungsten could be produced in a ductile state, were founded on the effect of mechanical working and of chemical purity on the ductility of some of the other unrelated elements. But even this hope seemed of doubtful fulfilment, owing to the apparently insuperable difficulties of mechanically working this particular material.

Mechanical working increases the ductility of some metals. Cast zinc, for example, undergoes a marked increase in ductility when subjected to ordinary wire drawing processes. Some special steels, also, which, as cast, are coarsely crystalline, have to be handled very carefully until they have undergone a certain amount of mechanical reduction, while from this point on they are very ductile.

Chemical purity also, is, in general, conducive to ductility and in some instances, slight amounts of impurity produce a marked effect. Some striking examples of this are the following:

Copper is very sensitive to the presence of bismuth, even 0.02 per cent of the latter rendering it brittle when hot, and 0.05 per cent brittle when cold. Sulphur is also a harmful impurity, and copper containing 0.25 per cent of it is only moderately malleable.

Gold is rendered brittle by 0.05 per cent of lead, bismuth, or tin, and is no longer malleable when it contains as little as 0.0003 per cent of antimony.

Nickel is rendered unsuitable for rolling by the presence of 0.1 per cent of either arsenic or sulphur.

Platinum is made hard and brittle by 0.03 per cent of silicon. Its ductility is also considerably lessened by the presence of small quantities of the other platinum metals.

Tin is brittle when cast at a temperature either too high or too low.

The analogy with iron is in some ways more interesting than the above, for both tungsten and iron take up carbon, and may be greatly hardened thereby. And iron is extremely sensitive to traces of sulphur, phosphorus, and arsenic.

Our early experiments in mechanically working tungsten

led to work on tungsten alloys, and on suspensions of tungsten powder in metals in which there was little or no alloying. One of the most interesting suspending media proved to be an alloy of cadmium, bismuth, and mercury. This amalgam is very pliable. For our purpose it has several other important characteristics. Upon heating to about 140 deg. cent., it becomes soft and plastic, and from this point it retains its plasticity over a considerable temperature interval. While the amalgam is in this state, it is possible to incorporate with it considerable quantities of many foreign substances, such as tungsten, in powdered form. (Such a mixture, containing about 30 per cent by weight of tungsten was exhibited at the meeting.) At room temperature, it is about as hard as lead, but, at a temperature of about 110 deg. cent., it can be readily pressed through a diamond die, and comes out as a silvery looking strong pliable wire. If this wire could be freed from everything but tungsten and still preserve its present strength and ductility, it would solve the tungsten filament problem. But such is not the case. Upon heating it by the passage of current, in a non-oxidizing atmosphere, the mercury first distils out then the cadmium and then the bismuth. Some shrinkage takes place as the foreign metals leave the filament, and the remainder is brought about by raising the temperature to white heat. Most of the ductility of the wire leaves with the mercury, and the remainder goes with the cadmium. This finished filament has been used in thousands of lamps, but these all lack ductility.

The above experience was duplicated when we tried copper as a binding agent for the tungsten, and again when nickel was used for this purpose. In each case there was a ductile stage in which the filament could be bent and otherwise manipulated, but not a trace of this ductility remained after the removal of the foreign element.

The above experiments gave us several new and valuable methods for producing tungsten filaments of the usual quality. But in so far as our ultimate goal, a ductile tungsten filament, was concerned, they were not promising. They were, however, in one respect, instructive, for in the case of all of the above and with many other foreign additions, we got a complete removal of the foreign elements, at least so far as our analytical tests showed, with the final high temperature treatment of the filament. This seemed to indicate that we either did not need to worry about contamination from such elements, or else that brittleness was due to traces of impurity so minute as to escape detection by our analytical methods.

To return now to the mechanical working of pure tungsten. This work received a great impetus by our discovery that an ordinary, dense, well sintered, tungsten filament can be easily bent and put into various forms, and otherwise manipulated at temperatures well below redness, and even below the temperature at which appreciable oxidation takes place. This helped us in two ways. First, it reduced the temperature at which mechanical working operations could be carried on, and, second, it gave a means of recognizing which of the mechanical and chemical processes involved in our experiments were bringing us nearer to the goal. Anything which reduced the temperature at which the metal could be permanently bent was, clearly, helping us.

We found that steps tending to the elimination of the last traces of certain impurities did greatly improve the resulting product. While it may be true that certain impurities present in small amount are harmless or even helpful, we know that certain other impurities are detrimental. We also found that a certain micrographic structure in the tungsten rod with which we start, was conducive to mechanical working and to ductility in the resulting product. Once arrived at the point where mechanical working was easy and where there was a certain amount of ductility in the product even when cold, the development became more rapid. It was aided by the construction of more refined apparatus, in the design of which care was taken to guard against the taking up of impurities during mechanical reduction processes, both from the atmosphere in which the work is carried on and from the surfaces of the tools.

Hand in hand with this improvement on the mechanical side has gone the work on greater chemical purity of the metal with which we start. One of the difficulties in purifying tungsten has been due to the fact, which has been pointed out by Smith and Exner and others, that tungstic acid is very prone to form difficultly separable complexes. Because of this tendency, especial care must be taken with regard to the purity of the reagents used, as otherwise recrystallization beyond a certain point does not result in corresponding purification.

The knowledge obtained from our various lines of research now makes it possible for us to prepare tungsten which can be mechanically worked without more difficulty than would naturally attend the manipulation of very fine wire.

The product which we now have is a perfectly pliable ductile wire, which has the strength of steel. (Specimens of ductile tungsten wire of various sizes were exhibited at the meeting.) It gives a lamp which is strong and whose filament retains its ductility throughout the life of the lamp.

The following data on the drawn wire, obtained from measurements made in the laboratory by Dr. Colin G. Fink, may be of interest:

Diameter (in inches)	Tensile strength (lb. per sq. in)	Specific gravity
0.150	——	19.30
0.005	490,000	——
0.0028	530,000	——
0.0015	600,000	20.19

The electrical resistivity at 25 deg. cent., expressed in microhms per centimeter cube, is, for the hard drawn wire, 6.2, and for the same annealed, 5.0.

The temperature coefficient of electrical resistivity between 0 deg. and 170 deg. cent is 0.0051 per degree centigrade.

The above values, with the possible exception of the temperature coefficient, are of course somewhat dependent on the early history of the wire from which they were determined.

The work which has been outlined above is the result of the close coöperation of about 20 trained research chemists, with a large body of assistants, in the research laboratory. These men were of course given, from the factory organization, all of the mechanical and electrical assistance they could use, and were assisted in no small measure by the staff of the incandescent lamp factory.

Steinmetz Assesses Energy Resources

IN this provocative paper, one of the nation's most respected electrical engineers analyzed energy alternatives and long-range trends. This may serve to remind us that informed engineers were concerned with conservation and forecasting an eventual need to utilize solar energy well before this became a public issue of the 1970's. Steinmetz's ability to formulate the problem in terms of ultimate limits and in a manner that could be grasped even by nonengineers is well illustrated by this paper. In another paper published in 1922 shortly before his death, Steinmetz again reflected on the energy problem and suggested that future biological engineers might develop plants that would more efficiently transform solar energy into stored chemical energy "thus in a closed cycle perpetually supplying unlimited energy from the sun's rays." (See a reprint of Steinmetz's paper entitled "The White Revolution" in *IEEE Spectrum*, pp. 90–95, April 1965.)

For a biographical note on Steinmetz, see Paper 20.

Presented at the 34th Annual Convention of the American Institute of Electrical Engineers, Atlantic City, N. J., June 27, 1918.

AMERICA'S ENERGY SUPPLY

BY CHARLES P. STEINMETZ

ABSTRACT OF PAPER

The gist of the paper is to demonstrate that the economical utilization of the country's energy supply requires generating electric power wherever hydraulic or fuel energy is available, and *collecting the power electrically, just as we distribute it electrically.*

In the first section a short review of the country's energy supply in fuel and water power is given, and it is shown that the total potential hydraulic energy of the country is about equal to the total utilized fuel energy.

In the second section it is shown that the modern synchronous station is necessary for large hydraulic powers, but the solution of the problem of the economic development of the far more numerous smaller waterpowers is the adoption of the induction generator. However, the simplicity of the induction generator station results from the relegation of all the functions of excitation, regulation and control to the main synchronous station. The economic advantage of the induction generator station is, that its simplicity permits elimination of most of the hydraulic development by using, instead of one large synchronous station, a number of induction generator stations and collecting their power electrically.

The third section considers the characteristics of the induction generator and the induction-generator station, and its method of operation, and discusses the condition of "dropping out of step of the induction generator" and its avoidance.

In the appendix the corresponding problem is pointed out with reference to fuel power, showing that many millions of kilowatts of potential power are wasted by burning fuel and thereby degrading its energy, that could be recovered by interposing simple steam turbine induction generators between the boiler and the steam heating systems, and collecting their power electrically. It is shown that the value of the recovered power would be an appreciable part of that of the fuel, and that organized and controlled by the central stations, this fuel power collection would improve the station load factor, give the advantages of the isolated plant without its disadvantages, and produce a saving of many millions of tons of coal.

I. The Available Sources of Energy

A. Coal

THE only two sources of energy, which are so plentiful as to come into consideration in supplying our modern industrial civilization, are coal, including oil, natural gas, etc., and water power.

Reprinted from *Trans. AIEE*, vol. XXXVII, pp. 985–991 (extract of a longer paper), June 27, 1918.

While it would be difficult to estimate the coal consumption directly, it is given fairly closely by the coal production, at least during the last decades, where wood as fuel had become negligible and export and import, besides more or less balancing each other, were small compared with the production. Coal has been mined since 1822, and in Fig. 1 is recorded the coal production of the United States, from the governmental reports. The annual production is marked by circles, the decennial average marked by crosses for every five years. Table I gives the decennial averages, in millions of tons per year.

TABLE I

AVERAGE COAL PRODUCTION OF THE UNITED STATES

(decennial average)

Year	Million tons per year	Per cent increase per year
1825	0.11	....
30	0.32	22.4
35	0.83	19.7
40	1.92	17.0
45	4.00	14.5
50	7.46	10.45
55	10.8	8.35
60	16.6	8.72
65	25.9	9.22
70	40.2	8.58
75	56.8	7.42
80	82.2	7.95
85	122	6.80
90	160	5.40
95	206	5.75
1900	281	6.96
05	404	6.60
10	532	....

In Fig. 1 the logarithms of the coal production in tons are used as ordinates. With this scale, a straight line means a constant proportional increase, that is, the same percentage increase per year, and in the third column of Table I are given the average percentage increase of coal production per year.

This Fig. 1 is extremely interesting by showing the great irregularity of production from year to year, and at the same time a very great regularity over a long period of time. Since 1870 the average production may be represented by a straight line, the values lying irregularly above and below the line, which represents an annual increase of 6.35 per cent and thus represents the average coal production[1] C by the equation

$$C = 45.3 \times 10^{0.0267\,(y-1870)} \text{ million tons}$$

or

$$\log C = 0.0267\,(y-1870) + 7.656$$

where y = year.

Before this time, from 1846 to 1884, the coal production could be represented by

$$C = 7.26 \times 10^{.0365\,(y-1850)} \text{ million tons}$$

or

$$\log C = 0.0365\,(y-1850) + 6.861$$

representing an average annual increase of 8.78 per cent.

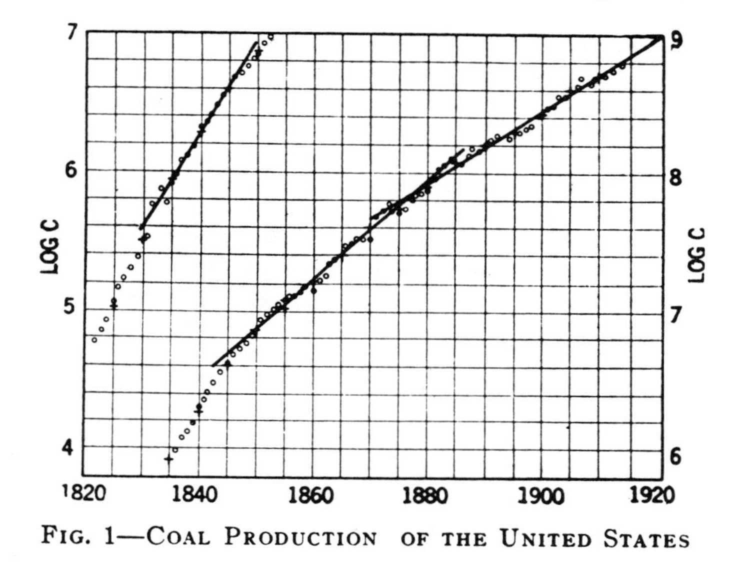

FIG. 1—COAL PRODUCTION OF THE UNITED STATES

It is startling to note how inappreciable, on the rising curve of coal production, is the effect of the most catastrophic political and industrial convulsions, such as the Civil War and the Industrial panic of the early 90's; they are indistinguishable from the constantly recurring annual fluctuations. It means, that the curve is the result of economic laws, which are laws of nature.

Extrapolating from the curve of Fig. 1, whieh is permissible, due to its regularity, gives 867 million tons as this year's coal consumption. As it is difficult to get a conception of such enormous amounts, I may be allowed to illustrate it. One of the great wonders of the world is the Chinese Wall, running

1. Soft coal and anthracite, and including oil reduced to coal by its fuel value.

across the country for hundreds of miles, by means of which China unsuccessfully tried to protect its northern frontier against invasion. Using the coal produced in one year as building material, we could with it build a wall like the Chinese Wall, all around the United States, following the Canadian and Mexican frontier, the Atlantic, Gulf and Pacific Coast, and with the chemical energy contained in the next year's coal production, we could lift this entire wall up into space, 200 miles high. Or, with the coal produced in one year used as building material, we could build 400 pyramids, larger than the largest pyramid of Egypt.

It is interesting to note that 100 thousand tons of coal were produced in the United States in 1825; one million tons in 1836; 10 million tons in 1852 and 100 million tons in 1882. The production will reach about 1000 million tons in 1920, and, if it continues to increase at the same rate, it would reach 10,000 million tons in 1958.

Estimating the chemical energy of the average coal as a little above 7000 cal., *the chemical energy of one ton of coal equals approximately the electrical energy of one kilowatt year* (24 *hour service*). That is, one ton of coal is approximately equal in potential energy to one kilowatt-year.

Thus the annual consumption of 867 millions of tons of coal represents, in energy, 867 million kilowatt-years.

However, as the average efficiency of conversion of the chemical energy of fuel into electrical energy is probably about 10 per cent, the coal production, converted into electrical energy, would give about 87 million kilowatts.

Assuming however, that only one half of the coal is used for power, at 10 per cent efficiency, the other half as fuel, for metallurgical work etc., at efficiencies varying from 10 per cent to 80 per cent, with an average efficiency of 40 per cent, then we get 217 million kilowatts (24 hour service) as the total utilized energy of our present annual coal production of 867 million tons.

B. The Potential Water Powers of the United States

Without considering the present limitation in the development of water powers, which permits the use of only the largest and most concentrated powers, we may try to get a conception of the total amount of hydraulic energy which exists in our country, irrespective of whether means have yet been developed or ever will be developed for its complete utilization. We therefore proceed to estimate the energy of the total rain fall.

Superimposing the map of rain fall in the United States, upon the map of elevation, we divide the entire territory into sections by rain fall and elevation. This is done in Table II, for the part of our continent between 30 and 50 degrees northern latitude.

TABLE II

TOTAL POTENTIAL WATER POWER OF UNITED STATES

In. rain fall	Ft. elevation	Area $m^2 10^{12} \times$	Avg. elevation m.	Avg. rainfall cm.	Kg-m. per m^2; $10^3 \times$	Kg-m. total $10^{15} \times$
>10	>5000	0.54	2100	12.5	263	142
	1000–5000	0.29	900		112	32.5
10–20	>5000	1.18	2100	37.5	787	930
	1000–5000	1.96	900		338	660
20–30	1000–5000	0.32	900	62.5	563	183
	100–1000	0.97	150		94	91
30–40	1000–5000	0.35	900	87.5	786	275
	100–1000	1.40	150		131	184
40–60	1000–5000	0.27	900	125	1130	305
	100–1000	1.03	150		188	194
						$\Sigma = 2996$
						3000

As obviously only the general magnitude of the energy value is of interest, I have made only few sub-divisions: five of rain fall and four of elevation, as recorded in columns 1 and 2 of Table II[2]. The third column gives the area of each section, in millions of square kilometers, the fourth column the estimated average elevation, in meters, and the fifth column the average rain fall, in centimeters. The sixth column gives the energy, in kilogram-meters per square meter of area, and the last column the total energy of the section, in kilogram-meters, which would be represented by the rain fall, if the total hydraulic energy of every drop of rain were counted, from the elevation where it fell, down to sea level.

As seen from Table II, the total rain fall of the North American Continent between 30 deg. and 50 deg. latitude represents 3000×10^{15} kg-m. This equals 950 million kilowatt years (24 hour service). That is, the total potential water power of the United States, or the hydraulic energy of the total

1. The lowest elevation, < 100 ft., is not included, as having little potential energy.

rain fall, from the elevation where it fell, down to sea level, gives about 1000 million kilowatts.

However, this is not available, as it would leave no water for agriculture; and even if the entire country were one hydraulic development, there would be losses by seepage and evaporation.

An approximate estimate of the maximum potential power of the rain fall, after a minimum allowance for agriculture and for losses is made in Table III, allowing 12.5 cm. rain fall for wastage, and 37.5 and 25 cm. respectively for agriculture where such is feasible.

TABLE III

AVAILABLE POTENTIAL WATER POWER OF THE UNITED STATES

Avg. rainfall cm.	Avg. elevation *m.*	Area m^2 $10^{12} \times$	Wastage cm.	Agricul-ture, cm.	Available rainfall cm.	Kg. m. per m^2 $10^3 \times$	Kg-m. total $10^{15} \times$
12.5	2100	0.54	12.5	....	....	....	....
....	900	0.29	12.5	....	....	....	....
37.5	2100	0.39	12.5	25	....	....	....
....	2100	0.79	12.5	....	25	525	415
....	900	0.98	12.5	25	....	....	....
....	900	0.98	12.5	....	25	225	220
62.5	900	0.21	12.5	37.5	12.5	112	23
....	900	0.11	12.5	....	50	450	50
....	150	0.97	12.5	37.5	12.5	19	18
87.5	900	0.35	12.5	37.5	37.5	337	118
....	150	1.40	12.5	37.5	37.5	56	78
125	900	0.27	12.5	27.5	75	674	182
....	150	1.03	12.5	37.5	75	112	116
							$\Sigma = 1220$

This gives about 1200×10^{15} kg-m. as the total available potential energy, which is equal to 380 million kilowatts (24 hour service). Assuming now an efficiency of 60 per cent from the stream to the distribution center, gives 230 million kilowatts. (24 hour service) as the maximum possible hydroelectric power, which could be produced, if every river, stream, brook or little creek throughout its entire length, from the Spring to the ocean, and during all seasons, including all the waters of the freshets, were used and could be used. It would mean that there would be no more running water in the country, but stagnant pools connected by pipe lines to turbines exhausting into the next lower pool. Obviously, we could never hope to develop more than a part of this power.

C. Discussion

It is interesting to note that the maximum possible hydraulic energy of 230 million kilowatts, is little more than the total energy which we now produce from coal, and is about equal to the present total energy consumption of the country, including all forms of energy.

This was rather startling to me. It means that the hope that when coal once begins to fail we may use the water powers of the country as the source of energy, is and must remain a dream, because if all the potential water powers of the country were now developed, and every rain drop used, it would not supply our present energy demand.

Thus hydraulic energy may and should supplement that of coal, but can never entirely replace it as a source of energy. This probably is the strongest argument for efforts to increase the efficiency of our methods of using coal.

A source of energy which is practically unlimited, if it could only be used, is solar radiation. The solar radiation at the earth's surface is estimated at 1.4 cal. per $cm.^2$ per min. Assuming 50 per cent cloudiness, this would give an average throughout the year (24 hours per day), of about 0.14 cal. per $cm.^2$ horizontal surface per min., and on the total area considered in the preceding table, of 8.3 million square kilometers of North America between 30 and 50 latitude, a total of approximately 800,000 million kilowatts (24 hour service), or a thousand times as much as the total chemical energy of our coal consumption; 800 times as much as the potential energy of the total rainfall.

Considering that the potential energy of the rainfall from surface level to sea level, is a small part of the potential energy spent by solar radiation in raising the rain to the clouds, and that the latter is a small part of the total solar radiation, this is reasonable.

Considering only the 2.7 million square kilometers of Table III, which are assumed as unsuited for agriculture, and assuming that in some future time, and by inventions not yet made, half of the solar radiation could be collected, this would give an energy production of 130,000 million kilowatts.

Thus, even if only one-tenth of this could be realized, or 13,000 million kilowatts, it would be many times larger than all the potential energy of coal and water. Here then would be the great source of energy for the future.

⋄⋄⋄ ⋄⋄⋄ ⋄⋄⋄

Report of a Pioneering Interconnection of Power Systems

THIS brief item from the *Electrical World* of 1914 discussed the revolutionary implications of the creation in the South of "the most extensive interconnected transmission system in the world." The pioneer innovator in this development was the Southern Power Company (later Duke Power Company) that had been organized in 1905 with the goal of supplying low cost electrical power to the rapidly expanding southern textile industry. Its success in achieving this goal was so apparent by 1910 that the Editor of *Electrical World* credited the Southern Power Company with "stimulating a whole population from a condition of former commercial and industrial apathy to an activity comparable . . . to that which characterized a new Western State." The Southern system of systems became an important precedent for later efforts to establish so-called "Giant Power" and "Super Power" grids in the northeast. The Editor's suggestion that a large artificial lake be created in the southern Appalachians as a "project of tremendous possibilities" was implemented dramatically by the TVA beginning in 1933. See a paper on the history of Giant Power by Thomas Hughes and a paper on the origins of TVA by Thomas McCraw, both in the PROCEEDINGS OF THE IEEE for September 1976. See also Paper 29 for a later assessment of the impact of the innovation by W. S. Lee, longtime chief engineer of the Southern Power Company.

The Great Southern Transmission Network

It will be startling news to some of our readers that there has quietly grown up in the South what is to-day by far the most extensive interconnected transmission system in the world. That splendid work was being done in the development of the water-powers of the southern Appalachian country has been already made evident, but the linkage of the various networks into what now approximates, and one day will become, a united whole is a comparatively new phase of the situation. Gradually in the South transmission enterprises grew and flourished until there are now seven great systems with a wide ramification of transmission and distribution lines. Of these, six, covering the States of North Carolina, South Carolina, northern Georgia and a large part of Tennessee, are now actually in physical connection through various traffic arrangements.

If the feat were desirable, one could operate a motor in Nashville, Tenn., by energy generated at Rockingham, N. C., over a circuit roughly 1000 miles long. Some day in the near future perhaps it will actually be done, if only as a unique accomplishment in long-distance transmission. The energy would first pass over the lines of the Carolina Power & Light Company at 100,000 volts to a point near Raleigh. Then it would turn westward and pass upon the lines of the Southern Power Company and follow them southwest at the same voltage until it reached the great generating plant at Tallulah Falls. Then it would pass on with unlowered voltage to Atlanta over the lines of the Georgia Railway & Power Company, and thence over another branch of the same system northward to the Tennessee boundary, where it would join the lines of the Tennessee Power Company and be passed on at 66,000 volts to Nashville. A little later, when the line between Nashville and Memphis is completed, the western terminus of the line carrying energy derived from the Cape Fear River, which runs into the Atlantic, would lie on the banks of the Mississippi.

The independent networks have naturally touched elbows and then joined hands. Operating at first in sections quite independent of each other, they have gradually spread out and worked together until actual interchange of energy has become the most natural thing in the world. The main point of the matter is not what has been done but what may be done. At present the local interchange of energy is a mere convenience; the connection exists, but it is not yet ready to be utilized in full. In the first place, the system of the Alabama Power Company will spread out and be connected with those of its neighbors, and then some far-sighted captain of industry must grasp the situation and make the most of it. To-day there are seven neighboring systems, covering five States. They are in a condition comparable with an equal number of large suburban towns in each of which has grown up its own electrical system along various lines of engineering and finance. The task before the great organizer is to unite the systems of the towns or the States into a coherent network capable of gaining all the advantages that lie in such union and of giving better and more extensive service than is possible with the present loose connection of the plants.

A glance at the map tells the story. A few hundred miles of tie line would provide main feeders from Tallulah Falls to Parksville, give an independent connection by way of Augusta between the systems of the Southern Power Company and the Georgia Power Company, and would tie in the lines of the Alabama Power Company at two points, one by way of the Georgia Power Company's lines and the other northward by the Tennessee Power Company system. This plan unites two watersheds, that of the Mississippi Valley and that of the Atlantic, giving all the advantages that come from what one may call the diversity-factor of flow. It would not make much difference at first whether there were actual financial union or not. The main thing is complete interconnection and friendly traffic relations whereby each plant on the system could be utilized to its greatest advantage and the waste of water that takes place through disconnected hydraulic effort could be averted. As a final engineering step, with the energy of five States behind the work, it ought to be possible to block up one of the gorges in the mountains and construct an artificial lake large enough to tide over the dry season and utilize the prodigious rainfall of the southern Appalachians. The country is without natural lakes but there are few districts where natural facilities for impounding the flow could be more easily or cheaply obtained. The work already done brings to view a project of tremendous possibilities.

Reprinted from *Elec. World*, vol. 63, p. 1201, May 30, 1914.

Description of an Integrated Power System of Steam and Hydroelectric Plants

IN this paper from the *Transactions of the AIEE* of 1929, one of the country's foremost hydroelectric power engineers noted the astonishing growth of the power industry since World War I, and explained how his own Company had responded to the challenge by combining five groups of power plants in a single system. He pointed out that this diversity enabled "the full and economical utilization of the available natural resources." Lee also made a striking generalization that "industrial expansion follows the transmission lines of a central system," a belief shared by the advocates of the TVA (a development that Lee bitterly opposed).

William S. Lee (1872–1934) was born in Lancaster, S.C. and received an engineering degree from the Citadel in 1894. As resident engineer with the Anderson Power Company, he directed construction of one of the South's first hydroelectric plants at Portman Shoals on the Seneca River. In 1898 Lee became resident engineer of the Columbus Power Company and was in charge of building what was described as "the first large dam built in the South" on the Chattahoochee River in Columbus. Lee then joined the Catawba Power Company as chief engineer. The Catawba Company became the nucleus of the Southern Power Company established in 1905 with Lee as chief engineer. In addition to being responsible for the design and construction of the system described in this paper, Lee designed several large hydroelectric plants in Canada during the 1920's. He served as President of the AIEE in 1931 and as President of the American Engineering Council in 1932. See "American Electrical Engineers. W. S. Lee," *Electrical World*, vol. 55, pp. 738–739, 1910 and the *NCAB*, vol. 24, p. 155. Also see the references given with Paper 28.

Economies in Central Power Service
as Illustrated by the Duke Power System

BY W. S. LEE
Fellow, A. I. E. E.

***Synopsis.**—This paper invites the attention of the engineering profession to the advantages and economies which may be obtained by coordinating both hydro and steam power stations of various kinds, and electric transmission and distribution lines serving a large industrial territory in a unified and centrally controlled system. For illustration, a general description of the generating and distribution system of the Duke Power Company, operating in the Piedmont section of the Carolinas, is given.*

EVERYONE who studies the statistics on electric power output in the United States will be amazed at the rapid growth of the electrical industry during the past decade and how, by the construction of larger and larger commercial electrical central stations and the merging of existing electrical properties, the increasing demand for electricity for lighting, power, and transportation was met.

Through this centralization, it was made possible to give uniform and dependable service to the customer, to maintain the same rates, and even lower them to domestic consumers in the face of increasing costs of labor and materials, to make the investments in properties of such character safe and attractive, to afford the support of an organization highly trained in the efficient planning, constructing, operating, and improving of utility systems, to make purchases in large quantities and to engage aggressively in building-up the electric light and power business. Other advantages are the grouping of hydro and steam power stations of various kinds to permit the full and economical utilization of the available natural resources of power.

The large generating plants deliver the electric energy to trunk transmission lines which feed the distributing systems of cities and towns and furnish power to the lines at industrial centers. In addition, interconnections of trunk lines of the transmission systems of large independent power companies operating in adjoining territories provide means for marketing wholesale power blocks from one utility to another, for interchange of power between systems and, in case of emergency, for stand-by service.

Due to their higher plant efficiencies and the lowering of construction costs, the number of large steam electric generating stations is rapidly increasing and water-powers of large size, which were considered uneconomical on account of the inability of marketing such large blocks of power within a reasonable period of time by local utilities, are being developed. Industrial plants abandon operation of their old inefficient power stations, and wasteful generating stations of small suburban electric properties are being shut down. An increasing number of manufacturing plants is being located away from the large centers of population, this trend towards decentralization being made possible by reason of the fact that high-tension transmission lines can be tapped at suitable locations away from metropolitan areas. Old inefficient water-power plants are being overhauled and existing hydro stations utilizing only a portion of the available water-power of the stream are being rebuilt and enlarged so as to lower the cost of the electric power generated.

A striking example showing the development of a local concern organized for the sale of electricity generated at a single water-power plant about 23 years ago to one of the largest central stations generating and distributing systems in the world is given by the Duke Power Company operating in the Piedmont section of the Carolinas. This company is now operating a number of hydro and steam electric plants having an aggregate installed capacity of 873,895 kv-a., and the power is fed into a transmission and distribution system consisting of approximately 4000 mi. of circuits, of which about 50 per cent are 100,000-volt steel tower lines. One hundred and sixty thriving industrial communities and many isolated cotton mills and factories are served by these lines, and in 1927 the total power generated and purchased by the Duke Power System amounted to 1,745,776,428 kw-hr. This compares with an average output of 19,000,000 kw-hr. of the old Catawba Station, which was the first plant operated and which was redesigned and reconstructed in 1925.

The generating system of the Duke Power Company may be subdivided into the following groups:

1. Hydro stations which use the fall and run of the river.
2. Hydro stations which use the fall and run of the river plus a small amount of storage taking care of the night and Sunday flow and, perhaps, of minor weekly fluctuations of the flow of the river.
3. Hydro stations which have a large storage capacity behind them and which are used as a valve, or outlet, to supply the deficiency in power of other stations during low-water seasons.
4. Steam plants designed to operate intermittently and during low-water periods, termed "stand-by plants."
5. Steam plants designed to operate twelve months of the year and capable of supplying a constant amount of power, termed "base plants."

Of special interest are the twelve hydro stations

1. Vice-president and Chief Engineer, Duke Power Co., Charlotte, N. C.

Presented at the Regional Meeting of Southern Dist. No. 4 of the A. I. E. E., Atlanta, Ga., Oct. 29-31, 1928.

Reprinted from *Trans. AIEE*, vol. 48, pp. 203–205, Jan. 1929.

located on the Catawba River the flow of which is almost completely regulated so that all these stations may be classed under Group 3. Beginning with the power plant farthest upstream, the names and installed capacities of these stations are as follows:

Name of hydro station on Catawba River	Rated capacity of station kv-a.
Bridgewater	25,000
Rhodhiss	21,875
Oxford	45,000
Lookout Shoals	23,400
Mountain Island	75,000
Catawba	75,000
Fishing Creek	37,500
Great Falls	30,000
Dearborn	56,250
Rocky Creek	30,000
Cedar Creek	56,250
Wateree	70,000
Total	555,275

The Bridgewater, Rhodhiss, and Oxford Stations, located in the upper region of the Catawba River, and the Catawba Station, located below the junction of the South Fork of the Catawba River, have large storage reservoirs behind them for the retention of the flood-waters and excess flow of the river.

The aggregate installed capacity of the six steam power plants is 244,313 kv-a., of which the Buck Station, using powdered fuel and having a rated capacity of 87,500 kv-a., is to be classed under Group 5. It should be added that the Duke Power Company now has under construction a large powdered fuel steam power station on the Catawba River near the Mountain Island hydro station, where an ample supply of condensing water is available. This station is located closest to the center of gravity of the entire load system and the initial installation will consist of two 60,000-kw. units, whereas the plans call for an ultimate installation of 480,000 kw.

The hydro stations to be classed under Groups 1 and 2 are operated in conjunction with the stand-by steam plants under Group 4, which arrangement will permit the most effective utilization of the stream flow under the conditions. The locations and rated capacities of these stations are as follows:

Name of hydro station	Name of river	Rated capacity of station kv-a.
Gaston Shoals	Broad	12,750
99 Island	Broad	22,500
Tuxedo	Green	6,250
Turner	Green	6,880
Portman Shoals	Seneca	9,500
Gregg Shoals	Savannah	2,250
Lake Lure	Rocky Broad	4,500
Saluda	Saluda	2,600
Idols	Yadkin	1,000
Plants of less than 1000 kv-a. capacity		6,077
Total		74,307

The steam power plants are located at strategical points of the system; combining Groups 4 and 5, they are:

Name of steam station	Rated capacity of station kv-a.
Buck	87,500
Eno	31,250
Greensboro	8,000
Greenville	8,000
Mount Holly	45,500
Tiger	37,500
Stations leased	26,563
Total	244,313

A summary giving the generating capacity of the Duke Power System, arranged by groups, discloses the following interesting facts:

Number of group	Total rated capacity of stations kv-a.	Per cent of total capacity of entire system
Hydro No. 1	7,777	0.89
Hydro No. 2	66,530	7.61
Hydro No. 3	555,275	63.54
Steam No. 4	156,813	17.95
Steam No. 5	87,500	10.01
All groups	873,895	100.00

If the steam power station of 150,000 kv-a. initial capacity now under construction is included, the ratio of steam to total generating capacity of the system will be increased from 27.96 per cent to 38.51 per cent.

To give an outline of the transmission and distribution system of the Duke Power Company, 100,000-volt double-circuit feeder lines from points of concentration of power generation tap the 100,000-volt double-circuit trunk lines running through the load centers of the Piedmont Section of the Carolinas. At these centers, the current is stepped down to 44,000 volts and delivered to the distribution system. The object of using the lower voltage is to reduce the cost of the substations serving the large power customers and public utility branches of the company; and wherever feasible, a generating station is tied-in directly to the 44,000-volt system to avoid double transformation.

At present approximately the following lines are in operation:

Voltage of circuit	Approximate length of circuit miles
100,000	2000
44,000	1900
13,200	100

The standard secondary delivery voltages are 575 and 2300.

Interconnection of the Duke Power Company's trunk lines is made for delivery of large blocks of power to it at the High Rock Station on the Yadkin River of the Tallassee Power Company, and a line is contemplated for connection with the hydro plant of the Lexington Water Power Company, now under construction on the Saluda River near Columbia, S. C. Interconnections for interchange of power and stand-by service are made with the trunk lines of the South-Eastern Power Company and the Carolina Power & Light Company.

In the process of development of the Duke Power Company effort was made to maintain the entire system in good operating condition and to obtain the best results by making improvements on existing plants wherever necessary, even to the extent of abandoning old plants in the interest of fuller and more efficient utilization of the properties. This latter policy was followed up with the fullest success in the case of the old Catawba Station representing an investment of more than one million dollars.

As already mentioned, this hydro plant containing eight rope-driven generators of a total capacity of 8000 kv-a. was built 23 years ago. The normal head at this station was 23 ft. and the over-all efficiency of the plant was about 70 per cent. The storage capacity of the pond was practically negligible and to obtain the benefit of the night and Sunday flow, the plant had to be operated continuously. Furthermore, the available uniform flow at this station was considerably augmented due to the regulating effect of the large storage reservoirs, since built on the Upper Catawba River. The question therefore arose whether additional machinery should be installed and the old plant overhauled or the dam be raised so that the water would back to the tail-race of the company's Mountain Island Station, 25 mi. further upstream, thus creating another large storage reservoir, and the building of an entirely new power house. By the adoption of this latter scheme, not only a reservoir of approximately ten billion cubic feet available storage capacity was created, making an additional regulated flow available at the existing stations on the lower Catawba River with a combined head of 250 ft., but the installation of the new power station could be increased from 8000 kv-a. of the old station to 75,000 kv-a. due to the higher head of 70 ft. and full utilization of the regulated stream flow

The map in Fig. 1 shows the outline of the states of North and South Carolina. There is shown by dots, each 10,000 spindles located within these two states. It will readily be noted that the industrial expansion follows the transmission lines of a central system.

Notwithstanding the great industrial development in the Carolinas during the last ten years, the prospects for further growth of the electric light and power business are very bright. The demand for power and light in the home, on the farm, and in industries is increasing steadily and new fields are being opened up constantly.

In anticipation of this ever increasing demand for electric energy, the Duke Power Company is going ahead with a definite program of extensions and improvements to its system to meet future service requirements.

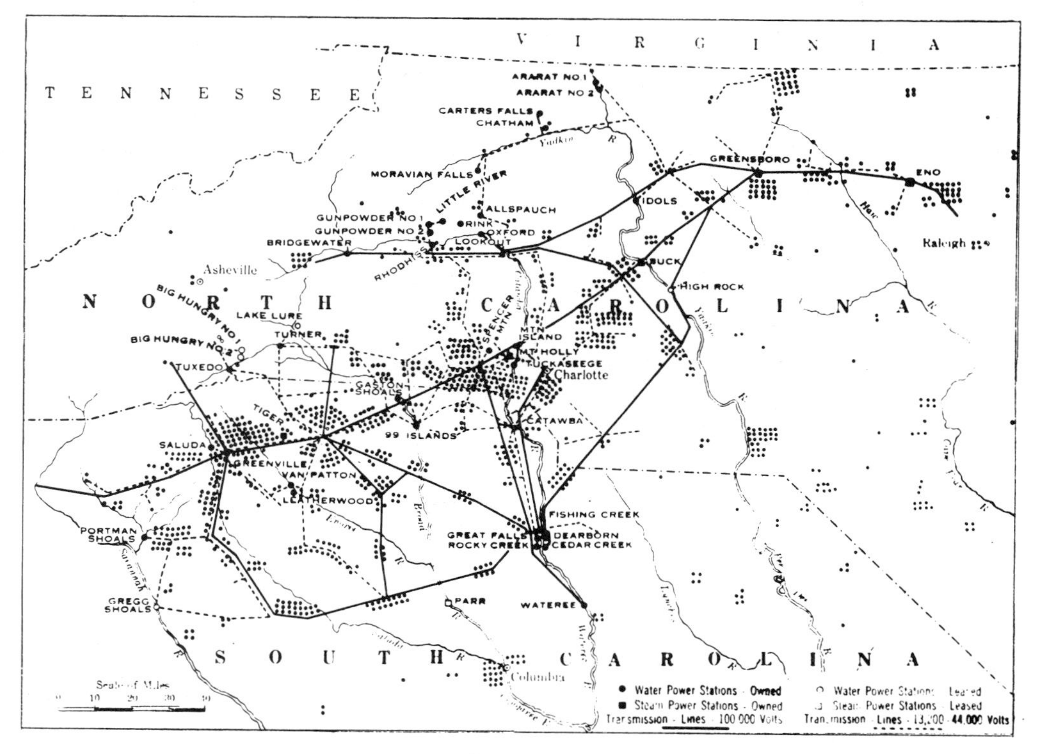

FIG. 1

Sporn's Assessment of the Power Industry in 1959

IN this paper published on the 75th anniversary of the AIEE, Philip Sporn gave an overview of developments in the American electric power industry during the past quarter of a century. Although he pointed to no single technical innovation that had been as revolutionary as earlier ones, the changes in scale and complexity had been substantial. The interpretation of developments in this period poses a challenging problem for electrical historians who have tended to concentrate on revolutionary innovations and heroic figures like Edison, Sprague, and Tesla. As Sporn suggests, much more was involved that "simply constructing transmission lines, substations and units of generating capacity." He also notes that, during the past 25 years, "electric energy has penetrated deeply into every phase of modern living." His optimistic forecast for the next quarter century that would extend to the AIEE centennial of 1884 antedated the energy crisis and environmental concerns of the recent past. For an interesting sequel that indicates that Sporn was right in predicting a greater role for electronic computers in the industry, see Hans Glavitsch, "Computer Control of Electric-Power Systems," *Scientific American*, pp. 34–44, November 1974. For a paper that outlines the impact of electrical appliances in the home with some surprising conclusions, see Ruth Schwartz Cowan, "The 'Industrial Revolution' in the Home: Household Technology and Social Change in the 20th Century," *Technology and Culture* vol. 17, pp. 1–23, 1976.

Philip Sporn (1896–   ) was born in Austria, but brought to the U.S. while a young child. He graduated from Columbia University in electrical engineering in 1917 and remained for a year of graduate work. After working for a brief time with the Crocker–Wheeler Company and a power company in Michigan, he joined the American Gas and Electric Company in 1920. It later became the American Electric Power Company and grew into the country's largest privately owned power system. Sporn rose to prominence in the Company and became its President in 1947. His technical contributions were in the areas that he describes in this paper such as improvements in circuit breakers and high-voltage transmission technology. He received the Edison Medal of the AIEE in 1945 and was elected to the National Academy of Engineers in 1965. See *Engineers of Distinction* (Engineers Joint Council, 1970), p. 394 and *John Fritz Medal. Biography of Medalist*, 10 pp., 1956.

Shasta Dam, 375,000 kw, Sacramento River, Calif., gross head 480 feet, 4 units. Photo courtesy U. S. Bureau of Reclamation

THE HISTORY of the electric utility industry in the United States in the 25 years since AIEE celebrated its Golden Anniversary in 1934 has been that of remarkable physical growth. The technical achievements of the industry during this period helped to foster this growth and were, in turn, stimulated and made possible by it. During this 25-year period, electric utility generation expanded by well over 700%, or at a rate almost 2½ times as fast as the increase in the real gross national product (GNP) after eliminating changes in the price level. The expanding use of electric energy in every aspect of American life has contributed to and, in turn, has been made possible by the rising standards of living and the increased productivity this country has enjoyed in this period.

In 1934, the electric utility industry had only 24.7 million customers, but by the end of 1958, the total had increased almost 2½ times to over 56 million. While this was partly the result of an expanding population, it was in large measure attributable to the expansion in electric service to a larger portion of the population so that well over 95% of the total residences in the United States, or all but the most remotely located, now have central station electric service available, compared with only 65% in 1934. Contributing to this result was the substantial completion of the program of rural electrification—undertaken initially by the private utilities and, later, more intensively by both the private utilities and the Rural Electrification Administration. At the end of 1934, almost 744,000 farms, or slightly less than 11% of the total number of farms in the United States, were served by the utility industry. By 1958, however, the extent of rural service was almost equal to that found in urban areas, with almost 95% of the total farms receiving utility service.

The 20.4 million residential customers in 1934 used an average of only 629 kwhr per customer; in 1958, the more than 46 million residential customers consumed 3,385 kwhr per customer. In the same period, average price per kwhr had fallen from 5.33 to 2.52¢, despite a more than doubling in the consumer price level.

Similar growth took place in the commercial and industrial consumption of electric energy. Electricity used per production worker man-hour rose from 4.07 kwhr in 1934 to more than 9 kwhr in 1958, excluding energy used by the aluminum industry and the Atomic Energy Commission. In the total, electric utility generation grew in these 25 years from a little over 87 billion to 641 billion kwhr, and the generating capacity of the industry expanded from 34 million to 140 million kw.

This growth in the past quarter century has also involved a change in the character of a considerable part of the industry from relatively small, more or less isolated systems into large integrated systems, many of these in turn being interconnected in larger pools of systems, with the advantages of mass production and

Philip Sporn is president, American Electric Power Service Corporation, New York, N. Y.

Reprinted from *Elec. Eng.*, vol. 78, pp. 542–545, 547–555, May 1959.

transmission of electric power at costs lower than could have been achieved by the smaller systems. This phase of the industry's development has been made possible in large part by a number of important technological achievements, and the large systems themselves have made possible the incorporation of cost-saving technological developments which would otherwise not have been made. For example, the savings in construction cost per kilowatt of capacity by building larger units would not have been possible without the large systems able to absorb them, and the savings in fuel costs now being achieved through the use of larger, more efficient units and the transmission of large blocks of power from plants built close to a source of fuel and condensing water would not have been possible without the need for large quantities of power in single systems.

Important forward strides in technology have taken place, as will be seen in the following discussion, in every phase of the industry's operations.

STEAM–ELECTRIC GENERATION

THE EXPANSION of electric utility capacity from 34 million to 140 million kw was more than a simple multiplication in kind. As a consequence, the generating plant, particularly the steam–electric plant, being installed on the utility systems of the United States today can hardly be classified as of the same species as those installed in 1934. The size of units, the steam temperatures, the pressures, and the efficiencies were all almost undreamed of, or even considered impossible in 1934. The growth of unit sizes in this period has been most notable. In 1934, the typical "large" turbine–generator unit delivered for power system installation was 40,000 to 50,000 kw. In 1937, a 40,000-kw unit was the largest single-shaft 3,600-rpm machine installed up to that time. Today, a tandem-compound single-shaft 3,600-rpm unit of 250,000 kw is operating, and one of 325,000 kw is on order. At the same time, cross-compound units have progressed from the 3,600/1,800-rpm combinations of 90,000 kw that were typical in 1939, to the 325,000-kw units now in operation and the 3,600/3,600-rpm 450,000-kw units now under construction and 500,000-kw units now on order.

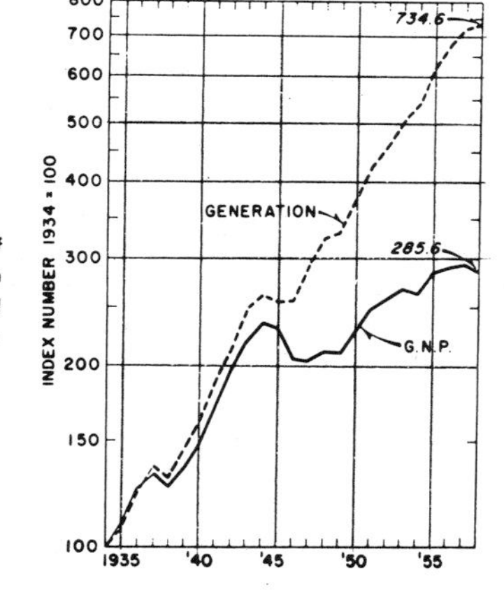

Comparative growth of electric utility generation and real gross national product, excluding price-level changes.

While these advances have been the result of many technical improvements, perhaps the most significant as far as the generators are concerned are the advances made in methods of cooling. In this development, the most important single step has been the use of hydrogen as the coolant, first at 0.5 psi (pounds per square inch), and later increased to 15, 30, and, finally, to 45 psi. The more recent development of inner cooling of conductors, both stator and rotor windings, has made possible the latest increases in unit sizes to what were considered impossible figures only a few years ago. On the turbine, the development of longer last-stage buckets—and even greater length is now under development —along with multiflow exhausts, has made possible the handling of the large volumes of steam required in units ranging in size from 325,000 to 500,000 kw.

Along with and contributing to the increased unit size have been the major advances in turbine and boiler technology in steam temperatures, pressures, reheat, and size and design of boilers. Steam temperatures of 900 F were first achieved in 1936, and in two steps this was advanced first to 925 F in 1937 and to 940 F in 1940. Further development was of necessity suspended during World War II, but in 1947 a major step was taken in the Atlantic City No. 7 unit, which advanced temperatures to 1,000 F. This was quickly followed by 1,050 F at Sewaren in 1948, 1,100 F at Kearny in 1953, 1,150 F for Philo No. 6 in 1957, and finally, the maximum to the present time, 1,200 F for Eddystone No. 1, now under construction. Although the economic validity of the very high temperatures of the last two steps remains to be established, it does represent a remarkable technical advance in steam temperatures of 300 F in the past 25 years which has left its mark on heat energy conversion technology.

Pressure technology, having broken through the supercritical barrier, may be said to have undergone even more notable advances than those in temperature. Pressures of 1,100 psi were achieved in the early 1930's, but developments were not extended appreciably beyond this until the 2,400-psi Twin Branch No. 3 unit was installed in 1940, largely as an advanced experimental prototype. Although World War II delayed further development, this 2,400-psi installation paved the way for the highly successful postwar boiler technology in the 2,000-psi range, which prevailed until the 4,500-psi supercritical unit was installed at Philo in 1957. This unit, having a capability of 107,000 kw and, parenthetically, occupying the space formerly occupied by a 40,000-kw unit of 1925 vintage, was developed as an experimental prototype which has led the way to construction of a number of larger-size units of a similar type. This has been extended to 5,000 psi, 1,200 F in the 380,000-kw Eddystone No. 1 unit referred to earlier. More particularly, it has provided the basis for the design of two 450,000-kw units on the American Electric Power (AEP) System, where, because of the relatively lower value of thermal savings

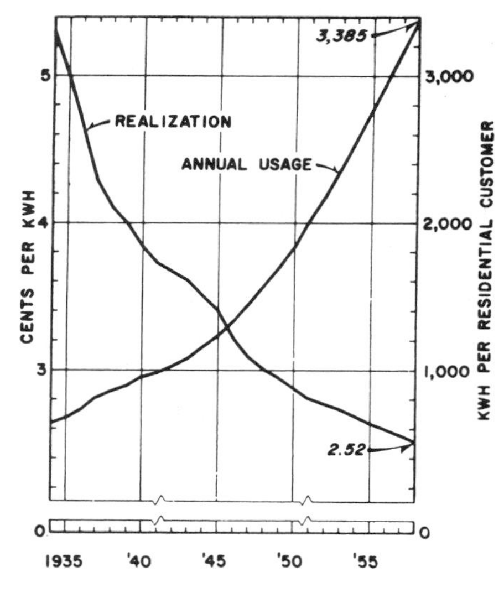

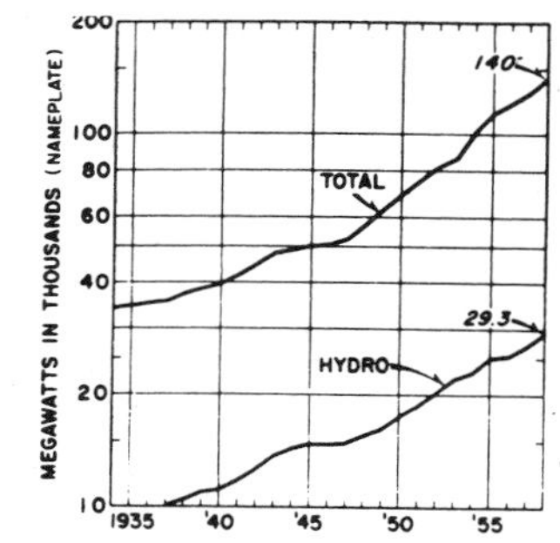

(Left) Increase in average annual kwhr use and decline in cost for residential consumer. **(Above)** Growth in electric utility industry total and hydro generating capacity. **(Right)** Thermal efficiency improvements. Declining heat rate, industry average, and best plant.

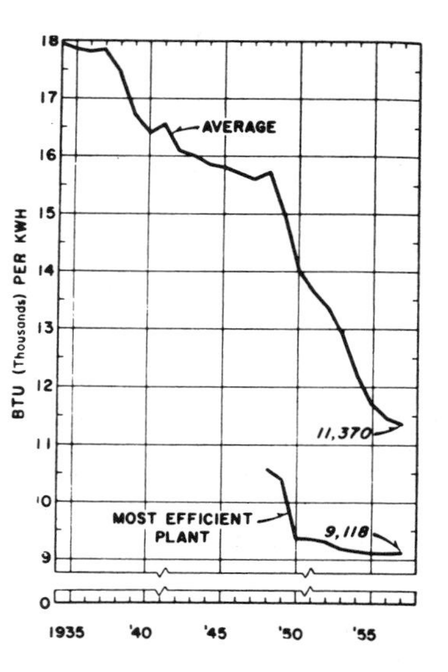

made possible only at the expense of higher costs for more expensive alloy materials required for higher temperatures, the temperature and pressure levels were backed off to 1,050 F and 3,500 psi. For similar reasons a second Eddystone unit is being designed for 3,500 psi and 1,050 F.

The reheat cycle which was initiated early in this period has now become general practice and has made possible substantial gains in thermal efficiency. Single reheat applications were developed as follows:

Date	*Initial Temperature*	*Reheat Temperature*
1940	940 F	900 F
1949	1,050 F	1,000 F
1953	1,050 F	1,050 F

More recently, the application of a double-reheat cycle has been developed. These include the experimental 1,150/1,050/1,000 F at Philo; the experimental 1,200/1,050/1,050 F for Eddystone No. 1, and the 1,050/1,050/1,050 F for the 450,000-kw units at Breed No. 1 and Sporn No. 5. These basic improvements in the heat cycle, along with many other refinements in boilers and turbines, have increased thermal efficiencies markedly. Compared with an average heat rate of 16,500 Btu per kwhr for units installed in 1934, the average for units installed in 1958 was close to 9,900 Btu per kwhr, and for one whole system, the AEP System, average over-all system heat rate in 1958 was reduced below 10,000 Btu per kwhr. This is all the more significant considering that it was only as recently as in 1950 that the Philip Sporn Plant was the first to achieve an average plant heat rate below 10,000 Btu per kwhr.

Along with improvements in thermal efficiency, the great increase in unit sizes has been one of the major factors in holding the cost per kw of capacity down to levels approaching those of a decade ago, despite an

Courtesy Commonwealth Edison Company

Modern turbine room of 1934, State Line Unit 1. 208,000 kw, triple cross-compound, 1,800/1,800/1,800 rpm. Main steam 600 psi, 730 F, reheat 500 F; optimum heat rate is 12,950 Btu; average annual heat rate 13,355 Btu.

Courtesy Tennessee Valley Authority

Modern turbine room of 1,600,000-kw Kingston steam plant. Nine 3,600-rpm, 1,800-psi tandem-compound units; four 150,000-kw, 1,000 F, reheat 1,000 F; five 200,000-kw, 1,050 F, reheat 1,000 F. Unit boiler–turbine layout, generator cooling H_2 at 15 and 30 psi.

almost doubling of construction costs in the postwar period. This is readily illustrated, as noted in the foregoing, by the space occupied by the Philo No. 6 Unit, almost tripling the capacity previously installed in a given area. In addition to the savings in construction costs made possible by increases in unit capacity, savings in operating expense have also been made possible through a reduction in the manpower requirements from upwards of 1 man per 1,000 kw 25 years ago to as low as 0.25 man per 1,000 kw today. Even more gratifying is the indication that this trend is not at an end, and that even lower manpower requirements per unit of capacity can be expected on the newer 450,000- and 500,000-kw units. The reduction in manpower requirements has been materially advanced by the use of centralized and largely automatized control arrangements along with the general adoption in the postwar period of the single-unit boiler–turbine combination for even the largest units, including the 450,000- and 500,000-kw units now under construction. The million-pound-per-hour steam boiler has now been replaced by the large steam generators rated at from 3 to 4 billion Btu-per-hour input. The increase in efficiencies has managed to offset a substantial part of the effect of the increases in fuel cost since the end of the war.

HYDROELECTRIC GENERATION

ALONG WITH THE LARGE EXPANSION of our steam–electric generation, progress has not been static in the further development of hydro resources not only in all areas of the United States where hydro potential is available, but also to a notable extent in Canada. An indication of progress in this country is given by the growth of installed hydro capacity during the past 25 years from a figure of 9,345,000 kw in 1934 to 29,318,000 kw in 1958, along with an increase in total hydro energy generation from 32,684 million kwhr to about 140,000 million kwhr in the same period.

Among other major developments are those completed and under way on the Niagara and St. Lawrence Rivers. At Niagara, the Lewiston Plant of 1,950,000 kw will incorporate the largest hydro units to date, 150,000 kw each, compared with Grand Coulee units of 108,000 kw each. This plant and its associated plant, the Tuscarora pump–hydro 240,000-kw plant, will make a total project of 2,190,000 kw, the largest single project to be developed in this country as well as in this hemisphere. Completion of this project is scheduled for 1962. The St. Lawrence development at Barnhart Island comprises essentially two plants in one, each of the same rating, the Robert H. Saunders–St. Lawrence Generating Station in the Canadian portion and the Robert Moses Generating Station on the American side. Both of these plants are now in full operation, each with a capacity of 912,000 kw.

Elsewhere in the United States, the principal areas of development have included the major projects in the Pacific Northwest, such as Bonneville, Grand Coulee, Cabinet Gorge, Ross, Brownlee, and many others; the Shasta Dam, along with numerous smaller projects, in California; Missouri River plants, such as Garrison and Fort Randall; Clark Hill and John Kerr in the Southeast; the numerous plants on the Tennessee and its tributaries involved in the TVA (Tennessee Valley Authority) complex; and finally, the New England area.

In Canada, a number of large projects on the order of 1 million kw or larger have been completed and others are under construction. These include the Ontario Hydro Electric Commission Adam Beck No. 2 Plant and pump–hydro installation at Niagara Falls, and the Robert H. Saunders Station on the St. Lawrence at Barnhart Island mentioned earlier. Quebec Hydro-Electric Commission projects include the 1.5 million kw Bersimis development and the Beauharnois Plant on the St. Lawrence River near Quebec. Aluminum Company of Canada has carried out an impressive group of hydro projects, including Kitimat in British Columbia with its 1,250,000-kw underground powerhouse, high-altitude transmission line, and the Lake St. John complex of plants totaling some 2 million kw on the Saguenay and Peribonka Rivers.

With many of the more favorable hydro sites already developed, a trend toward more economical methods of dam construction has made a significant contribution to extending the economic feasibility of remaining sites. In particular, two types of construction may be cited, first the rock-fill construction in which the waterproofing is obtained by means of a deck on the upper face made of asphaltic cement to give it the necessary degree of flexibility without cracking. The other type consists of a combination rock and earth fill, the latter comprising a core of well-packed clay soil in between rock fill on both upstream and downstream sides.

Considerable interest also has been evidenced in pump–hydro developments although total capacities involved so far are not great. Projects during the past 25 years include the Hiwassee Plant of TVA, the Colorado Big Thompson project at Estes Park, the Tuscarora and Adam Beck projects on the Niagara River previously mentioned, and a new project under development at Smith Mountain, Va. The trend in later developments has been toward the use of a single unit to serve as both a generator–turbine combination and a motor–pump combination as compared with separate units for each function.

Although a heavy program of hydro construction has been carried out in this country, it is significant to point out that generation has grown even faster so that the proportion of United States over-all power requirements contributed by hydroelectric plants has shown a considerable decline over the past 25 years. With the one exception of the Pacific Northwest area, this is true even in areas of large hydro potential such as in California, as well as in the TVA system where more than two thirds of the total power requirements are now generated by steam. Likewise, the Hydro Electric Power Commission of Ontario is basing future growth of generating capacity principally upon steam. The extent to which the expansion of steam–electric genera-

Courtesy Westinghouse Electric Corporation

Railway car mounted 50,000-kva 3-phase, 138 kv—69/34.5 kv mobile transformer at Lima, Ohio.

Courtesy Indiana & Michigan Electric Company

200,000-kva 345/138-kv 3-phase autotransformers at Olive Station, Ind. Reduced insulation design 1,050-kv basic insulation level.

Courtesy Detroit Edison Company

Large 3-phase step-up transformer, 370,000 kva, 17/129 kv, at St. Clair Station, Detroit, Mich. General Electric FOA design.

tion has outstripped the growth of hydro resources in the United States is indicated by the drop in percentage of total generated hydraulically from 37.5% in 1934 to 21.8% in 1958.

TRANSFORMERS

In keeping with the growth of generator unit sizes as well as with over-all growth of systems, power transformer sizes have been increased to capacity ratings hardly visualized 25 years ago. This has come about by improvements in design from the standpoint of insulation, efficiency in the use of materials, improvements in the quality and characteristics of materials, and finally, by drastic changes in shipping methods, all of which have made possible larger and larger capacities within practical space limitations.

One of the first steps in this direction was the one-step reduction in basic insulation level (BIL) for high-voltage transformers, 115 kv and above, made possible by more effective co-ordination of insulation strength and lightning arrester protective characteristics. An example of this was the use of a 550-kv BIL, reduced from 650 kv, for 138-kv transformers, tried out successfully as early as 1934, and gradually adopted as standard practice thereafter.

Other important steps were the development in 1941 of grain-oriented steel, permitting a 1/3 increase in core flux density, and the introduction of the FOA design, combining forced-oil circulation and forced-air cooling. Further reduction in oil and material requirements was obtained by the use of special tanks designed to fit core and coils more closely than previous straight-walled tanks.

A bold step to insure the adequacy of these transformer designs to withstand exposure to lightning surges under field conditions was the introduction in the early part of this 25-year period of the practice of using impulse tests for controlling the quality of transformers in regular production as well as for research purposes on new development models. This has paid off in reducing transformer failures in this country caused by lightning to very small proportions.

With the solid background of research and technology in transformer design and construction, including the various developments leading to concentration of larger capacities in smaller dimensions, the manufacturers were able to take in their stride the design and construction of 345,000-volt transformers when these were first required in 1953. The first group of these transformers, designed with a basic insulation level of 1,175 kv, 1½ steps below the full 1,550-kv level, were 150,000-kva 3-phase autotransformers, 345 kv to 138 kv with 37,500-kva tertiary windings. Subsequent installations of larger 345-kv transformers include 3-phase 200,000-kva autotransformers similar to the original 150,000-kva units and a 3-phase 275,000-kva 345-kv generator step-up transformer.

Consistent with the general adoption of the unit arrangement for turbine, generator, and boiler installations, it is significant that available sizes of single-

Courtesy Commonwealth Edison Company

100,000-kva 3-phase 22-kv ±38-degree phase-shifting transformer, at State Line Station in Illinois. Westinghouse design.

unit transformers have kept pace with the increase in sizes of generators and turbines. These transformers have progressed from 315,000-kva 3-phase units in 1955, to the largest to date of 380,000-kva 3-phase, both for generator step-up purposes. As in the case of large capacity generators, the introduction of forced cooling which has made possible the development of improved techniques in cooling, in this case primarily in the physical arrangement of conductors and insulation, has been an important factor in making these high-capacity ratings possible.

A further step in reducing the relative size, cost, and weight of transformers, particularly in the higher voltage ratings, is an additional lowering of BIL which is now being tried out, taking advantage of improvements in accuracy of protective levels of lightning arresters. For 345-kv transformers, this has been a full 2-step reduction in BIL from 1,550 kv to 1,050 kv compared with the 1,175-kv level or 1½-step reduction based in earlier designs. At lower voltages, such as 138 kv, 230 kv, and others, a similar 2-step reduction in BIL is being tried out in some installations, including a number of 138-kv transformers with 450-kv BIL, two steps below the full insulation level of 650 kv. Although the largest 345-kv transformers now under construction using the 1,050 BIL are the 275,000-kva, 3-phase step-up units for the Breed 450,000-kw generators, manufacturers today indicate willingness to undertake building such units in ratings as high as 550,000 kva or even higher.

TRANSMISSION SYSTEMS

DEVELOPMENTS IN GENERATION, which have resulted in large capacity units and in an increasing number of both steam and hydro stations with more than a million kw of capacity, would not have been possible under the geographical limitations of water, land, and availability of fuel without major expansion of transmission systems capable of moving large quantities of power economically to centers of load. To a considerable extent and particularly on systems where transmission distances were not excessive, the necessary expansion of transmission capacity has been obtained at voltage levels already in use, such as 115 kv, 138 kv, and 161 kv, not to mention 230 kv, which was already in use on a number of systems 25 years ago, having been initiated in California in 1923.

While there exist a number of examples of major expansions in transmission capacity at existing voltage levels, these are exceptional. In general, the movement of increasingly large quantities of power at these lower voltages has become more and more burdensome from the standpoint not only of the multiplication in number and cost of transmission lines required, but also from the standpoint of the serious right-of-way problems involved, particularly in areas of expanding populations and suburban residential build-up. For many systems, a satisfactory solution has been the adoption of 230-kv transmission, more than 15,000 circuit-miles of which is now in operation in the United States alone. It has been a popular voltage, not only for superposition on systems up to 115 kv and even 138 kv, but also for bulk power transmission on many hydro developments involving long distances, both in this country and in Canada.

A notable exception to this pattern was occasioned by the Hoover Dam project in the early 30's, involving transmission of a large block of power some 275 miles to Los Angeles. For this purpose, 230 kv was considered inadequate and new ground was broken in developing a 287-kv transmission system including all related equipment. Although this system, including a third line added in 1939, was successful in operation, expansion elsewhere during the next decade continued at the 230-kv level.

In the meantime, other systems still using 115- or 138-kv transmission were recognizing the need for a higher transmission voltage which to them did not appear could be met by either 230 or 287 kv. This very

Courtesy Ohio Power Company

345-kv transmission, conventional double-circuit construction, two ground wires, single conductor 1,414 MCM, 1.75-inch diameter, expanded ACSR, eighteen 5¾-inch suspension insulators.

345-kv transmission, double-circuit tower design with delta spacing and bundled conductors: two 795-MCM ACSR on 16-inch spacing, two ground wires, eighteen 5¾-inch suspension insulators.

Courtesy Ohio Power Company

conclusion was reached by the AEP System and led to the establishing in 1947, in co-operation with several manufacturers, of the Tidd high-voltage test project in Ohio. There, various aspects of transmission at voltages ranging from 265 kv to 525 kv were studied extensively over a period of several years. The resulting evaluation of the information obtained finally led to the adoption of 345 kv as the new backbone transmission voltage for the AEP System. At the present time, this voltage is being utilized or planned for installation not only on the AEP System but also in several other areas of the United States and in Canada. In addition to a total of some 2,000 circuit-miles now in operation at this voltage on the AEP and Ohio Valley Electric Corporation systems, it is now being used by Bonneville Power Administration in the Northwest, by the Commonwealth Edison Company, by Ohio Edison Company, and by the British Columbia Electric Company in Canada. Although this is the highest transmission voltage in the United States or Canada at the present time, studies are under way on a number of systems of 460 kv and even higher voltages to handle possible long-term requirements.

SYSTEM CONTROL, PROTECTION, AND COMMUNICATION

THE DEVELOPMENT of today's far-flung interconnected transmission networks obviously involved much more than simply constructing transmission lines, substations, and units of generating capacity. The solution of a great many technical problems and the development of many types of specialized equipment were required in order to mold these basic elements into effectively controlled, economically operated systems capable of supplying adequate and reliable transmission and distribution service and doing so under severe weather and other abnormal conditions. Among the problems involved were circuit interruption at new orders of magnitude of short-circuit current; protective relay schemes of greater accuracy, speed, and dependability; control, communication, and telemetering facilities for successful operation of individual systems; parallel operation of interconnected systems; and many other related problems.

The development of circuit breakers with increased operating speed and interrupting capacity has been one of the outstanding accomplishments. In 1934, the maximum interrupting ratings available up to 230 kv was 2.5 million kva and 8-cycle interrupting time was a recently brought-out improvement over the 15 cycles or more which prevailed shortly before. Spurred by the transmission and stability requirements of the Hoover Dam project in 1936, a radical advance was made in a special design of a 3-cycle 2.5-million-kva breaker for operation at 287 kv. For other duties at 230 kv and below, however, several years had to elapse before new designs with both increased interrupting capacities and higher speeds became available. In the development of these new breaker designs with much higher ratings, major field short-circuit tests supplementing laboratory development work played an important part. These included 138-kv tests up to 3.5 million kva at Philo in 1944, 230-kv tests at 7.5 million kva at

High-capacity high-altitude 300-kv line. Extra-heavy construction single-circuit line over 5,300-foot Kildala Pass, B. C., Canada. Conductor 2.32-inch ACSR to carry 600,000 kw per circuit.

Courtesy Aluminum Company of Canada

Courtesy Long Island Lighting Company

Modern dispatching center. System operations control center, Hicksville, N. Y. Underground location.

Grand Coulee in 1948, 138-kv tests up to almost 7 million kva at Philip Sporn in 1954, and finally 345-kv tests also at Philip Sporn in excess of 13 million kva in 1957. Contrasted with 2.5-million-kva ratings in 1934, circuit breakers are available today in ratings of 15 million kva at 138 kv, 20 million kva at 230 kv, and 25 million kva at 345 kv. It is notable also that design improvements have produced these high-capacity oil circuit breakers in even smaller-dimensioned tanks and lesser oil requirements than those of the 2.5-million-kva breakers 25 years ago.

In addition, substantial advances have been made in the use of mediums other than oil for breakers. For example, more than 100 breakers using compressed air as the interrupting medium are now in operation or on order in the United States, ranging in ratings from 2.5-million-kva 5-cycle opening for 115-kv designs to 15-million-kva 3-cycle ratings for 230 kv and 300 kv, and 25-million-kva 3-cycle ratings for 345 kv. At the present time, development of circuit breakers using sulfur hexafluoride gas as the interrupting medium is under way and shows promise not only of higher interrupting capacities, but also of superior performance from a maintenance standpoint.

Along with circuit breaker developments, substantial improvements have been made in protective relaying schemes for transmission systems. An early improvement was the speeding up of carrier-current differential relaying to 1-cycle operation, compared with the four cycles or more previously available. Other improvements include the introduction of a modified type of impedance relay which, combined with directional-comparison schemes with much higher current settings, permits carrying heavier overloads or swings under nonfault load conditions without unwanted tripping.

One of the developments in the transmission and protection art, which is not only outstanding in its importance but dramatic in its conception and performance, is the successful ultrahigh-speed reclosing of transmission lines by which a faulted line is opened at both ends simultaneously to clear the fault and returned to normal operation by reclosing, all in a fraction of a second. In view of the sometimes ponderous dimensions of the high-voltage switches controlling the opposite ends of a transmission line many miles in length and the high arc energy involved in a short circuit caused by a lightning flashover, the accomplishment of this opening and reclosing sequence in 1/3 to 1/4 of a second is an above-the-ordinary technical achievement. Because lightning flashover, in spite of much progress in lightning research made during the past 25 years, remains the greatest menace to transmission reliability not only throughout a great part of the United States and Canada, but in other important areas of the world as well, the significance of this technique can hardly be overemphasized, particularly in view of the increasing dependence placed upon continuous electric service in all phases of any modern industrially developed society.

Beginning with the first trial installation in 1936 on a 138-kv line with 8-cycle breakers, the use of ultrahigh-speed reclosing has been extended until, at the present time, it has become standard practice for a large portion of the industry, particularly in moderate to severe lightning areas. With the availability of 3-cycle breakers and modern carrier relaying, the art today has advanced to over-all reclosing times as low as 15 cycles (1/4 second) on 138-kv lines and 22 cycles on 345-kv lines. Performance records of 90 to 95% successful reclosure are being obtained in areas where lightning is the principal cause of line faults.

In the field of control of power system generation, frequency, etc., the operation of large complex systems and particularly interconnected groups of systems would have been greatly hampered if not rendered impracticable without developments in techniques and equipment for automatic control of frequency and tie-line loading which are in widespread use today. The need for such techniques became apparent even earlier than the beginning of this 25-year period and early applications of automatic control were made at that time on several systems, including an interconnected group of some 30 companies operating in parallel in Ohio, Indiana, western Pennsylvania, and in adjacent states to the south, all with a then-combined generating capacity of around 5 million kw.

Co-operative efforts among the operating organizations of these systems were undertaken to develop and improve automatic control techniques, and this co-operation has continued with the growth of the systems and the addition of new members to the interconnection pool until today the group comprises 104 companies operating in 29 states, with a total combined generating capacity of more than 60 million kw. Initial attempts at automatic frequency control by placing the burden on a single centrally located generating plant were quickly modified to spread the burden of frequency regulation to other stations. This approach has now expanded to the point where every important generating station in the interconnected systems group is assigned an appropriate share or "band" of generating capacity to be raised or lowered in the event of a system frequency deviation above or below a normal frequency band. The basic function of modern frequency and tie-line control is, of course, that of regulating the total generating output in each individual system in such a way that the total generation will match continuously the total system load including scheduled loads on interconnection tie-lines.

Obviously, the successful development and application of these automatic control functions along with many other requirements of modern system operation has necessitated the accompanying development of adequate communication systems. One of these functions is that of economic dispatching of generation on individual systems. This process of scheduling plant loadings for best over-all economy has been brought to varying stages of development, in some cases by continuous manual supervision from a central dispatching center using the incremental slide rule in conjunction

Courtesy Ohio Power Company

Courtesy Ohio Power Company

Courtesy Westinghouse

Mobile space radio equipment in electric utility line truck.

Microwave repeater station with steel tower.

EHV oil circuit breaker, 345-kv 25-million-kva interrupting capacity, 3-cycle opening, high-speed reclosing.

with an incremental transmission loss computer to obtain maximum economy, and in other cases by incremental loading computers installed to carry out the economic dispatching automatically. While great progress has been made, this function of system control is still in a state of development and experimentation, with various systems being proposed and a few of them under trial.

Here again adequate communication is of outstanding importance. Carrier current over power lines has been used for many years for telephone communication, relaying, telemetering, load control, etc., and is still the reliable and much-utilized medium in this field. Equipment, including transmitters and receivers, has been greatly improved in performance, particularly with respect to signal-to-noise ratio.

In many systems, however, carrier-current channels have become inadequate, both as to the number available and the quality of circuits provided. This has led to the increasing use of microwave communication systems for which suitable equipment and license authorizations became available some 10 years ago. These systems today provide high-quality voice transmission, free from atmospheric and man-made interference, and can handle a large number of channels for all communication requirements, including telemetering, system control, etc. At present, some 700 microwave stations are in service on transmission systems in this country alone. With reliability and technical performance equal to or better than any other alternative means of communication, it is expected that these systems will be greatly expanded to meet continuing growth in communication requirements.

Another form of communication, developed almost entirely within the last 25 years, is the mobile radio system by which communication is maintained between fixed centers of operation and mobile units in the field. These systems have been of tremendous assistance in the efficient deployment and supervision of manpower for all functions of utility work including construction, system operation, and both routine and emergency maintenance. Beginning with a few experimental installations 25 years ago, commercial equipment has since been developed and greatly improved, with changes from AM to FM and increased sensitivity, selectivity, etc. A recent count shows some 9,000 fixed stations and 125,000 mobile units plus a number of portable units in service today.

ROLE OF COMPUTERS IN THE POWER INDUSTRY

THE USE OF NETWORK ANALYZERS for power system planning had begun to a limited extent at the beginning of this 25-year period in 1934 when some three or four such analyzers had been made available. Today, some 35 of these are in use, several of the more recent designs being equipped with a large number of automatically controlled generators and other improved features to facilitate analysis of load and voltage problems on today's complex and expanding systems. In addition to their use throughout the power industry as an indispensable tool for power system planning, these analyzers in some cases have also served as effective teaching mediums in engineering schools.

Much more recently, the application of digital computers to power system planning problems has made rapid strides. For example, the powerful high-speed IBM (International Business Machines Corporation) *704* computer has been very successfully programmed for the calculation of power flow in networks, as well as for other problems in network analysis such as stability problems, and now seems capable and perhaps destined to take over much of the job of network analyzers. The computer solution is, in fact, superior from the standpoint of economy and accuracy for load-flow problems and has been proved capable of solving other problems, such as extremely complex stability analyses which have been entirely beyond the capability of the network analyzer or any other practicable method of solution.

In system operation, digital computers are being used for certain calculations in connection with economic dispatching of power system generation as previously described. They have been of particular value in the calculation of transmission loss factors. Digital computers are being more and more extensively applied, of course, in the commercial and accounting phases of the industry and, in some cases, have done a unique job in calculating distribution transformer loading from customer billing records. Altogether, the use of digital computers appears to have opened up an extensive new field for analysis not only in system planning problems, but for many other types of decision-making problems as well.

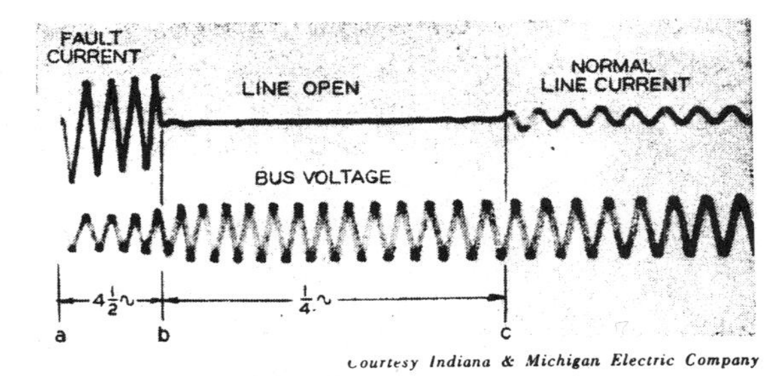

Courtesy Indiana & Michigan Electric Company

Oscillogram of ultrahigh-speed reclosing operation: (a) short-circuit current initiated (lightning flashover); (b) short-circuit current interrupted 4½ cycles later; (c) circuit reclosed after 14-cycle dead time and 18½ cycles after initial fault. Circuit restored with normal load current indicated.

Extrahigh - voltage air-blast circuit breaker, 345-kv 25-million-kva interrupting capacity, 3-cycle opening, high-speed reclosing, Lima, Ohio.

Courtesy Brown Boveri Company

Courtesy General Electric Company

Air-blast circuit breaker, 138-kv 10-million-kva interrupting capacity, 3-cycle opening, high-speed reclosing, New York, N. Y.

Courtesy U. S. Atomic Energy Commission

Portsmouth Project, uranium diffusion plant. Peak demand (1957) 2,137,000 kw; annual consumption (1957) 17,504 million kwhr. Energy supplied by Ohio Valley Electric Corporation.

DISTRIBUTION

JUST AS the great expansion in generation over the past 25 years would not have been possible without paralleling developments in transmission, similarly the efficient and economical delivery of this energy to ultimate consumers would have been severely handicapped if it had not been accompanied by the extensive and important developments in distribution which have taken place.

One of the major advances, as in the case of large power transmission, has been the development of higher distribution voltages. For example, 2,300-volt delta circuits have practically disappeared by conversion to 4,000-volt Y circuits with common neutral. At the same time, a very extensive growth has taken place in the use of the distribution voltages in the 7.2/12-kv Y class, beginning largely as a rural distribution voltage but subsequently developing into extensive use today as a highly economical voltage for urban distribution as well.

Courtesy Olin Mathieson Chemical Corporation

Aluminum reduction plant. View of potline showing 38 out of 168 pots (electrolytic cells) in a single line. Total plant load—five reduction potlines plus rolling mill—360,000 kw. Aluminum production 180,000 tons per year.

Courtesy Kaiser Steel Company

Strip mill, Kaiser steel plant, Fontana, Calif. Strip mill demand, 35,000 kw. Total plant demand, approximately 90,000 kw. Total plant annual usage, 500 million kwhr.

Although nowhere near as extensive as 12-kv Y distribution, a substantial beginning has been made in the use of 14.4/24.9-kv Y in a number of rural areas where it has proved to be practical and economical. It has also proved economical in some areas where existing 13.8-kv delta systems have been converted to 24-kv Y.

A start, at present principally in the stage of study and discussion, has been made in the use of a still higher distribution voltage, 19.9/34.5-kv Y. Trial installations at this voltage are now under way in the Northwest.

Major improvements in service reliability have been brought about by the use of high-speed circuit reclosures, not only for line-sectionalizing service, but also for use as substation feeder breakers. By this means, circuit clearing is being accomplished at three times the speed of conventional circuit breakers, along with fast restoration of service for transient fault conditions.

In the area of distribution transformers, very important developments have taken place in design improvements, giving better electrical characteristics, reduced physical size, and greater reliability. The use of grain-oriented steel, strip-wound cores, and more efficient insulating materials has practically doubled the capacity available within a given dimension, and at the same time losses and impedance values have been reduced. In one case, as an example, even a fairly recent 167-kva distribution transformer design has been replaced in exactly the same tank size with a 250-kva unit.

While underground distribution in concentrated urban areas was fairly prevalent 25 years ago, particularly in larger cities, this has been greatly expanded during this period, both in large metropolitan areas and in moderate-sized cities and towns.

The use of shunt capacitors, both switched and unswitched, has expanded to a very great extent during

Courtesy Indiana & Michigan Electric Company

Electrically heated school building. Penn Township High School near Mishawaka, Ind.; 51 classrooms; total connected load 4,816 kw.

the past 25 years, stimulated greatly by improved manufacturing methods. This has resulted in the development of progressively larger unit sizes, increasing from 15 to 25 kva, 25 to 50 kva, and most recently to 100 kva, all at the prevailing distribution voltages up to 15 kv. At the same time, in contrast to the rising costs for other equipment, the cost of capacitors in dollars per kva has actually been lowered.

Another important development, extensive in its effect, has been the almost explosive increase in the use of synthetics for cable conductor and other insulating requirements. These include polyethylene, neoprene, polyvinyl chloride, Butyl, and many others. Greatly improved characteristics in aging, resistance to chemicals, sunlight, flammability, and many others have been realized.

The use of aluminum has superseded to a very large extent the use of copper for many distribution purposes, including service drops where it is becoming almost universal practice. Aluminum is also used as a sheathing for paper-insulated cable.

Finally, in secondary voltage practice considerable use is being made of the higher-level 277/480-volt systems. It has not yet been introduced as an official standard, but its coming as an approved practice in the not-too-distant future is clearly foreshadowed.

UTILIZATION

IN THE PAST 25 YEARS, and more especially in the period since World War II, electric energy has penetrated deeply into every phase of modern living. It is essential not only to power our industry but to perform in commerce, on the farm, and in the home the many tasks that have now come to be regarded as essential elements in our day-to-day living. The applications

Courtesy Ohio Power Company

Modern electrified farm in Ohio has 160 acres in corn, wheat, pasture, and hay for hogs and steers. Farm equipment: food grinder, mixer, elevator, pig brooder. Home equipment: water system, two ranges, water heater, clothes dryer, deep freezer. Approximate demand 40 kw.

Courtesy Indiana & Michigan Electric Company

Modern electrically heated home, Fort Wayne, Ind., with 10-kw baseboard heating plus range, water heater, dryer. Average annual consumption 18,000 kwhr.

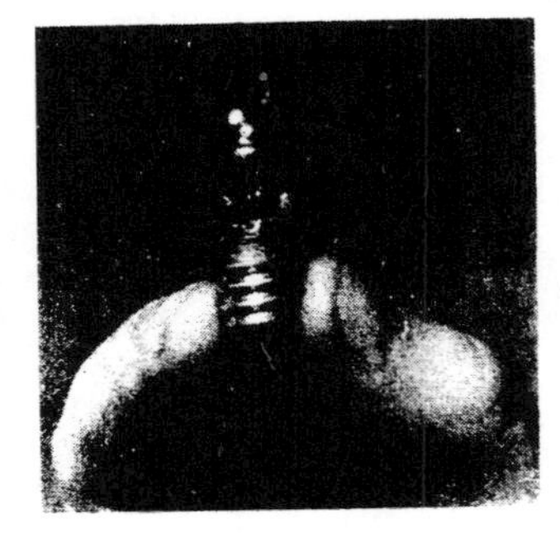

Neon light, 1/4 watt. One of the smallest loads on a central station power system.

Courtesy Uris Brothers

Modern office building, at 750 Third Ave., New York City; 34 stories, 602,000 square feet rentable area. Total connected load: lighting 3,680 kw, motors 2,400 hp. Maximum demand (Summer 1958) 3,260 kva.

of electric energy range widely both in function and in magnitude; it would be difficult to conceive our society without the availability of an adequate supply of electric energy. There is little doubt that the requirements for electric energy and its applications to more and more functions will continue to expand. The growing requirements for new metals such as titanium and magnesium along with the rising demands for aluminum, all of which require electrical processes for their production, and the processing of low-grade ore such as taconite, will expand the need for electric energy in industry markedly. And beyond this, the need to expand our productivity to provide the rising standard of living that we have come to expect for a growing population in which the hours of work are falling, and the proportion of the population of labor-force age is declining, will also require the the application of substantial and increasing quantities of electric energy.

Similar considerations apply on the farm and in commerce. In the home, our rising standards of living are intimately associated with the expanding use of electric energy for the many devices that reduce the work required to maintain the home—to cook, to clean, to preserve foods, and to provide entertainment and information. In both commerce and in the home, year-round weather conditioning is becoming increasingly important. The shopping center protected from the weather and maintaining constant temperatures throughout the year through electric cooling and heating is just at the early stages of development, and the all-electric home to provide similar comforts is also on the verge of rapid development and extension. All of this indicates a continuation of the rapid long-term increase in electric energy requirements which is likely to continue for a long time—certainly until the centenary edition of this publication appears.

But this is not a result that can be expected to develop automatically in the natural course of events. A great deal of effort in research, development, and in utilization of new concepts and discoveries will be required on the part of many technicians, engineers, and technologists associated with the industry. Particularly in utilization, much remains to be done to develop further those devices which show promise of contributing importantly to our national productivity, welfare, and well-being. To cite just one example, an important step forward would be the development of an efficient heat storage system which would make possible the combination of an electric heat pump and solar heating system.

Although the industry in the past quarter century has made very substantial technological strides which have given the country perhaps the finest series of systems for making available to its economy an abundant and highly economical supply of electric energy, many technological challenges loom up for the quarter century ahead. In the field of atomic energy only a beginning has been made, and there still remains the finding of a solution to the many difficult problems which will make possible economic generation of electric energy by means of nuclear fission, and to make it possible for the atom to pick up some of the burden of supplying the country's ever-increasing heavy energy requirements. Much remains to be done to improve even further the efficiency of generation, transmission, and distribution and to extend the field of application of electric energy. The history of how this has been accomplished, and of how challenges have been met and responsibilities discharged by the industry and by its technicians, engineers, and technologists should make most exciting and stimulating reading in 1984.

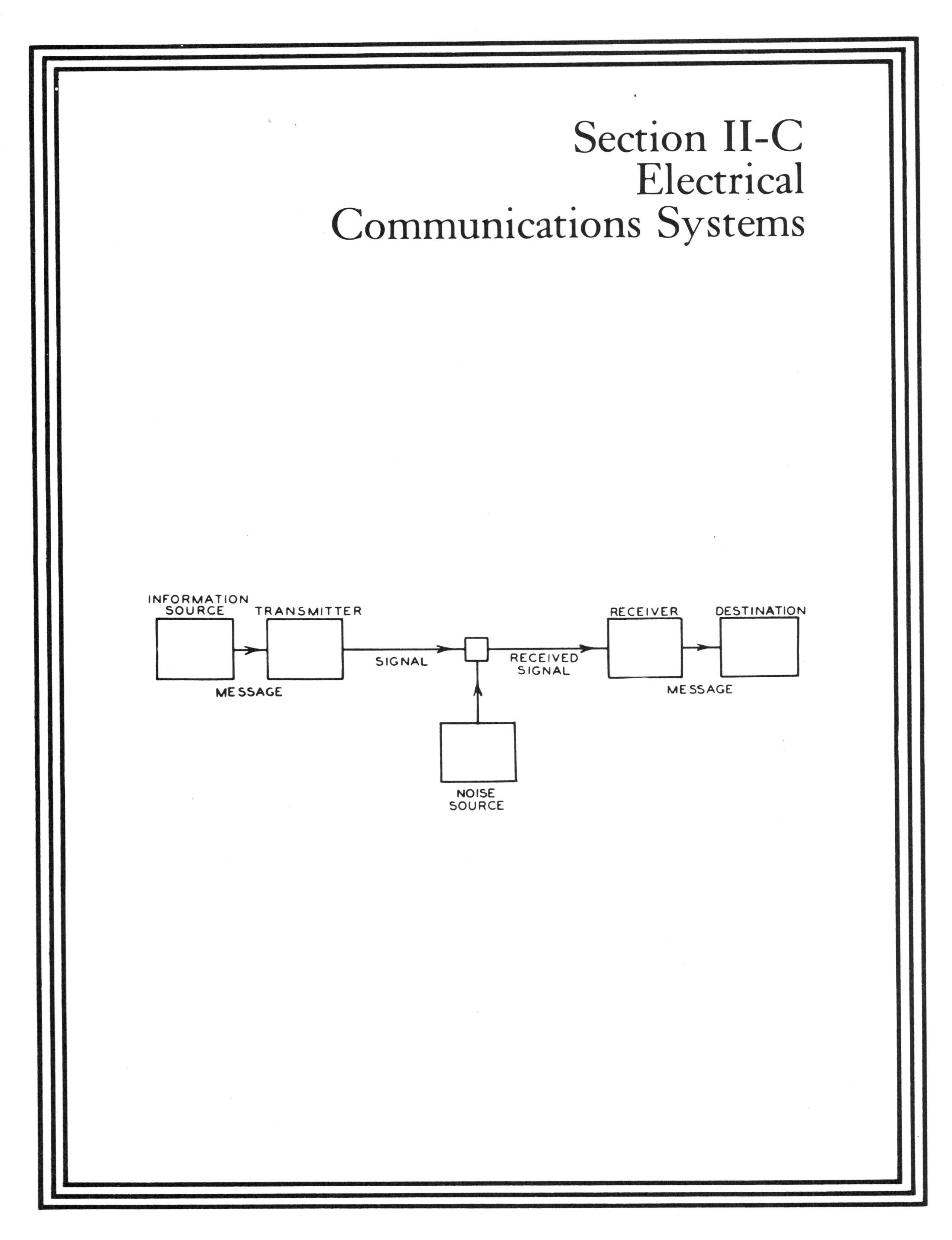
Section II-C
Electrical
Communications Systems
INFORMATION SOURCE
TRANSMITTER
SIGNAL
RECEIVED SIGNAL
RECEIVER
DESTINATION
MESSAGE
MESSAGE
NOISE SOURCE

Pupin on Loaded Telephone Conductors

IN this brief paper, one of the most famous pioneers in electrical engineering education and invention described the loading coil that brought him considerable wealth as well as prestige. By the time this article appeared, he had already received approximately $200,000 from the Bell Company for the patent rights and would ultimately receive almost half a million dollars before the patents expired. One might argue that the payment was not excessive since Bell's control of loading enabled the Company to maintain an effective monopoly on long-distance circuits and a substantial advantage in urban cable circuits as well. More than one million loading coils were installed by 1925 and more than 14 million were made as late as 1974 for use by the Bell System. According to an estimate published in 1926, the loading coil innovation had saved the Bell Company approximately 100 million dollars in capital costs. The loading coil is an early example of an innovation that required an understanding of the Maxwell–Heaviside electromagnetic theory and proficiency in mathematical analysis by the inventor. Both Pupin and a rival claimant to priority in the invention of the loading coil, George A. Campbell, received advanced training in mathematical physics. See *A History of Engineering and Science in the Bell System. The Early Years (1875–1925)*, edited by M. D. Fagan (Bell Telephone Laboratories, Inc., 1975), pp. 241–252. Also see James E. Brittain, "The Introduction of the Loading Coil: George A. Campbell and Michael I. Pupin," *Technology and Culture*, vol. 11, pp. 36–57, 1970.

Michael I. Pupin (1858–1935) was born in Idvor in what was then part of Austria–Hungary. He came to the U.S. in 1874 and worked as a farm laborer and in a cracker factory before entering Columbia where he graduated in 1883. He received a fellowship that enabled him to study in Europe and where he received a Ph.D. degree from Berlin University in 1889. Pupin then returned to teach at Columbia in a newly established electrical engineering program. He remained at Columbia until his retirement in 1931 and was also quite active in both the AIEE and IRE. His autobiography, entitled *From Immigrant to Inventor*, originally published in 1922, became a best seller and was awarded a Pulitzer Prize. He was one of the first in the U.S. to experiment with X-rays in 1896, but his greatest fame resulted from the loading coil that he patented in 1900. See the *DSB*, vol. 11, p. 213 and the *NAS*, vol. 19, pp. 307–323.

A Note on Loaded Conductors.

By M. I. Pupin, Ph.D.

In my publications on wave propagation over non-uniform conductors it was shown that such conductors act approximately like their corresponding uniform conductors when the number of loads per wave length is sufficiently large. Sixteen loads per wave length was shown to be so close an approximation that experimentally no difference could be detected between the non-uniform conductor and its corresponding uniform conductor. It was also shown, experimentally, that with a small number of, say, four loads per wave length the resemblance between the two is small, and the efficiency of transmission over the non-uniform conductor is very much diminished in spite of the increased inductance. In other words, when the number of loads per wave length is small

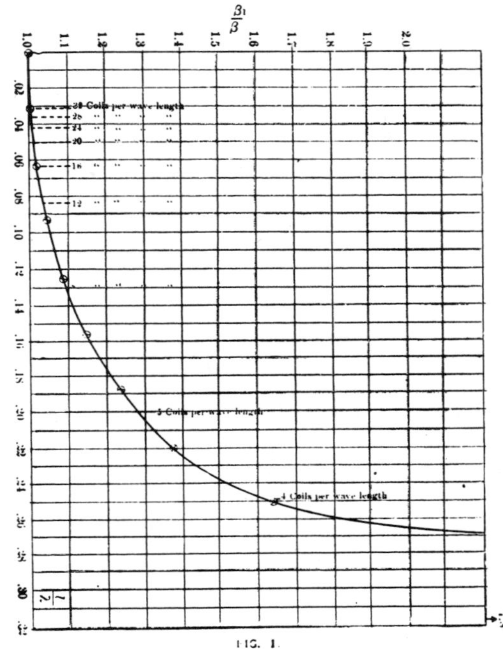

Fig. 1.

the loaded conductor becomes an inefficient transmitter and, therefore, the loading is poor.

It is proposed to show in this note, numerically, the action of a poorly loaded conductor by comparing, numerically, its propagation constants to those of the corresponding uniform conductor. Numerical calculation can be readily made by formulæ deduced in my two papers referred to below. The two types of loaded conductors will be considered separately. The first type was discussed in Section II. of my paper on "Propagation of Long Electrical Waves," *Transactions* of the American Institute of Electrical Engineers, Vol. XV. (March 22, 1899). The second type will be found in a paper read by me before the American Mathematical Society, Dec. 28, 1899, and published in the *Transactions* of that society, Vol. 1, No. 3. The behavior of these two types of conductors will be found to be the same under all conditions which can arise in ordinary practice.

A. Loaded Conductor of the First Type.

In this case wave propagation is expressed in terms of a complex angle $\theta = \frac{1}{2}(\alpha_1 + i\beta_1)$ which is defined by the following equation:

$$h = -4\, sin^2\, \theta \; * \qquad (1)$$

which can be also written

$$sin\, \theta = sin\, \frac{1}{2}(\alpha_1 + i\beta_1) = \frac{1}{2}(\alpha + i\beta) \; ** \qquad (2)$$

This equation determines completely the propagation constants α_1 and β_1 of the loaded conductor in terms of the propagation constants α and β of the corresponding uniform conductor.

For numerical calculations the real and the imaginary parts can be easily separated, thus

$$\left.\begin{aligned} cos\, h\, \frac{\beta_1}{2}\; sin\, \frac{\alpha_1}{2} &= \frac{\alpha}{2} \\ sin\, h\, \frac{\beta_1}{2}\; cos\, \frac{\alpha_1}{2} &= \frac{\beta}{2} \end{aligned}\right\} \qquad (3)$$

or what is the same thing

$$\left.\begin{aligned} cos\, \frac{\alpha_1}{2} &= \frac{1}{2}\sqrt{\frac{1}{2}\left\{\sqrt{[4-(\alpha^2+\beta^2)]^2+16\beta^2}+[4-(\alpha^2+\beta^2)]\right\}} \\ sin\, h\, \frac{\beta_1}{2} &= \frac{1}{2}\sqrt{\frac{1}{2}\left\{\sqrt{[4-(\alpha^2+\beta^2)]^2+16\beta^2}-[4-(\alpha^2+\beta^2)]\right\}} \end{aligned}\right\} \qquad (4)$$

B. Loaded Conductor of the Second Type.

In this case wave propagation is expressed in terms of a complex angle $\psi_s = \frac{1}{2}(\alpha_1 + i\beta_1)$ which is defined by the following equation:

$$sin\, \psi_s = sin\, \frac{1}{2}(\alpha_1 + i\beta_1) = \frac{1}{2}\, \frac{l}{k+1}(\alpha_s{}^1 + i\beta_s{}^1)^{(1} \qquad (5)$$

that is, provided the distance between the loads is sufficiently small in comparison with the wave length on the conductor before it is loaded. This provision imposes no serious limitation since this wave length is five to ten times longer than the wave length on the corresponding uniform conductor. Equations (5) and (2) are identical, so that the second type of loaded conductor does not differ from the first.

In preparing the numerical tables from which the curves in Fig. 1 and Fig. 2 were plotted much labor is saved by considering separately the values of $cos\, \frac{\alpha_1}{2}$ and $sin\, h\, \frac{\beta_1}{2}$ in the following three intervals:

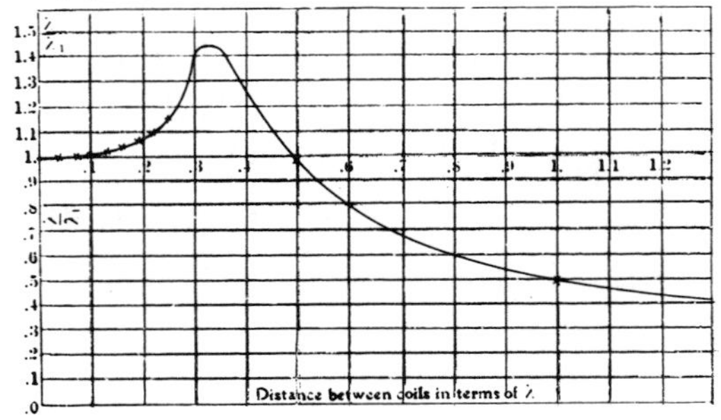

Fig. 2.

First interval: The quantity $4 - (\alpha^2 + \beta^2)$ is considerably greater than zero.

Since β^2 is very small in comparison with unity equations (4) reduce to

$$\left.\begin{aligned} cos\, \frac{\alpha_1}{2} &= \sqrt{1-\left(\frac{\alpha}{2}\right)^2} \\ sin\, h\, \frac{\beta_1}{2} &= \frac{\frac{\beta}{2}}{\sqrt{1-\left(\frac{\alpha}{2}\right)^2}} \end{aligned}\right\} \qquad (6)$$

Second interval: The quantity $4 - (\alpha^2 + \beta^2) = 0$ very nearly. Equations (4) reduce to

$$\left.\begin{aligned} cos\, \frac{\alpha_1}{2} &= \sqrt{\frac{\beta}{2}} \\ sin\, h\, \frac{\beta_1}{2} &= \sqrt{\frac{\beta}{2}} \end{aligned}\right\} \qquad (7)$$

Third interval: The quantity $4 - (\alpha^2 + \beta^2)$ is considerably smaller than zero.

* Propagation of Long Electrical Waves; *Transactions* of the American Institute of Electrical Engineers, Vol. xv, p. 144.

**Same paper, p. 151.

[1]M. I. Pupin, Wave Propagation Over Non Uniform Electrical Conductors; *Transactions* of the American Mathematical Society, Vol. I, No. 3, p. 276.

Reprinted from *Elec. World and Eng.*, vol. XXXVIII, pp. 587–588, Oct. 12, 1901.

Equations (4) reduce to

$$\left.\begin{aligned} \cos\frac{a_1}{2} &= \frac{\frac{\beta}{2}}{\sqrt{\left(\frac{a}{2}\right)^2 - 1}} \\ \sinh\frac{\beta_1}{2} &= \sqrt{\left(\frac{a}{2}\right)^2 - 1} \end{aligned}\right\} \quad (8)$$

A brief explanation of the symbols is desirable. Let l be the distance between the loads on the loaded conductor of the second type. It will be the length of the corresponding uniform conductor which is equivalent to one coil of the loaded conductor of the first type.

Let λ be the wave length on the corresponding uniform conductor. Let λ_1 be the wave length on the non-uniform conductor.

Then $a = \frac{2\pi l}{\lambda}$ = angular distance on the corresponding uniform conductor,

$a_1 = \frac{2\pi l}{\lambda_1}$ = angular distance on the non-uniform conductor.

$$\beta^1 = l\,\beta$$
$$\beta_1^1 = l\,\beta_1$$

where β^1 and β_1^1 are the attenuation constants on the non-uniform and its corresponding uniform conductor, respectively.

In the following table the ratios of the wave lengths $\frac{\lambda}{\lambda_1}$ and of the attenuation constants $\frac{\beta_1^1}{\beta^1}$ of the loaded conductor and the corresponding uniform conductor are given for various distances between the loads, these distances being measured in terms of the wave length on the corresponding uniform conductor. The column headed by $\frac{l}{\lambda}$ contains these distances. The curves in Fig. 1 and Fig. 2 were plotted from the above table.

TABLE.

$\frac{l}{\lambda}$	$\frac{\lambda}{\lambda_1}$	$\frac{\beta_1^1}{\beta^1}$
.0313	1.001	1.005
.0626	1.009	1.02
.094	1.015	1.047
.125	1.029	1.088
.1565	1.045	1.15
.1788	1.071	1.24
.22	1.093	1.38
.25	1.15	1.64
.318	1.443	7.8
.5	.986	40
1.0	.5	34.5

These curves were calculated for the conductor described in my paper read before the American Institute of Electrical Engineers May 18, 1900. For this conductor

$$\beta_1^1 = .0141$$

It is evident that up to about four coils per wave length they will hold true for any loaded conductor that may occur in practice.

Observe the rapid increase in the attenuation constant of the loaded conductor as soon as the number of coils per wave length becomes less than six. It is dangerously high at four coils per wave length and may be anything if that number is between 4 and π.

Observe also that the wave lengths of the loaded conductor is a definite quantity only as long as the conductor is properly loaded. On a poorly loaded conductor the wave length may be either smaller or greater than the wave length on the corresponding uniform conductor.

McMeen Defines the Telephone Engineer

In this paper an engineer of the Western Electric Company proposed a definition of the telephone engineer and defended the telephone engineer's claim to legitimacy as an engineering professional. He sought to counter the impression that telephone engineers did not "deal with exact quantities" or practice the accuracy and precision commonly expected of engineers. In support of his position, McMeen cited the work of Pupin (see Paper 31), although a critic might have noted that Pupin was an engineering educator and not a practicing telephone engineer. McMeen neglected to mention a major reason for the lack of appreciation of the work of the telephone engineer. It had long been a policy of the Bell Company to not publish the results of the research of their engineers that might be of value to independent competitors. The editor of *Electrical World* published a strong attack on this policy early in 1904, calling it "short sighted, imprudent and immoral." Prior to that time, technical papers by telephone engineers at Bell had been unusual in engineering journals.

Samuel G. McMeen (1864–1934) was born in Eugene, Ind., and studied for a time at Purdue, but did not graduate. He worked for a number of Bell-affiliated telephone companies in the Midwest from 1884 to 1902 when he joined the Western Electric Company. In 1904 McMeen and Kempster B. Miller formed a consulting engineering firm. McMeen frequently served as an expert witness in patent litigations and was coauthor with Miller of a popular book entitled *Telephony*. See *Who's Who in Engineering* (1922–1923), p. 823 and an obituary in *Electrical Engineering*, vol. 53, p. 1330, 1934.

A paper read at a meeting of the Chicago Branch of the American Institute of Electrical Engineers, May 19, 1903.

CONCERNING THE TELEPHONE ENGINEER.

BY S. G. McMEEN.

Mr. Lockwood's contribution concerning the telephone switchboard, and the discussion brought out by it, form so valuable a portion of the too meager literature of the telephone that the writer makes bold to act as spokesman for those who would record appreciation.

There are two principal values in all statements of fact; one, the record of the truth for immediate information and subsequent reference; the other, the correcting of false impressions. The paper and discussion referred to have gone far in exhibiting both of these values, and at the risk of paralleling what has been said so well and honestly, this item of additional discussion is presented.

The telephone engineer's profession is so varied in its contact with commercial and social conditions that its scope is apparently not well understood even by those practising similar professions. Indeed, the very existence of the telephone engineer is in some quarters denied. Let us look at his work from some of the several sides to see if we may generalize about it.

To many persons, the telephone instrument on the wall is the telephone system, and an art based on so little a thing seems a simple one. Yet there are involved, in even the instrument on the wall, the problems of apparatus-design for electrically transmitting speech at the highest efficiency; for the least and greatest distances without change of local equipment and of receiving speech and signals from distant stations.

Granting that the substation telephone set stands for too narrow a view of the whole matter, and that at least a connecting

Reprinted from *Trans. AIEE*, vol. XXI, pp. 81–83, May 19, 1903.

line is necessary, it is seen that here is involved a problem of alternating current transmission. Note, also, that this line, in such a system as Mr. Lockwood has described, must do in the best manner several kinds of things such as carrying currents for speech and signaling, and some of these things must be done simultaneously. Owing to conditions of topography and degree of congestion of population, the line may be buried or erected with a wide variety of surroundings. As its insulation, capacity and conductivity have vital bearing on its efficiency. and as they also govern its cost, a responsibility rests with the engineer who controls the design of these points. For example: "If by decreasing the effective electrostatic capacity between wire and wire of an underground telephone cable by *b* microfarads, the efficiency of speech transmission is increased by *d* per cent., at a cost per mile per year of *c* dollars, is the expense warranted?" Instances might be multiplied to illustrate the importance of knowledge and judgment in connection with this costly and important part of a telephone system.

To the telephone instruments and systems of lines must be added the central office equipment, so well described in the principal paper under discussion. Of the switchboard and its appurtenances which make up this equipment, three important features have been pointed out. First, the switchboard is an aggregation of a great many parts; secondly, the parts must inter-operate with complete harmony; and thirdly, the aggregation is often of such size as to compare favorably with the big work of an engineer of any other craft.

In connection with the equipment in general in its relation to the lines in general. The location of central offices, their sizes, now and hereafter, the number, size and character of underground ducts and cables, together with all other elements, must be considered and solved for one conclusion—the giving of the best telephone service at the least cost, now and hereafter. And it is not to be overlooked that the factors upon which the solution depends are subject to change so that such change must be foreseen and discounted if possible

It may be replied that all other engineers must take account of rates of interest, prices of material and labor, density of population, etc., and the possible shifting of values. Granted. But a telephone system must be planned and begun it is believed, a longer time before its development to its ultimate size than most other works of magnitude. Because the speaking current of the telephone is very small, usually alternating, and so not susceptible of easy and exact measurement, there has grown up the idea that the telephone engineer does not deal with exact quantities, and does not use in his work that accuracy of observation and precision of execution common to the engineering profession. This impression is not correct, for two reasons. First, the speech-transmission current is far better understood to-day than formerly and is determined with a very great degree of accuracy, as also are the many other currents having to do with the operations attending speech-transmission. Second'y, the telephone engineer requires the same observing and reasoning powers which distinguish engineers in general from the men about them in other occupations.

As an instance of the possibility of mathematical and physical accuracy, witness the work of Prof. Pupin in the matter of distortionless circuits. An intelligent mathematical reading of the specifications of either of his patents upon the elimination of the hurtful effects of the electrostatic capacity of an alternating current transmission line is recommended as furnishing conclusive proof of the claims here advanced. Considering the art as caused by and based upon a single device—the telephone itself—it is evident that the great usefulness of the present system is largely due to the cooperation of a great mass of additional circuits, methods and apparatus.

The destiny of the art is to place all the peoples of the earth in speaking relation with one another, it is also destiny that the telephone engineers shall specialize. Yet the assertion is ventured that no other branch of engineering demands so thorough a retention of the general lore of the art by the specialist within the art as does this.

Upon the basis of these thoughts, may we not define as follows:

A telephone engineer is one who practises the design and construction of telephone apparatus, telephone circuits and telephone operating methods, and the correlation of all three into systems of telephone communication.

If the definition be approved, let us then claim a place for him among the engineers of more classical reputation, giving him credit for his accomplishment and our best wishes for his future.

The First AIEE Discussion of Wireless Telegraphy

THE meeting reported in this paper from the *Transactions of the AIEE* of 1897 featured a brief demonstration and discussion of the Marconi wireless apparatus. The demonstration seems not to have been an unqualified success due to difficulties with the coherer detector. By the time of a topical discussion on "The Possibilities of Wireless Telegraphy" published in the *Transactions* of 1899, the discussion led by Reginald Fessenden and Michael Pupin was much more informed. This was followed by the dramatic session of January 1902 when Marconi himself was on hand to report his success in transmitting across the Atlantic. For a recent and authoritative interpretation of the early history of radio and especially Marconi's contributions, see Hugh G. J. Aitken, *Syntony and Spark—The Origins of Radio* (New York, 1976). See also Alexander A. McKenzie, "The Three Jewels of Marconi," *IEEE Spectrum*, pp. 46–52, December 1974. Also see John F. Ramsay, "Microwave Antenna and Waveguide Techniques Before 1900," *Proceedings of the IRE*, vol. 46, pp. 405–415, 1958.

WIRELESS TELEGRAPHY.

December 15, 1897.

MR. W. J. CLARKE:—Mr. Chairman and gentlemen: Having been requested this evening to give you an exhibition of the Marconi apparatus, I am pleased to be able to say that I am in a position to do so. I think perhaps it might be well to give you a short explanation of the apparatus which we use. In the first place, we have here on the front of the base, a Marconi coherer consisting of a small glass tube, fitted with silver plugs connected to platinum wires at the end. These plugs are very close together, about the centre of the tube, the intervening space being filled in with nickel filings. I have found that it is entirely unnecessary to either exhaust or seal the tube. I also find that we can use almost any kind of metal for the filings. When the cohesion is examined under the microscope, I find that what we have to provide for is the proper size of the filings, and not the proper kind of metal. This coherer is placed in series with this relay which is of about 1200 ohms resistance, and made extremely sensitive, more so than the ordinary Morse relay. In series with it, also, are two cells of small dry battery placed in the base. These are arranged so that we can use either one cell or two, the object of this being that when the cells are new, we can use only one in order to prevent corrosion of the filings where they cohere. As the cells grow older we simply throw our *a* switch, and use two cells. This sounder is a 20-ohm instrument. It is placed in multiple with this cohering apparatus consisting of a 20-ohm vibrator placed in the base. The local battery consists of three cells of dry battery arranged as with the main battery, so that we can either use one, two or three of the cells. This Morse key is simply used in order that we may see that the apparatus is in proper condition. Pressing this key short-circuits the coherer. Our transmitter consists of an eight-inch induction coil. A much smaller one, though, is all that is necessary for a short distance. This coil has its secondary terminals connected with two brass balls, each an inch and a half in diameter. These balls are brought into close proximity, in fact in most cases touching two large brass balls, four inches in diameter. The large balls are securely cemented in the ends of a rubber tube, and the distance between them on the inside of the tube is about one-thirty-second of an inch. The space between the balls in the tube is filled with the purest quality of vaseline oil, and the moment that we close the circuit through the primary of the coil the electric waves generated strike the coherer, the filings cohere, and the resistance is sufficiently reduced to operate the relay. Sometimes we have trouble with this particular instrument, not on account of the system being imperfect, but on account of the fact that it is the first piece of portable apparatus

Reprinted from *Trans. AIEE*, vol. XIV, pp. 607–613, Dec. 15, 1897.

that I know of built in this country, and of course you understand that the first piece of apparatus is liable to imperfection. I think now that everything is in proper shape, and I will proceed to the other room and close the circuit of the primary of induction coil, and I think you will see the receiver respond, and after doing this we will close the doors and work the receiver through the glass doors.

[Apparatus shown in operation.]

THE CHAIRMAN:—[Dr. Kennelly]. Gentlemen, you have witnessed an interesting exhibition, and I am sure that many must be desirous of asking Mr. Clarke some questions upon the difficulties he has had to encounter in making this instrument, and the various matters he has found necessary to take into account. I am sure that if he is as ready to answer as he has been ready to describe the apparatus that he will respond.

MR. GANO S. DUNN:—I should like to ask whether the circuit remains closed while the shower of sparks is passing, and also, if Mr. Clarke would describe the action of the coherer.

MR. CLARKE:—I would like to say that with proper adjustment, the circuit remains closed as long as the primary circuit of the induction coil is closed, and as long as the vibrator of the induction coil is in action. Unfortunately the adjustment which we have on our decohering apparatus in this instrument is not sufficiently under control, so that I cannot get the fine adjustment necessary for transmitting intelligible Morse signals, but this is simply a question of proper adjustment. Now with regard to the coherer, as I said before, it simply consists of two conductors separated by a very short interval, this space being filled with metallic filings. The filings are of such a size and the distance apart of the conductors is such, that while the filings are lying loose, the resistance through the tube is very high indeed. But the moment that the filings cohere, the resistance is very much reduced, so much so that the current in series with the relay is able to pass to a sufficient extent to operate the relay. I may say that I have been so very busily engaged in getting a smaller and less expensive set of apparatus ready for the market, that I have not had an opportunity to experiment very largely with the question of the distance, but I expect during the next week to be able to accomplish something in that direction.

MR. DUNN:—I have done a good deal of telegraphing and am familiar with the frequency requisite to get all the signals in. For instance, when the operator is sending very rapidly, his "e" dot, by the time it reaches the other end may be only a very small fraction of a second. I noticed the frequency of your coil was readily observable, I could almost count it by beats, and I wanted to inquire how long one discharge of the coil would cause the filings to cohere, and hold the circuit closed.

MR. CLARKE:—The coherer, when in proper shape, in which this hardly is, is so very sensitive that it is not necessary to have a vibrator on the coil at all. It is simply necessary to have a Morse key. The coherer responds the moment you close the circuit, only in this case, of course, it simply coheres and decoheres instantly, and the circuit of the sounder is closed and opened.

MR. DUNN:—If when you close your key on the induction coil that causes the filings to cohere, and then decohere immediately, your sounder here would make its down click, and be followed by its up click when the key in the other room has not risen.

MR. CLARKE:—Yes.

MR. DUNN:—Then you would have to use some different system from the ordinary Morse for sending.

MR. CLARKE:—No, because when we use the vibrator on the coil, and everything is in proper adjustment, the filings cohere as long as our coil is working, and decohere the moment we stop it by opening our key.

MR. DUNN:—The vibrator frequency is about 800 a minute.

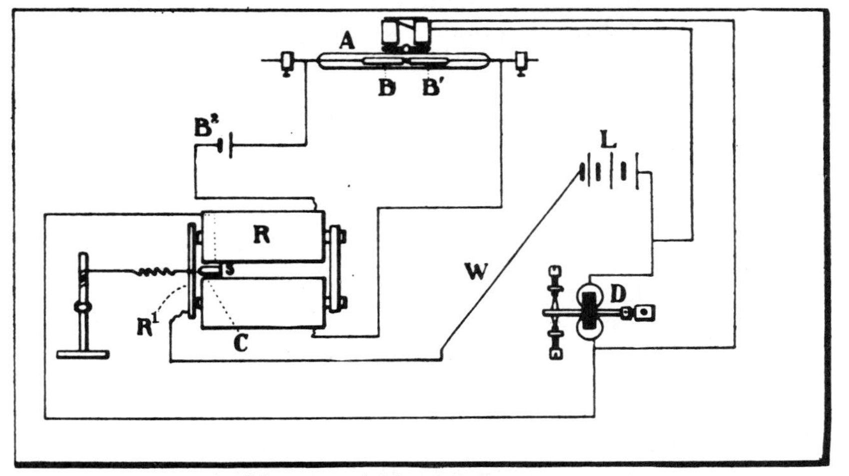

Diagram of Marconi Apparatus.

MR. CLARKE:—Well, I may say that the coil we are using is not the coil that we use for quick signaling. We use a Tesla coil for that purpose, a very high-frequency coil.

THE CHAIRMAN:—I might suggest for the benefit of some who may not be familiar with the subject, that if you sketch the outline of the connection on the board it may help.

MR. CLARKE:—I will do that, I might say that at some future time I would be very glad to show you the stereopticon diagrams on the sheet, of the different classes of this apparatus we are making up. It is a little difficult to explain it on the blackboard, but I will do the best that I can. Here we have the coherer A somewhat enlarged. These B B' are the silver plugs. The filings are between the ends of those plugs. The main battery B^2 is connected to one end of the coherer. The other terminal of the battery is connected to one terminal of the 1200-ohm relay R. The other terminal of the relay is connected to the coherer. Now when the apparatus is in its normal condition, the relay is so adjusted and the resistance of the filings is such, that the current from the battery will not pass to a sufficient extent to operate the armature of the relay, but when the waves from the transmitter strike the filings and they cohere to each other, and also to the silver plugs, as examination under the microscope shows that they do, the resistance is so reduced as to allow sufficient current to pass from this battery to operate the relay. The moment the relay operates, it pulls up its armature, making the contact C. The moment this contact is made, the current from the local battery L traverses the wire W to the armature of the relay R[1], from the armature across the contact to the screw S, back to the sounder D and back again to the battery. Now once the filings have responded to the waves and cohere, they will remain cohering unless we have some means of decohering them. In order to accomplish this, we have a vibrating hammer which strikes the tube. This vibrator is placed in multiple circuit with the sounder. They are both of comparatively high resistance.

MR. EDWARD DURANT:—I would like to ask what is the greatest distance at which you can get communication?

MR. CLARKE:—I may say that I have not tried the instruments at very long range. I have not had either the time or the opportunity. But I understand from the reports coming from the other side, that Mr. Marconi has had no difficulty in transmitting intelligible signals twelve miles.

MR. MACGREGOR:—I should like to ask one or two questions. The first is,—does it make any difference as to what the relative position of the tubes containing the filings is? Is the position you have it in all essential?

MR. CLARKE:—I have not found that it is.

MR. MACGREGOR—The second thing I want to ask is, whether there is any fixed relation between these two pieces of apparatus. In other words, would any other induction coil produce the same effect as this one, or are these two instruments related so as to form a pair? If intelligence transmitted from one piece of apparatus could be read by any one of a hundred machines, this method of telegraphy would have its disadvantages.

MR. CLARKE:—In the first place, I have not found using the short ranges over which I have tried the apparatus, that it makes any difference in what position either the receiver or transmitter is placed. In regard to their being any tune, as we may call it, between the transmitter and receiver, I may say that Marconi claims that it is absolutely necessary to have the transmitter and the receiver in what we might call synchronism. In order to accomplish that, he takes two strips of copper, each about half an inch in width, and attaches them to the end of his tube, one on each end. These strips of copper he claims have been shortened or lengthened in accordance with the frequency of our transmitting apparatus. In order to ascertain what is the proper length, he takes a piece of glass and pastes upon it a strip of tin foil about half an inch in width, and two or three feet in length. He divides this strip in the centre with a very sharp knife. He sets his transmitter in a dark room and places this testing apparatus in front of it, at a few yards distant. He operates his transmitter, and when he does so, the same as in the experiments of Hertz, he notices small sparks passing across between the divided tin foil. He moves his tester further from the transmitter until the sparks entirely disappear. Then he cuts off, say half an inch, from each end of his tin foil and immediately the sparks reappear. He goes further, until the sparks disappear again, and he keeps on cutting down each end of his tin foil, until he finds that he is working in the other direction, and that if he cuts it any more, he simply has to go back closer to the coil in order that the sparks at this point will reappear. Then he finds he has got the proper length. Accordingly, he cuts what he calls the wings of the coherer nearly the same length as these two pieces of tin foil. I may say that in the comparatively small distance over which I have tried the apparatus, I have not found it necessary to have any wings on the tube at all, or that it is necessary to have the apparatus in any kind of synchronism, I find that some recent writers in the London *Electrician* bear me out in this. I see it stated that in one of the exhibitions given by Mr. Marconi, or some of his associates, that in order to show how readily the waves would go through a piece of iron, the coherer was placed in an iron box, and in order to place it there, it was necessary to remove the wings, but the coherer responded just the same.

Dr. Sam'l Sheldon:—I would like to ask if it is essential that you should have silver for the two electrodes.

Mr. Clarke:—The only object, as far as I can see, is simply to keep the points of contact clean.

Dr. Sheldon:—Then the object of a vacuum would be to prevent the silver from oxidizing.

Mr. Clarke:—Yes, and I may say in this connection too, that experiments have been tried with putting the filings in hydrogen gas, but it was found that they speedily became too bright and clean, so much so that they cohered all the time. If the filings are examined under a microscope and the waves are acting upon them, it will be noticed that the moment they decohere you will see little bright spots where the particles of metal have been touching each other. Those bright spots rapidly disappear.

Mr. C. T. Child:—I would like to ask if that decoherer is in the nature of an electro-magnetic vibrator.

Mr. Clarke:—Yes.

MR. CHILD:—How do you shield the coherer from the waves sent out from that instrument?

MR. CLARKE:—We do not find that it affects them in the least.

MR. CHILD:—I should think that those waves would affect it.

MR. CLARKE:—Well, Mr. Marconi claims that it does. I have not found any effect at all. Perhaps it is because I have guarded against that by cutting down the sparks at the contact with resistance coils, and also placing a resistance across the terminals of all the magnets.

A MEMBER:—I fancy that the resistance must be very low with those metal filings there, so low that the variation must be very small as we find at once in telephone experiments where we use metal filings. May I ask what the resistance is of that coherer?

MR. CLARKE:—I may say that although I have the facilities, I have not had the time to make a thorough test in that direction and I can only guess at what the resistance really is. I expect very shortly to devote very considerable time in making a thorough test.

THE CHAIRMAN:—It is very high, isn't it?

MR. CLARKE:—I hardly think so, because you will see we have a resistance in the relay of 1200 ohms, and one cell of dry battery is all that we require to operate it. In fact if we use silver filings it is difficult to get them to decohere enough to interrupt the flow of current. I should have stated that silver is the only metal I have found that is too sensitive.

MR. WINTRINGHAM:—How so?

MR. CLARKE:—If we use silver filings we find it almost impossible to adjust the apparatus so that the circuit will remain open and then close when the waves strike it. Silver filings are so very clean and bright, that they form a very low resistance across the plugs in the tube, much lower than nickel or any other metal that I have tried. I have not tried gold or any metal of that kind. I do not know how they would act.

THE CHAIRMAN:—If there are no further questions, I feel sure that you will endorse my recommendation that a vote of thanks be extended to Mr. Clarke for his kindness in exhibiting the apparatus, and in so patiently answering all our questions prompted by a very natural curiosity. I need hardly put that to vote. I think I am justified in extending that vote of thanks.

MR. ALBON MAN:—Will the gentleman answer one question? The origin of waves seems to be provided for between the large balls in the rubber tubes. Why are not other waves generated between the small balls and the large ones, or are they generated, and do they have the same effect as the ones in the tube? I understand that sparks pass between the terminals of the small balls and the large ones.

MR. CLARKE:—Yes.

MR. MAN :—Are they not a source of electrical waves?

MR. CLARKE :—In regard to that I can hardly say ; I have not experimented far enough. Mr. Marconi claims that it is necessary not only to use two balls four inches in diameter in the tube, but also use the other balls of the same size. Now I find that over the short distance that I have worked the apparatus, it is not necessary even to have the small balls at all. All that is necessary is to connect the terminals of the coil to the large balls, and I have used a coil of only one inch spark capacity. I feel quite positive that a coil of that size is all that is necessary for transmitting this distance.

MR. MAN :—I would like to know also if the diameter of the large ball and its mass, is a necessity to the transmission of the waves, or the production of the waves. Has the mass of the balls anything to do with the force of the waves?

MR. CLARKE :—I can only say in regard to that, that I have followed Mr. Marconi's experiments very closely. I have only so far constructed a transmitter with the four inch solid brass balls. Marconi claims that it is absolutely necessary to have solid balls for long distance transmission, and that with hollow balls, the distance over which we transmit will be less than half. My own impression is, that hollow brass spheres filled with lead, or other inexpensive metal, would be just as good, and we are going to experiment with that shortly, and also with the smaller spheres, on account of the larger spheres being very expensive and very heavy.

MR. MAN :—Hertz found his electrical waves produced as readily from small balls, or even any kind of a spark as from larger masses. It was for that reason that I asked the question.

MR. CLARKE :—Knowing the results of the experiments of Hertz, I am of the opinion myself that the size and mass of the ball does not make the difference that is claimed for it, and for that reason we are going to experiment in order to determine that accurately.

MR. HAMBLET—I move a vote of thanks to Mr. Clarke for the very interesting exhibition that he has given us this evening.

[The motion was carried, and the meeting adjourned.]

Kennelly Calculates the Properties of the Kennelly–Heaviside Layer

IN this short note from the *Electrical World and Engineer* of 1902, Kennelly deduced the probable height and conductivity of a layer of the upper atmosphere that might serve to reflect wireless telegraph signals and greatly extend their range. It is interesting that Marconi's successful transmission across the Atlantic had taken place about three months before Kennelly's paper appeared. The great British electromagnetic theorist, Oliver Heaviside, published a similar analysis later the same year in the *Encyclopedia Britannica*, leading to the common designation of the layer as the "Kennelly–Heaviside Layer." See a bibliography on the K-H layer in the PROCEEDINGS OF THE IRE, vol. 19, pp. 1066–1071, 1931.

Arthur E. Kennelly (1861–1939) was born in India where his father was employed as harbor master in Bombay. Kennelly received his early education in England, but did not attend a university. At the age of 14 he became an office boy at the Society of Telegraph Engineers in London and took advantage of the opportunity to read books on telegraphy and electromagnetics at the Society's library. Kennelly worked as a telegrapher and on cable-laying ships before accepting a position as technical assistant to Thomas Edison at his West Orange Laboratory in 1887. In 1894 Kennelly joined Edwin Houston in organizing a consulting engineering firm. He became an Engineering Professor at Harvard in 1902, a position he retained until 1930. He also taught part time at M.I.T. during the years 1913–1924. He served as President of both AIEE (1898) and the IRE (1916). Kennelly was a very prolific author and published approximately 400 papers. He was the author or coauthor of 28 books. He had a great talent for simplifying difficult concepts or mathematical methods and expressing them in terms that the average engineer could grasp. See the *DAB*, vol. 22, pp. 357–359; the *NCAB*, vol. 13, pp. 452–453; the *DSB*, vol. 7, pp. 288–289; and the *NAS*, vol. 22, pp. 83–119.

On the Elevation of the Electrically-Conducting Strata of the Earth's Atmosphere.

By A. E. Kennelly.

According to the measurements of Professor J. J. Thomson ("Recent Researches in Electricity and Magnetism," p. 101), air at a pressure of 1-100 mm. of mercury has a conductivity for alternating currents approximately equal to that of a 25 per cent aqueous solution of sulphuric acid. The latter is known to be roughly 1 mho-per-centimeter, so that a centimeter cube would have a resistance of about one ohm. Consequently, air at ordinary temperatures, and at a rarefaction 76,000 times greater than that at sea level, has a conductivity some 20 times greater than that of ocean water, although about 600,000 times less than that of copper.

If we apply the ordinary formula for finding the elevation corresponding to a given air-rarefaction, we find that if the air had a uniform temperature of 0 deg. C., the height of this stratum of air with a rarefaction of 76,000, would be

$$18.39 \log 76{,}000 \text{ kilometers above the sea,}$$
$$\text{or } 89.77 \text{ kilometers,}$$
$$\text{or } 55.77 \text{ miles.}$$

If the air had a uniform temperature of —50 degs. C. this elevation would be reduced 18.3 per cent. or to 73.3 kilometers (45.5 miles). The temperature of the earth's atmosphere has only been measured within a range of a very few kilometers above the surface of the sea, and consequently the materials are not at hand for any precise calculation of the height of electrically conducting strata. It may be safe to infer, however, that at an elevation of about 80 kilometers, or 50 miles, a rarefaction exists which, at ordinary temperatures, accompanies a conductivity to low-frequency alternating currents about 20 times as great as that of ocean water.

There is well-known evidence that the waves of wireless telegraphy, propagated through the ether and atmosphere over the surface of the ocean, are reflected by that electrically-conducting surface. On waves that are transmitted but a few miles the upper conducting strata of the atmosphere may have but little influence. On waves that are transmitted, however, to distances that are large by comparison with 50 miles, it seems likely that the waves may also find an upper reflecting surface in the conducting rarefied strata of the air. It seems reasonable to infer that electromagnetic disturbances emitted from a wireless sending antennæ spread horizontally outwards, and also upwards, until the conducting strata of the atmosphere are encountered, after which the waves will move horizontally outwards in a 50-mile layer between the electrically-reflecting surface of the ocean beneath, and an electrically-reflecting surface, or successive series of surfaces, in the rarefied air above.

If this reasoning is correct, the curvature of the earth plays no significant part in the phenomena, and beyond a radius of, say, 100 miles from the transmitter, the waves are propagated with uniform attenuation cylindrically, as though in two-dimensional space. The problem of long-distance wireless wave transmission would then be reduced to the relatively simple condition of propagation in a plane, beyond a certain radius from the transmitting station. Outside this radius the voluminal energy of the waves would diminish in simple proportion to the distance, neglecting absorption losses at the upper and lower reflecting surfaces, so that at twice the distance the energy per square meter of wave front would be halved. In the absence of such an upper reflecting surface, the attenuation would be considerably greater. As soon as long-distance wireless waves come under the sway of accurate measurement, we may hope to find, from the observed attenuations, data for computing the electrical conditions of the upper atmosphere. If the attenuation is found to be nearly in simple proportion to the distance, it would seem that the existence of the upper reflecting-surface could be regarded as demonstrated.

Reprinted from *Elec. World and Eng.*, vol. XXXIX, p. 473, Mar. 15, 1902.

Alfred Goldsmith's Reflective View of World Communications

A distinguished founder of the radio engineering profession in the U.S. gave this considered assessment of the current state and probable future of world communications systems in 1921. He listed what seemed to him to constitute the requirements of "an ideal system" that "should include every person on the globe." Goldsmith then outlined a reasonable approximation to the ideal that might be achieved through a synthesis of guided and unguided communications networks. Much of the discussion in this paper would still be appropriate, although his ideal system has yet to be realized. For a recent reflective analysis along similar lines, see W. L. Everitt's paper in the special issue of the PROCEEDINGS OF THE IEEE for September 1976.

Alfred N. Goldsmith (1888–1974) was born in New York City and graduated from the City College of New York in 1907. He was awarded the Ph.D. degree from Columbia University in 1911. Goldsmith joined the Wireless Institute and was one of three who arranged the merger of the Wireless Institute and the Society of Wireless Telegraph Engineers to form the Institute of Radio Engineers in 1912. (See Paper 15.) He was Editor of the PROCEEDINGS OF THE IRE from its inception until 1954. He was a consultant to the General Electric Company during World War I, and later worked as Research Director at the Marconi Wireless Company of America, and in the same capacity for the Radio Corporation of America after it was formed in 1919. Goldsmith remained with RCA until 1931 when he became an independent consultant. See Carl Dreher, "His Colleagues Remember 'The Doctor'," *IEEE Spectrum*, pp. 32–36, August 1974, and an obituary in the same issue, pp. 114–115.

JOURNAL
OF THE
AMERICAN INSTITUTE OF ELECTRICAL ENGINEERS

Vol. XL DECEMBER, 1921 Number 12

World Communication

BY ALFRED N. GOLDSMITH

Director, Research Department, Radio Corporation of America

IT is not intended, in this brief presentation of the broad outlines of the important subject of world communication, to do more than touch on the technical methods whereby communication is now being carried on or by which it is expected that communication will be carried on in the future. The subject matter is altogether too voluminous to permit more than a very general summary of the methods now in vogue or to be expected, and the setting forth of a general perspective of the problems involved and their relations. The writer must also indicate that his special interest in the field of radio communication has caused him to lay the main emphasis in this paper thereon, although, in a perfectly balanced paper on local and long-distance communication, the major portion would certainly deal with wire communication. While the writer has endeavored to minimize this lack of proportion in the paper, the reader is nevertheless advised to regard the paper as generally descriptive of the topics treated, but he should not judge the relative importance of the types of communication considered as proportionate to the space devoted to them.

Before proceeding further, the writer takes pleasure in acknowledging his indebtedness to Messrs. John L. Merrill and B. H. Reynolds of All America Cables, Incorporated, to Mr. A. H. Griswold, and Mr. O. B. Blackwell of the American Telephone and Telegraph Company, to Mr. Donald McNicol, to Messrs. C. H. Taylor, W. A. Graham, W. A. Winterbottom, P. Boucheron of the Radio Corporation of America, and to Mr. C. M. Yorke of the Western Union Telegraph Company, for the illustrations, information, and advice which they have so kindly furnished in connection with the preparation of this paper.

It must doubtless have occurred to every person who carries on business of importance, and indeed even to that ubiquitous individual, "the man on the street," that communication by direct transportation of the communicant is a most clumsy though necessary method. The carriage of the ponderous bulk of the human body from one point to another, merely to enable word-of-mouth communication, is doubtless an ineluctable necessity in many cases at present, and yet it is far from satisfactory. What is desired is the communication of intelligence, not the painfully long and tedious motion of the individual in whom the intelligence is embodied. To be sure, ordinary speech from man to man is a form of communication which, within a restricted range and for a given tongue, meets all ordinary requirements. But it becomes unsatisfactory as soon as we become removed from anyone to whom we wish to transmit our ideas. What we all desire is means so that we can get into immediate touch with all of those with whom we do business or with whom we take our pleasures. Our interest in communication then, is roughly dependent on the distribution of these people. Evidently we are the most interested then in transmission within a comparatively restricted area. In view, however, of the wide range of modern business and the widening range of personal travel, long-distance communication has also become of great interest to us.

It will be found that there are two well-defined methods of communication, each having its particular characteristics. Broadly, these may be termed "guided communication" and unguided communication." Ordinary speech, (unless it is particularly closely directed by a megaphone) is of the unguided variety. In other words, the energy radiated passes out freely in *all* directions enabling any one on the surface of a sphere, having the supposed aerially suspended speaker at its center, to hear, that is, to receive the messages equally well. As a general rule, methods of communication utilizing free wave transmission are entirely unguided, partly because of the difficulty of radiating a sharply defined stream or "searchlight beam" of radiation and partly because of the difficulty of preventing its subsequent diffusion or spreading in all directions through diffraction, refraction, reflection, and passage through diffusing media (for example, fog). As already hinted, the best rough approximation to a directed beam based on free wave propagation is the modern searchlight, which has indeed been used for military signaling and for light telegraphy and light telephony (photophone transmission). In no

Lecture delivered at the 372d Meeting of the A. I. E. E., New York, N. Y., November 17, 1921.

Reprinted from *J. AIEE*, vol. XL, pp. 885–889 (extract of a longer paper), Dec. 1921.

case of communication by free electromagnetic waves do we find anything like the definiteness of destinations and number of parallel channels which are achieved in wire communication, where we have remarkable results in the compression of a great number of separate telephone conversations within the same cable sheath.

Ordinary telephony and telegraphy, and "guided radio," "wired wireless," or "carrier current telephony and telegraphy," (as it is variously called) are all good examples of guided forms of communication. Here we have electromagnetic waves very closely guided in and by material conductors (wires) to a definite point of destination, and without sidewise radiation of the lengthwise propagated energy except where the distributed line constants suffer abrupt change.

Both guided and unguided electromagnetic communication depend on the transmission of electromagnetic waves through space. In unguided transmission, which is generally known as radio, the wave-spread out over a wide area and in this way may be picked up by a large number of receiving stations. In guided communication, generally known as wire transmission, the conducting wires form, as it were, electrical paths, which guide the waves to the exact point to which it is desired to send them.

Both of these methods have their advantages and disadvantages. In general they are not competitors of each other for overland working but supplement each other.

Modern radio communication is unguided, and hence can be received equally well at all points equidistant from the transmitting station. Here and there partially directed radio beams have been produced, principally for military purposes or for harbor radio beacons or the like. There are distinct possibilities of future developments along this line of directional radio transmission, naturally, by preference on the shorter wave lengths and over the shorter distances, though again within the limits imposed by the mechanism of wave radiation and propagation and in no case approximating even roughly to what can be accomplished by wire methods.

Guided communication differs from the unguided methods in that the signaling energy is directed along a well-defined path terminating, as a general rule, exclusively at the true recipient of the message. Of course, the true recipient of a *telegraph* message is the final receiving operator; the person to whom it is addressed receives only a letter transcript of the message, carried to him by a low-speed projectile method of communication, namely a messenger boy. But between the sending and the receiving operator on a metallic-return buried-cable telegraph circuit there exists what is to all intents and purposes a unique and completely definite path. In strictest accuracy this statement would have to be modified because of the leakage of current from telegraph conductors, the induction of currents in nearby conductors even from circuits made up of twisted pairs of conductors, and the radiation of electromagnetic waves from a telegraph circuit during the transient conditions. Most of these effects are, in general, inappreciable and may be neglected for the moment, although their effects in long telephone circuits require careful consideration.

It appears that both guided and unguided communications have noteworthy characteristics and real spheres of usefulness. The advantages of guided communication are the much greater ease of securing secrecy of communication and freedom from "tapping," the higher efficiency of energy transmission, the definite and known nature of the destination of the message, and the comparative ease of establishment of a compact many-channeled communication network.

It is interesting to note that the *efficiency of energy transmission* by unguided systems of communication is practically nil. The approximate ratio of radiated power to receive power for fair transoceanic communication via radio is of the order of one hundred trillions to one. And by no means all of the power available in the antenna or radiating system of the transmitter is actually radiated. The efficiency of the mere act of radiation may be from roughly 50 per cent down to a per cent or two. The efficiency of transmission of energy over even a long telephone circuit or submarine cable is incomparably greater than the value given for unguided communication having a highly disadvantageous law of attenuation or diminution of signal amplitude with distance. Nevertheless, the *efficiency of communication* may be very high for unguided signaling, both as a system for transmitting intelligence and as an economical means for such transmission. Thus, the cost of accurately sending a message across the ocean via radio compares very favorably with the cost of transmission by submarine cable. That is, there is an approximate balance between the high cost of the long submarine cable and its comparatively inexpensive terminal stations and the high cost of the radio transmitting station, the low cost of the radio receiving station, and the zero cost of the medium of transmission. The balance is such as to indicate that the question of the intrinsic efficiency of energy transmission does not properly enter into a consideration of the real value of a communication system. It must not be inferred that the efficiency of energy transmission is without importance in comparing different types or systems of unguided communication with each other but only that it is not a suitable criterion for judging the commercial usefulness of an unguided system of communication as compared with a guided system.

It appears that unguided communications, such as radio, are preeminently fitted for broadcast service and will find a wide application in this direction. At present both the United States and Germany are using radio for the dissemination of information, and practically all the larger countries use it for broadcasting time signals, hydrographic information and the like.

The principal problems connected with radio communication, selected as a typical and effective form of unguided communication, are those connected with the avoidance of atmospheric disturbances of reception and those connected with the mutual interference of various transmitting stations.

It appears that there originate in the atmosphere, in a fashion not entirely understood, a number of forms of electric disturbances which tend to interfere with radio signals. Under favorable conditions, and with the most modern forms of receiver for eliminating such disturbances, their presence would hardly be suspected; on the other hand, in the middle of a lightning storm in the vicinity and with unsuitable receivers, all communication will be interrupted. As a result of determined research, there have been produced forms of receivers which markedly reduce the effect of such electric strays, and enable good communication to be fairly continuously maintained. It is to be expected that the normal further evolution of radio receivers, the increase in transmitter power, and the skilful handling of traffic will lead to *continuous* commerical communication regardless of these undesired intruders.

The use of the ether as a common channel for all radio messages is not an unmixed blessing. At times it leads to situations paralleling those encountered by the urban dweller who has many musically inclined neighbors, all of whom simultaneously persist in utilizing the available air for powerful acoustic radiation. The gratification of the listener is markedly diminished under such conditions of "interference" or "jamming" as it is termed by the radio operator. Monofrequent radiation and audio-frequency selectivity suggest themselves as the first type of solution in the acoustic case mentioned. For radio communication, we have aimed at as nearly monofrequent or "monochromatic" radiation as possible from the transmitting station, and the use of radio-frequency selection at the receiving station. This method is, on the whole, very effective and enables a considerable number of mutually non-interfering radio channels to be in simultaneous operation. In addition to frequency selection, in its several most recent forms, we have available directional selection. This latter involves the use of special antenna systems at the receiving station which respond most powerfully to signals coming from one or more directions. While the development of directional reception is still in its earlier stages, it is possible to reduce interference considerably by this means. Directional reception has other practically useful applications in the guiding of ships in times of fog or storm, using either the "radio compass" or the "radio beacon." In the former case, the directional receiver is located on land, and the ship is located during its transmission by triangulation between two or more land stations. In the latter case, the directional receiver is on the ship and its location is determined by triangulation on board ship after finding the bearing of two or more known land "beacon" stations which send out their indentification signals more or less steadily.

The establishment of the radio analog of "order wire circuits" is entirely possible, and leads to no particular difficulties. On the other hand, the development of effective radio ringing methods, which enable dispensing with the constant attendance of an operator, for example on board ship, is still in its earlier stages. This is not a serious drawback at present since the transoceanic circuits run nearly continuously and therefore require the constant attendance of operators in any case, and in the marine service it is very desirable to have an operator on duty at all times to pick up faint distress calls and hydrographic information (storm warnings and the like). However, the development of effective and reliable call signal systems may become of more importance for unguided communications as time goes on.

While two-way communication is readily possible through direct speech, that is, by the use of unguided air waves, true duplex communication is not so easily obtained. The essential element in full duplex communication is that there shall be two streams of signals flowing in opposite directions simultaneously and without mutual interference. Thus there are required two transmitters and two receivers, which influence each other only in pairs. In radio communication, full duplex communication is readily obtained and is, in fact, used largely in the transoceanic service. In the marine services, on the other hand, the users are generally content with two-way communication wherein the two parties to the communication take turns at sending and receiving alternately. The traffic-carrying capacity of a circuit is more than doubled by the substitution of full duplex for two-way service since the correction of errors is then more readily effected without delay and the coordination of traffic on both sides more easily carried out. It is interesting to note that not only duplex but even multiplex radio communication is possible on a single frequency or wave length by the use of audio or superaudio frequency modulation. That is, continuous waves of a certain length are modulated or controlled at a frequency lower than the wave frequency and the reception is accomplished with double selectivity; namely, an initial selectivity to the actual wave frequency followed by a selectivity toward the lower or modulation frequency. By thus building a lock which requires the simultaneous use of two keys to open it, we can increase the selectivity of receiving stations. It is not believed, however, that the ultimate total traffic-carrying capacity of the ether is increased in this way, even though it avoids admirably certain present-day difficulties and has a real sphere of usefulness.

If we revert to the broadest aspects of the problems of world communication, we find that an ideal system can be imagined which would presumably meet all reasonable requirements. This would be a continuously operative interference-free person-to-person network. It should include every person on the globe and clearly carry speech, which, it is hoped, would be in some lingua franca or universal language. Thus baldly stated, the problem seems frankly insoluble in terms of agencies now available. Yet it is not difficult to imagine equivalent systems which would constitute something better than a first approximation to this remarkable plan. Let us imagine a very large number of telephone substations located at all convenient points. Most of these substations will be fixed, and connected by wire line to the central stations or exchanges. However, there may be in addition movable substations which will, of course, be generally connected by radio

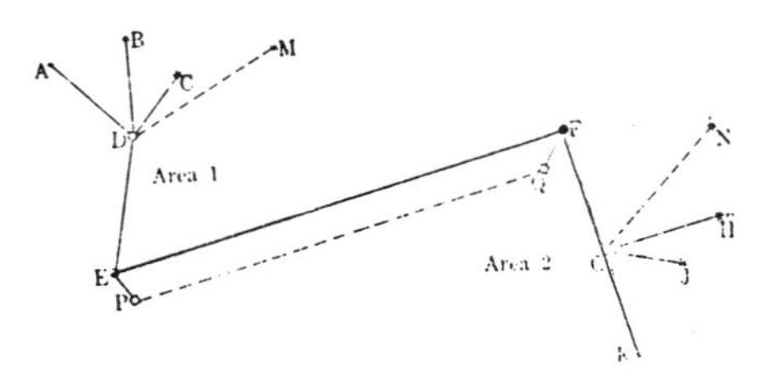

Fig. 1—Schematic Communication Network Embracing Wire and Radio Systems

to the central offices. As examples of such movable substations we may take ships, railroad trains, and aircraft equipped with radio telephone sets, or in special cases individuals supplied with highly portable radio telephone outfits. We may expect that even individuals, such as forest rangers and other groups of professional patrolmen, would make good use of such portable substation equipment. The central offices will be connected with each other, directly or indirectly, by wire systems where wires can be more effectively and economically maintained, and by radio where wires cannot be maintained. In Fig. 1 is shown schematically the layout of such a system including, however, only two of the communication areas.

We will suppose that person A in area 1 desires to speak to person H in area 2. Substation A is connected by wire to central office D of area 1. Thence the speech currents travel to the "long-distance" exchange at E. From E the speech travels by wire or radio to F, the "long-distance" exchange nearest to G, the central office of area 2. If the radio route is used, the radio stations P and Q, controlled by E and F respectively are employed. After passing to G, the speech is sent directly to H, the called station. At various points along their path, the speech currents are amplified as necessary. In case a movable substation desires to share the facilities of such a communication system, it will do so by establishing radio telephone communication with a radio station at the local central office or a radio station directly connected thereto. Thus if an occupant of the railroad train M provided with a radio telephone set, desires to use the telephone system shown, he does so by establishing communication with D by radio. A movable substation at N could similarly share the benefits of the system by connecting with central office G by radio. So that we can even imagine direct communication between stations M and N, both in motion, through this communication plan. It will be seen that communication with moving substations, and telephone communication across oceans, will probably be on a radio basis in this system, and that all the remaining portions of the communication (which constitute the major portion of the communication network in practically all cases) will be by wire. The particular system described illustrates the judicious and useful combination of guided and unguided systems of communication, though it represents an extreme case and a very small percentage of the total telephone traffic handled.

The great bulk of communication is now carried on by wire methods. There is no question but that wire transmission must always carry the great bulk of communications, particularly in the well settled and highly developed parts of the world where its methods permit the setting up of reliable and economical paths for carrying tremendous numbers of messages without mutual interference.

Radio, however, has an important field to fill, as an auxiliary of the wire system in extending service to ships at sea, to aeroplanes, perhaps even to automobiles, and in general to moving bodies to which wires cannot be connected. It has a pioneering field in undeveloped countries, or wherever natural barriers make it difficult to set up the wire system. It has perhaps its most profitable and commercially useful field in giving service between continents separated by wide reaches of ocean.

The possibilities of guided transmission can be well illustrated by two examples: First, the transcontinental line between New York and San Francisco, which is today giving high-grade commercial service between the cities of our eastern and western coasts. The second is the 1200 pair subscribers' cables, now being placed by the telephone companies, which permit of 1200 simultaneous conversations being transmitted through a sheath of less than 2.5 inches, inside diameter.

Two other illustrations which will likewise indicate the possibilities of radio are: First, the radio telegraph service from the eastern coast of this country to all of the countries in Europe. Second, the radio broadcasting stations, which have been set up and which permit all of those within range of these broadcasting stations who have the necessary apparatus to pick up the news and entertainment which is sent out from the stations.

For the case of long-distance wire communication in certain of the more important countries, such systems as those required in the previous plan already exist to a considerable extent. For the corresponding international transoceanic network, involving radio telephone trunk circuits, the surface has not yet been scratched, although the recent remarkable demonstrations by the American Telephone and Telegraph Company of telephone service from Catalina Island to Havana indicate the extent to which this problem is engaging the attention of investigators and firing the imagination of far-seeing executives of communication companies. In the case mentioned, telephone communication was established between two points thousand of miles apart, separated by portions of two oceans, and across a continent. The speech was sent from Catalina Island across a stretch of the Pacific Ocean via radio, thence by wire line across the United States and down to Key West, Florida, and finally through the submarine telephone cable to Havana. The transmission difficulties were equal to those in bridging the stretch from London to Pekin, or from New York to Buenos Aires, as pointed out by Dr. J. J. Carty. It is likely that for a long time mankind will find that a very approximate approach to the system of world communication previously described will satisfy the most ambitious engineers and traffic experts. The omission of the portable substations *M* and *N* would work no great hardship at this time.

Fig. 2—Long-Distance Line Telephone Repeater Installation A. T. & T. Co., Salt Lake City, Utah

particularly if the number of conveniently located substations such as *A*, *B*, *C*, and so on is continually increased into a practically all-embracing system.

The subscribers' local telephone stations (such as *B* and *J* of Fig. 1) already exist to a considerable extent in some parts of the world. Thus in the United States, there are 12,600,000 stations of this type in existence, interconnected by 25,400,000 miles of wire, and carrying more than eleven billion messages annually. This is one telephone per nine persons of the population, roughly, and approximately four telephones per square mile for the entire area of the country including its uninhabitable portions. Obviously we are approaching an era of truly universal communication service, when a modern country can show a telephone network such as that of the Bell Telephone System connecting over 70,000 cities, towns and rural communities. Some of the component elements of such a vast communication system deserve attention. Starting from long-distance exchanges, such as that in New York, the lines radiate in all directions. Careful "loading" of the lines prevents distortion of the speech. At intervals the attentuated line currents are automatically amplified in "repeater stations," such as that at Salt Lake City shown in Fig. 2. These land line circuits may be automatically extended through submarine telephone cables. Or radio telephone transmitters may be used as extensions of the land line system. Thus Catalina Island is connected to the California mainland and the Bell telephone system through a radio telephone link.

Fig. 3—Amplifier and Modulator Equipment of Radio Telephone Station, Arlington, Va.

◆◆◆ ◆◆◆ ◆◆◆

Alexanderson Discusses the Alexanderson Radio System

AN eminent radio pioneer whose radio alternator had been the basis for the creation of the Radio Corporation of America presented this reflective paper in 1920. He credited radio with continuing the "emancipation of the human spirit" that had been started by the printing press by making the communication of ideas "independent of brute force that might be used to isolate one part of the world from another." Alexanderson stressed that his system was based on the extension of the principles of power engineering to radio and obviously regarded a radio station as similar to a central power station. They were similar even in load diversity as the peak load for the Radio Central Station would occur at different times of the day and year for Europe and South America.

Ernst F. W. Alexanderson (1878–1975) was born in Sweden and graduated in engineering from the Royal Technical University in Stockholm in 1900. He then studied for a year at the Engineering College in Charlottenburg, Germany before coming to the United States in 1901. After working for a few months for the C & C Electric Company in New Jersey, he joined the General Electric Company in 1902. During his 50-year career with G.E., Alexanderson received approximately 350 patents for inventions, ranging across the whole spectrum of electrical engineering. He began work on a high-frequency radio alternator in 1904, a project that culminated in a 200-kilowatt alternator that was installed at New Brunswick in 1918. This station became the prototype for the worldwide chain that was underway when his paper was presented in 1920. Alexanderson became the first Chief Engineer of RCA and held the position until the system was completed. He then resumed his career as engineer–inventor at G.E. and made contributions to television, power electronics, and control systems during the last half of his career. (See Paper 61.) See the *NCAB*, vol. A, pp. 30–32. See also my paper on Steinmetz and Alexanderson in the PROCEEDINGS OF THE IEEE for September 1976. Also see James E. Brittain, "E. F. W. Alexanderson: Reflections on the Remarkable Career of an Engineer–Inventor," published in *The Bent of Tau Beta Pi*, Summer 1976.

CENTRAL STATIONS FOR RADIO COMMUNICATION*

By

Ernst F. W. Alexanderson

(Chief Engineer, Radio Corporation of America; Consulting Engineer, General Electric Company)

Radio achievements are often referred to as belonging in the realm of mystery, and it is indeed wonderful that we are now able to speak with a voice that carries thru empty space across the oceans. Whenever knowledge conquers a new force of nature for the use of humanity, it ceases to be a mystery, but the pursuit of this knowledge makes an even greater appeal to the imagination.

The development of the steam engine was a triumph of the engineering art of the last century, but it was not the engine itself but the steamship and the locomotive that interested humanity.

The telephone and cables no less than the steam engine have introduced a new era in human affairs. They have, to a degree, conquered space and time, but only with certain serious limitations.

An ocean cable runs only from one landing place to another and it can be cut in time of war; its use can be censored by its owners and controlled by military and naval power. When, on the other hand, you send a radio message, it may reach all parts of the world. Depending upon whether it has been sent in code or in plain language, it may be a confidential, private message or a press message intended for the world at large, but nobody can directly prevent the electromagnetic waves themselves that carry the message from reaching their destination. It is thus not exaggeration to say that the emancipation of the human spirit that was begun by the invention of the printing press has found its fulfilment in radio communication. Radio makes the transmission of ideas from man to man and from nation to nation independent not only of any frail material

*Received by the Editor, October 12, 1920. Presented before a joint meeting of The Institute of Radio Engineers and The New York Electrical Society, New York, November 10, 1920.

Reprinted from *Proc. IRE*, vol. 9, pp. 83–90, Apr. 1921.

carrier such as a wire, but above all it renders such communication independent of brute force that might be used to isolate one part of the world from another.

These are the ideal aims which inspire the engineers engaged in the development of the radio technique. This is also the explanation why some of the foremost lawyers, executives, financiers, officials, and statesmen of this country have found incentive in the human aspects of the radio technique to devote a great deal of their time and thought to its promotion and development on a world-wide scale.

The interest that is evidenced by all concerned in this subject has become much more serious since it has been established that the laws and forces with which we are dealing are within the control of our knowledge, so that engineers can now proceed with the design of a radio communication system with practically the same deliberate accuracy as in the design of an electric power transmission from a water fall to a railroad.

This audience is constituted of members of a society of electric power and light engineers as well as members of THE INSTITUTE OF RADIO ENGINEERS; so I shall take the opportunity to trace the close connection which now exists between electric power engineering and modern radio engineering and will demonstrate, as the specific subject of this paper, how the development of the Central Station for radio communication is as logical and inevitable as was the development of the central electric power station.

The entry of the Corporation with which I have been connected for the last twenty years upon the field of radio communication has been a gradual growth and a natural consequence of its general activities in power engineering. The engineers specializing in alternating current technique were in a natural position to take up the problem of designing alternators and transformers in the radio technique. These differ from the one used in the power technique principally in the fact that the number of alterations per second is about one thousand times as great. This speeding up of the performance one thousand times involved many new problems, but the most remarkable fact to record is that *the generally established principles of the alternating current power technique could be applied to the radio technique almost without change.* It meant that the magnetic properties of iron which had been reduced to an exact science by Steinmetz thirty years ago had to be studied again at radio frequencies; but it was found the Steinmetz laws of hysteresis eddy currents, and skin effect were as accurate at two hundred thousand cycles per second as at twenty-five cycles.

It was furthermore found that the established conceptions of phase displacement, power factor, and leading and lagging currents were as applicable and useful in the high frequency as in the lower frequency technique.

It is true that radically different methods had to be devised for measuring power factors of a fraction of one per cent from the methods used for measuring power factor of 50 to 100 per cent, but the new methods of investigation verified the well-known principles.

The starting point of this development work was the time when Fessenden brought to the General Electric Company the problem of generating alternating currents for radio transmission. In doing so Fessenden realized that a practical solution of this problem could be worked out only by an organization of specialists.

Some of the problems that presented themselves in the evolution of the radio power plant were:

The design of a dynamo-electric machine or alternator generating electric power in the form of alternating currents of frequencies one thousand times as great as those used for motors and lights.

The development of magnetic amplifying devices capable of translating telephone and telegraph currents into corresponding modulations of the radio frequency energy flowing from the power plant into the radiating antenna.

The development of a regulator so sensitive as to hold the speed of an ordinary induction motor constant within a few hundredths of one per cent, this being necessary in order to maintain the proper phase relations in a load circuit working at one-third of one per cent power factor.

Improvement of the tuning of the antenna so as to transform as large a part as possible of the generated energy into electromagnetic waves.

The realization of Fessenden's vision, the radio power plant of to-day, became thus the result of the combined effort of leading electrical and mechanical engineers. Among these, it is sufficient to mention Mr. W. L. R. Emmet, the creator of the giant electric power stations of today.

The radio power plant which resulted from this was shown to Senator Marconi during a visit to Schenectady, and because of his interest in its performance, it was transferred to the Marconi Radio Station at New Brunswick, where it had no sooner been installed than it was taken over for war service by the Navy. Further enlargements and developments of this installation which were undertaken by the Navy resulted in the plant which is now owned and operated by the Radio Corporation of America.

Here we had arrived at a point where two schools of engineering pursuing different aims, with widely different modes of thought, had been brought before a common problem. The one had been thinking in terms of power factor, kilowatts, and phase displacement, the other in terms of wave length, decrements, and tuning.

A third school of knowledge was at that time brought into contact with this technique and added new impetus to it. As soon as such scientists as Drs. Coolidge and Langmuir began to study the Fleming value and the remarkable little device invented by Dr. Lee De Forest and known as the audion, the foundation was laid for the vacuum tube technique which has so profoundly influenced the art of radio communication.

These scientists tell us that electricity is not the mysterious "power fluid" that we may have imagined flowing smoothly in our wires, but miniature plants or comets of condensed material electricity of definite charge and mass shooting across a miniature universe inside of a glass bulb and following orbits that can be calculated as accurately as the orbits of the stars.

Keeping in mind the origin of the modern art of radio communication in these three widely separate realms of knowledge, power engineering and electro-physics, we may now proceed to examine the essential parts. We find then—first, a modern electric power plant working at very high frequency; second, a network of wires a mile (1.6 km.) long, supported on tall masts; third, on the opposite side of the ocean a little glass bulb full of shooting stars. The question is: what does really happen?

Does the electricity generated by our alternator emanate from the antenna and flow in an undulating stream thru the air or thru the water or thru both? If we search for it in an airplane, we find it, and if we submerge ourselves in a submarine and search for it, we find it, and yet we are told it is not so.

Does the little electron, as an individual, take a leap off the aerial wires and, after devious paths, find its home in the glass bulb on the other side of the ocean? We are also told that it does not.

If I knew exactly what really does happen, and should try to tell you, then sooner or later somebody would claim that I was altogether mistaken. Therefore, I will only try to tell you how I imagine that it happens, wondering if any of you will see the same mental picture of the process that I see.

We were once told by the physicists that all space was filled by a fine substance that was called ether, and that the light and heat that radiated from the sun was a wave motion in the ether. The physicists now tell us that there is no ether, but still they say that light is a wave motion. Be this as it may, for the purpose of visualizing what takes place in radio transmission, it is convenient to cling to the theory of the ether.

We are familiar with other forms of wave motion—the air waves that carry sound to our ears and the water waves on the ocean. Thus the carrier of the radiated electric energy must not be likened to the flowing stream of water, or to the wind or to a bullet shot from a gun, but likened to a wave in a uniform medium where each particle of the medium oscillates around a stationary base line while the wave rolls forward.

The distance that a wave can travel in an absorbing medium before it fades out to a definite extent is proportional to its length. We may therefore introduce the idea of wave length, which is the distance from the crest of one wave to the next. The long swells of ocean travel for hundreds of miles, whereas a pebble dropped on a still surface of water produces a ripple that fades away in a short distance.

In radio communication it has been observed that the distance over which reliable communication can be maintained is about 500 times the length of the ether wave that is used. It may be more than a mere coincidence that the distance to which a sound wave travels in air, and a wave on the surface of water will travel before it fades out, is also about 500 wave lengths. The average wave length of sound of spoken words is abont one foot (0.3 m.), and we know that if we speak loudly our voices will carry a distance of about 500 feet (150 m.). The exceptions to this rule that will occur to anybody are also significant. We know what distances voices will carry over a lake in a quiet evening. We also know what extraordinary distances radio signals will carry sometimes in a quiet night. These are "excepttions that prove the rule," and the rule refers only to reliable communication under normal conditions.

A radio transmitting system is designed for the purpose of producing waves in the ether which we call electromagnetic waves, and for controlling the rate at which the waves are produced in such a way that a train of successive waves will carry the meaning of articulate speech or telegraphic code. If we wish to send a message a long distance, we must select a long wave. The distance to Europe is 5,000 kilometers (3,200 miles). If this distance is to be bridged by 500 wave lengths, each wave length must be at least 10 kilometers (six miles), or, as it is usually expressed, a wave length of 10,000 meters.

We can produce water waves by rocking a boat. If we rock a canoe rapidly we get a short wave, but if we rock a larger boat more slowly we get a correspondingly longer wave. To rock the boat requires energy, but in order to produce a wave of suitable length, the energy must act thru an intermediate member which has suitable size, proportions and period of oscillation.

In radio transmission the energy is furnished by the radio frequency power plant, but, in order to transform this energy into waves, there is required the intermediate member which makes contact with a large volume of the medium which carries the wave motion. This medium is the ether and corresponds to the water or the air in the more familiar forms of wave motion. The member that transforms the energy to the ether is the antenna. The waves used for trans-Atlantic communication are as a matter of fact 10,000 meters long, or even longer. The antenna corresponds to the hull of the rocking boat or the sounding board of the piano.

The analogy with water waves may be carried still further. The wave is a successive displacement of the medium, and the initial displacement produced by the member acting upon the medium is proportional to its volume. The water displacement of the boat corresponds to the effective volume of the antenna. The maximum voltage at which the antenna can be operated corresponds to the maximum angle to which the boat may be rocked before it ships water. This is the voltage at which the surrounding air breaks down under the electrostatic pressure. In electrical units, the displacement in the ether is expressed in meter-amperes. This is really a measure of volume as is apparent from the consideration that the amperes charging current at the limiting voltage is proportional to the two horizontal dimensions. The third dimension or the height appears directly in the product, and is expressed in meters.

The height of the antenna is the most expensive of the three dimensions by which we may create electric displacement in the ether. The tendency in stations designed for greatest economy is, therefore, towards structures of moderate height and great length, whereas, the tendency in the past, when dynamic efficiency was the principal consideration, was towards towers of great height. The unit of performance on the old basis was kilowatts consumed by the antenna. The unit on the new basis is "ether displacement." This modern measure of antenna radiating capacity is the number of meter-amperes of "ether displacement" that can be produced at the voltage which is limited by the breakdown of the air.

The antennas of the stations of New Brunswick and Marion which are now used in trans-Atlantic service are each one mile (1.6 km.) long. In the new Radio Central Station, which is being built by the Radio Corporation on Long Island, there will be ten or twelve antennas, each a mile and a quarter (2 km.) long. This station is intended to communicate efficiently with all parts of the world. When very long distances are to be spanned, correspondingly long waves will be used. For efficient transmission of these long powerful waves, an antenna will be needed that makes contact with a large volume of ether. This will be accomplished by combining several of these antennas into one unit. At other times the same antennas will be used for the simultaneous transmission of several messages over shorter distances.

The shifting of radiation power which has been referred to is made possible by the use of the multiple tuned antenna which has been described in a previous paper before THE INSTITUTE OF RADIO ENGINEERS.[1] The New Brunswick and Marion antennas are now tuned so that each acts as six single antennas operating in multiple. The combining of several such groups in multiple is only a further extension of the same principle.

When two such antenna groups are connected in multiple, the loss resistance is reduced to one-half. Hence the efficiency of the antenna is increased so that a given power produces more radiation. Still more important is, however, the fact that more power may be utilized at this increased efficiency, and so the net result is that the amplitude of the radiated wave is doubled, which means that four times as much energy is radiated.

The economical factors that point to the radio central station as the practical solution of the problem of long distance communication are practically the same as those that created the central

[1]See PROCEEDINGS OF THE INSTITUTE OF RADIO ENGINEERS, volume 8, number 2, pages 279–282.

electric power station. Broadly speaking, they provide for the utilization of the plant investment and operating force to the utmost by shifting the equipment from one service to another and combining it to meet various demands.

New York is a natural communication center and the service must extend to Europe, South America, and westward. Another Radio Central Station at Hawaii is being equipped to serve as a relay for all points on the other side of the Pacific Ocean.

While it is winter on the northern hemisphere, the radiating power to Europe can be much reduced, but this is the season when the South American traffic requires a maximum radiation because of summer conditions then existing on the southern hemisphere. The New York Radio Central Station can then divert some of its radiating power from the European to the South American circuits. There will also be daily fluctuations in traffic load which will occur at different hours due to the difference in geographic longitude. Thus the peak load of European traffic will occur at different times than the South American and Western traffic. The central station equipment can be utilized so as to take advantage of this.

The realization of trans-Atlantic radio telephony for commercial purposes is another object of the Radio Central installation. Trans-atlantic telephony will, no doubt, be something of a luxury for the immediate future. The radiation intensity needed for telephony is much greater than for telegraphy, and a plant designed purely for telephony might prove very expensive. However, the flexibility of the Radio Central, where any number of antennas can be combined when desired to produce a more efficient radiation, will make an extra powerful transmitter available when needed, while the plant may be used in a more economical way at other times for telegraphy.

SUMMARY: There are considered the mechanism of radiation and reception in radio communication. The design of the transmitting equipment is compared with the design of the usual alternators and power plants of electrical engineering. The main problems encountered are described, and an account is given of the solutions obtained. The development of the Radio Central Station for telegraphy and telephony is discussed, its arrangements described, and its usefulness indicated.

Gherardi and Jewett on Telephone Repeaters

IN this extract from a longer paper on the history of efforts to devise repeater amplifiers for telephone circuits, two leading telephone engineers credited the recently developed vacuum-tube repeater with starting "what bids fair to be a revolution in the entire scheme of telephonic transmission." They argued that the innovation could only have been achieved by "a great unified engineering and research department" and was "far beyond the limits of any single individual or any limited organization." The same observation might have been made about the earlier loading coil innovation. (See Paper 31.) For more details on early mechanical and electronic repeaters used by the Bell System, see *A History of Engineering and Science in the Bell System: The Early Years (1875–1925)*, edited by M. D. Fagen (Bell Telephone Laboratories, 1975), pp. 253–277.

Bancroft Gherardi (1873–1941) was born in San Francisco and was the son of a Naval officer who had served aboard the U.S.S. Niagara during the Atlantic Cable expedition of 1858. (See Paper 9.) Gherardi graduated from the Polytechnic Institute of Brooklyn in 1891 and then studied at Cornell where he received a Master's degree in engineering in 1894. He then joined the Metropolitan Telephone and Telegraph Company which later became the New York Telephone Company, a subsidiary of the Bell Company. He worked on the design of loaded lines and later was involved in the transcontinental telephone project that became operational in 1915. He became Chief Engineer of A.T.&T. in 1919 and remained in that capacity until his retirement in 1938. Gherardi was President of the AIEE in 1927 and was elected to the National Academy of Sciences in 1933. See the *DAB*, vol. 23, pp. 298–300 and *NAS*, vol. 30, pp. 157–177.

Frank B. Jewett (1879–1949) was born in Pasadena, Calif. and graduated from the Throop Institute of Technology (later California Institute of Technology) in 1898. He then enrolled at the University of Chicago where he was awarded the Ph.D. degree in 1902. Jewett taught physics at M.I.T. for two years before being persuaded to join the engineering staff of the Bell System by George Campbell. (See Paper 39.) In 1906 Jewett succeeded Campbell as head of the Electrical Department. Jewett arranged to hire a young physicist from the University of Chicago, H. D. Arnold, to work on the telephone repeater problem which became urgent in 1912 when a decision was made to attempt transcontinental telephony. When the Bell Telephone Laboratories were formally organized in 1925, Jewett was selected to head them with Arnold as Research Director. He was President of AIEE in 1922 and President of the National Academy of Sciences from 1939 to 1947. See the *DSB*, vol. 7, pp. 107–108 and the *NAS*, vol. 27, pp. 239–264.

Presented at a joint meeting of the American Institute of Electrical Engineers, and the Institute of Radio Engineers, New York, October 1, 1919.

TELEPHONE REPEATERS

BY BANCROFT GHERARDI AND FRANK B. JEWETT

ABSTRACT OF PAPER

In this paper the authors have endeavored to set forth briefly but clearly the history of the research and development work which has led up to the final production of successful telephone repeaters. The various forms of amplifiers which have been suggested are described and their possibilities and limitations pointed out. The essential properties of repeater networks together with the necessary line conditions for successful repeater operation are described and illustrated. Tandem operation of repeaters is discussed as is also the use of repeaters in four-wire circuits. Illustrations are given showing a few of the more important repeater installations now in regular commercial service in the United States.

INTRODUCTION

DURING the past ten years developments in the design and construction of telephone amplifiers, or telephone repeaters as they are more universally called, and in the art of their application to all forms of telephone circuits have progressed to such an extent as to justify the presentation to the American Institute of Electrical Engineers of a comprehensive paper which shall cover not only the earlier efforts but the present state of development.

As indicated below, the idea of one or more repeaters inserted in a line for the purpose of reinforcing from some local source of energy the weakened current from the sending station is older than the telephone itself. In telephony innumerable attempts by a large number of investigators have gone on continuously almost from the inception of the telephone in an effort to extend the range of telephonic speech through the utilization of energy applied to the telephone line at a point or points between the transmitting and receiving stations. While the net result of all this work is, of course, the state of the art as we now have it, a survey of the earlier developments in the light of present-day knowledge discloses in striking fashion the fact that, unknown to the investigators, the early attempts

Reprinted from *Trans. AIEE*, vol. XXXVIII, pp. 1287–1289 (extract of a longer paper), Oct. 1, 1919.

were destined not to succeed where success was measured by the development of a practical device which would give satisfactory results on a regular two-way telephone circuit.

We know now that the final successful development and application of the telephone repeater had to wait not only for the slow accumulation of comprehensive knowledge concerning all the factors which govern the successful transmission of speech electrically over wires and of the intricate relation of circuits and apparatus to produce desired results with the maze of frequencies involved in speech, but also on developments in physical science which were in themselves quite foreign to the specific realm of telephony. For these reasons the work of the earlier investigators, ingenious though it was, was for years foredoomed to failure by factors over which they had no control and of which, in many cases, they had no knowledge even. In the light of what we now know, much of this early work appears almost impossibly successful and attests to the ingenuity and resourcefulness of the men who conducted it.

Another thing which the final successful development and extensive application of the telephone repeater indicates with striking clarity is the fact that a complete solution of the problem was made possible only by the existence of a great unified engineering and research department, such as that maintained by the Bell System. The elements involved and the multiplicity of detail to be worked out and correlated were far beyond the limitations of any single individual or any limited organization. The best that mathematics, physics, chemistry, engineering, manufacturing and operation could supply were needed for the solution. Further, if the solution were to be obtained in a reasonable time it was essential that the efforts of the experts in all of these fields must be directed in a cooperative attack on the main problem. With the growth of fundamental research groups of highly trained scientists in the engineering department of the Bell System the means for attacking the elusive repeater problem in comprehensive fashion became available and progress toward success became certain and rapid. The loose ends of discrete and desultory researches were gathered in, past and present work scrutinized and the attack directed from a foundation of certain knowledge not theretofore available.

The wonderful success which has attended the work and

which has started what bids fair to be a revolution in the entire scheme of telephonic transmission should be a source of gratification to American electrical engineers, to whom more than to any others have been due the other wonderful telephone developments of the past.

In discussing the earlier work on telephone repeaters and that which led up to the first successful device, the authors have not attempted even to enumerate the large number of investigators, much less to allocate credit among them. The names mentioned are those of the men who stand out most conspicuously and the fact that many of them are employees of the Bell System is not remarkable, since it was the largest single agency consistently at work on all problems concerning telephone development. In a complex art which has been vigorously worked for nearly fifty years, it is inevitable that great numbers of patents should have been taken out on all conceivable kinds of devices and systems. Many of these are inherently worthless, while others contain elements of value, although not in themselves disclosing complete workable arrangements. While, therefore, no single invention has solved the repeater problem in its entirety, it is interesting to note that all of the inventions which make possible the successful telephone repeater of today are the work of Americans.

No paper dealing with the development and successful application of telephone repeaters would be complete without mention of the fact that throughout all the later years, during which progress has been certain and rapid, the principal credit for the broad engineering and commercial results obtained is due to Colonel J. J. Carty, past President of this Institute and for many years Chief Engineer and now Vice President of the American Telephone and Telegraph Company. To Colonel Carty more than to any one else is due the credit of having brought about that coordination of all of the elements mentioned above, which was vital to the ultimate success of the undertaking.

◆◆◆ ◆◆◆ ◆◆◆

Colpitts and Blackwell on the Carrier Multiplex System

THIS is an extract from a much longer paper describing an important innovation that had been recently introduced by the Bell System. The first commercial carrier system with four channels had been placed in operation in 1918 between Baltimore and Pittsburgh. As pointed out by Colpitts and Blackwell, the carrier concept dated back to the 19th century, but did not become technically feasible until the development of devices and circuits such as the thermionic tube and Campbell's band filter (see Papers 37 and 39). For more details and a bibliography on carrier systems, see the original paper in the TRANSACTIONS OF THE AIEE for 1921 or an edited reprint published in the PROCEEDINGS OF THE IEEE in April 1964, pp. 340–359. Also see *Engineering and Science in the Bell System: The Early Years (1875–1925)*, edited by M. D. Fagen (Bell Telephone Laboratories, 1975), pp. 277–290.

Edwin H. Colpitts (1872–1949) was born in New Brunswick, Canada and graduated from Allison University in 1893. He later received the B.A. and M.A. degrees in physics from Harvard in 1896 and 1898. Colpitts joined the engineering staff of the American Bell Company in Boston in 1899 and worked with Campbell on early loading coil experiments. In 1907 he became a research engineer with the Western Electric Company in N.Y. and worked on telephone repeater and radio telephone systems, as well as carrier multiplex. He became an assistant Vice President in the Department of Development and Research of A.T.&T. in 1924 and later was Vice President of the Bell Telephone Laboratories from 1934 until his retirement in 1937. Perhaps his best known invention was the Colpitts vacuum-tube oscillator. See the *NCAB*, vol. E, pp. 269–270 and an obituary in *Electrical Engineering*, vol. 68, p. 460, 1949.

Otto B. Blackwell (1884–1970) was born in Massachusetts and graduated from M.I.T. in 1906. He was hired by Jewett (see Paper 37) as an engineer with the Bell Company in 1906 and remained with the Bell System until his retirement in 1949. He was a Vice President of the Bell Telephone Laboratories from 1936 to 1944 and was awarded the Edison Medal by the AIEE in 1950. See H. S. Osborne, "The Edison Medalist," *Electrical Engineering*, vol. 70, pp. 186–188, 1951 and *Who Was Who in America*, vol. 5, 1969–1973.

Presented at the 9th Midwinter Convention of the American Institute of Electrical Engineers, New York, N. Y., February 17, 1921.

CARRIER CURRENT TELEPHONY AND TELEGRAPHY

BY E. H. COLPITTS
Western Electric Co.

AND O. B. BLACKWELL
American Telephone & Telegraph Co.

ABSTRACT OF PAPER

This paper briefly outlines first the history of the development of carrier multiplex telegraphy and telephony. The fundamental principles underlying particularly the newer developments of the art are then discussed. Consideration is likewise given to the propagation characteristics of open wire lines, including those containing intermediate lengths of cable. Commercial types of apparatus and actual installations are then described and a brief statement made as to further applications of the art.

THE carrier method of multiplexing telephone and telegraph lines is technically one of the most interesting and important of the developments which have been perfected in the art of electrical communications during the past few years. In this paper we are giving a brief sketch of the development of this art, an explanation of the principles on which it is based, and a description of the applications which have been made in the plant of the Bell Telephone System.

In a carrier multiplex system, a number of separate telephone, telegraph or signaling messages are superimposed simultaneously on a single electrical circuit by employing a separate alternating current, usually called a "carrier current," for each of the separate messages. This carrier current is made to vary in accordance with the variations of current representing the telephone, telegraph or signaling message. The different carrier frequencies which are superimposed on a circuit must differ sufficiently in frequency so that they may be separated from each other at the terminals by the use of proper electrical circuits. Each carrier

Reprinted from *Trans. AIEE*, vol. XL, pp. 205–207 (extract of a longer paper), Feb. 17, 1921.

may be of either audible or ultra-audible frequency, but its frequency must be higher than the highest frequency represented in the message to which it corresponds. These currents are known as carriers, since in a sense, they may be said to "carry" the telephone, telegraph or signaling currents by which they are controlled.

The underlying principles of such systems are old in the communication art and indeed go back to the date of the invention of the telephone itself, for it will be recalled that it was Bell's experiments with the vibrating reed type of multiplex telegraph system which led to his discovery of the telephone. A short history of the art during the forty odd years that have elapsed between the conception of its possibilities by the early communication pioneers and the present realization of their hopes is given below under the heading "Historical."

In looking back over the early history of the carrier art it is now clear that the development of successful multiplex carrier systems had necessarily to await not only the evolution of the fundamental ideas for carrier operation, but also the development of radically new types of apparatus and the developments in electrical wave transmission over wires which have characterized the recent progress of long-distance telephony.

The telephone and telegraph systems which are described in this paper are in daily use over long toll circuits in the Bell telephone system. The telephone installations in service furnish simultaneously as many as four two-way telephone conversations over each circuit in addition to the telephone and telegraph facilities normally afforded by the circuit. The telegraph systems in service are arranged to furnish as many as ten duplex carrier telegraph circuits over each circuit in addition to the telephone and telegraph facilities normally afforded by the circuit. These figures do not indicate the maximum numbers of facilities which it will be found economical to employ ultimately, but cover the facilities furnished by the systems which are now commercially employed.

The increased circuit facilities obtained in this way

are, in general, up to the high standards set for the best grade of long-distance circuits. They are relatively stable and are maintained by the regular telephone plant personnel. The carrier circuits, both telephone and telegraph, are so designed that as circuits they fit in completely with the more usual circuit facilities of the telephone system. They may be connected with each other and with ordinary circuits, and, in general, present much the same degree of practicability and flexibility of operation as do the more usual forms of circuits.

While the development has thus succeeded in making available to the communication art new types of circuit facilities, these facilities can only be made to meet the high standards required in a public service plant by the use of correspondingly high-grade equipment which, unfortunately, is correspondingly expensive. Indeed, the cost of these systems, at least at present, is such as to make their use economical only over relatively long toll circuits. For short-distance toll service, and for local exchange service, the equivalent facilities can be provided more cheaply by the older methods.

••• ••• •••

Campbell on the Electric Wave Filter

IN this brief extract from a longer paper on the theory of the wave filter, its inventor defined the device that had been discovered by an extension of his work on the theory of loading coils (see Paper 31). The great importance of the wave filter that Campbell had patented in 1917 had become apparent by 1922, as indicated by the Editor's introductory note to Campbell's paper. The broad advance in communications technology and the stimulating exchange of ideas and innovations among power, electronics, wire, and wireless engineers during the first quarter of the 20th century may be readily seen from Papers 35–40. See Anatol I. Zverev, "The Golden Anniversary of Electric Wave Filters," *IEEE Spectrum*, pp. 129–131, March 1966.

George A. Campbell (1870–1954) was born in Hastings, Minn. and graduated from M.I.T. in 1891. He received a master's degree from Harvard in 1893 and then received a fellowship that enabled him to devote three years to graduate study in Europe. He was awarded the Ph.D. degree from Harvard in 1901. Campbell began his career as research engineer with the Bell Company in Boston in 1897. His conception of the loading coil was made in 1899, and he derived a general formula for loaded lines that became known as Campbell's Equation. Priority for the invention was later awarded to M. I. Pupin on the basis of a disclosure date two days earlier than Campbell's. In addition to his wave filter, Campbell worked on the theory of antennas, transients, repeater circuits, and measurement theory. F. B. Jewett stated in 1935 that Campbell's achievements "entitle him beyond question to rank first among his generation of theoretical workers in electrical communication." Campbell was awarded the Edison Medal of the AIEE in 1940. See the *NCAB*, vol. 45, pp. 120–121 and *The Collected Papers of George Ashley Campbell* (A.T.&T., N.Y., 1937).

Physical Theory of the Electric Wave-Filter

By GEORGE A. CAMPBELL

NOTE: The electric wave-filter, an invention of Dr. Campbell, is one of the most important of present day circuit developments, being indispensable in many branches of electrical communication. It makes possible the separation of a broad band of frequencies into narrow bands in any desired manner, and as will be gathered from the present article, it effects the separation much more sharply than do tuned circuits. As the communication art develops, the need will arise to transmit a growing number of telephone and telegraph messages on a given pair of line wires and a growing number of radio messages through the ether, and the filter will prove increasingly useful in coping with this situation. The filter stands beside the vacuum tube as one of the two devices making carrier telegraphy and telephony practicable, being used in standard carrier equipment to separate the various carrier frequencies. It is a part of every telephone repeater set, cutting out and preventing the amplification of extreme line frequencies for which the line is not accurately balanced by its balancing network. It is being applied to certain types of composited lines for the separation of the d.c. Morse channels from the telephone channel. It is finding many applications to radio of which multiplex radio is an illustration. The filter is also being put to numerous uses in the research laboratory.

The present paper is the first of a series on the electric wave-filter to be contributed to the Technical Journal by various authors. Being an introductory paper the author has chosen to discuss his subject from a physical rather than mathematical point of view, the fundamental characteristics of filters being deduced by purely physical reasoning and the derivation of formulas being left to a mathematical appendix.—*Editor.*

THE purpose of this paper is to present an elementary, physical explanation of the wave-filter as a device for separating sinusoidal electrical currents of different frequencies. The discussion

Reprinted with permission from *Bell Syst. Tech. J.*, vol. I, pp. 1–3 (extract of a longer paper), Nov. 1922.

will be general, and will not involve assumptions as to the detailed construction of the wave-filter; but in order to secure a certain numerical concreteness, curves for some simple wave-filters will be included. The formulas employed in calculating these curves are special cases of the general formulas for the wave-filters which are, in conclusion, deduced by the method employed in the physical theory.

All the physical facts which are to be presented in this paper, together with many others, are implicitly contained in the compact formulas of the appendix. Although only comparatively few words of explanation are required to derive these formulas, they will not be presented at the start, since the path of least resistance is to rely implicitly upon formulas for results, and ignore the troublesome question as to the physical explanation of the wave-filter. In order to examine directly the nature of the wave-filter in itself, as a physical structure, we proceed as though these formulas did not exist.

It is intended that the present paper shall serve as an introduction to important papers by others in which such subjects as transients on wave-filters, specialized types of wave-filters, and the practical design of the most efficient types of wave-filters will be discussed.[1]

Definition of Wave-Filter

A wave-filter is a device for separating waves characterized by a difference in frequency. Thus, the wave-filter differentiates between certain states of motion and not between certain kinds of matter, as does the ordinary filter. One form of wave-filter which is well known is the color screen which passes only certain bands of light frequencies; diffraction gratings and Lippmann color photographs also filter light. Wave-filters might be constructed and employed for separating air waves, water waves, or waves in solids. This paper will consider only the filtering of electric waves; the same principles apply in every case, however.

In its usual form the electric wave-filter transmits currents of all frequencies lying within one or more specified ranges, and excludes currents of all other frequencies, but does not absorb the energy of these excluded frequencies. Hence, a combination of two or more wave-filters may be employed where it is desired to separate a broad band of frequencies, so that each of several receiving devices is sup-

[1] I take pleasure in acknowledging my indebtedness to Mr. O. J. Zobel for specific suggestions, and for the light thrown on the whole subject of wave-filters by his introduction of substitutions which change the propagation constant without changing the iterative impedance.

plied with its assigned narrower range of frequencies. Thus, for instance, with three wave-filters the band of frequencies necessary for ordinary telephony might be transmitted to one receiving device, all lower frequencies transmitted to a second device, and all higher frequencies transmitted to a third device—separation being made without serious loss of energy in any one of the three bands.

By means of wave-filters interference between different circuits or channels of communication in telephony and telegraphy, both wire and radio, can be reduced provided they operate at different frequencies. The method is furthermore applicable, at least theoretically, to the reduction of interference between power and communication circuits. The same is true of the simultaneous use of the ether, the earth return, and of expensive pieces of apparatus employed for several power or communication purposes. In all cases the principle involved is the same as that of confining the transmission in each circuit or channel to those frequencies which serve a useful purpose therein and excluding or suppressing the transmission of all other frequencies. In the future, as the utility of electrical applications becomes more widely and completely appreciated, there will be an imperative necessity for more and more completely superposing the varied applications of electricity; it will then be necessary, to avoid interference, to make the utmost use of every method of separating frequencies including balancing, tuning, and the use of wave-filters.

◆◆◆ ◆◆◆ ◆◆◆

Jansky Founds a New Science—Radio Astronomy

In this extract from a longer paper, Karl Jansky published his first account of the "static of unknown origin" that marked the birth of a new science. His conclusions were still very tentative, although his associate, George Southworth, later wrote that July and August of 1932 had seen the turning point that led Jansky to his astonishing suggestion that the static was coming from the Milky Way. Jansky proposed this interpretation of his data in 1933. Professional astronomers seemingly showed little interest in this phenomenon until after World War II when the newly developed radar technology proved adaptable in radio astronomy. Jansky's discovery seems to be a classic example of the role of serendipity in science. It has been speculated that Jansky would have been awarded a Nobel Prize but for his untimely death in 1950.

Karl G. Jansky (1905–1950) was born in Norman, Okla. where his father was Dean of Engineering at the University. The family moved to Madison, Wis. in 1908 and Jansky received a degree in physics at the University of Wisconsin in 1927. He taught at Wisconsin for a year before joining the Bell Telephone Laboratories in 1928 as a radio research engineer. He was assigned to study sources of interference to radio telephony and designed the short-wave recording system described in this paper. His observations were made at Holmdel beginning in 1931 and lasting until 1938. He was troubled by poor health during most of his career and died just as the importance of his work was gaining recognition. See George C. Southworth, "Early History of Radio Astronomy," *The Scientific Monthly*, vol. 82, pp. 55–66, 1956; C. M. Jansky, Jr., "The Beginnings of Radio Astronomy," *American Scientist*, vol. 45, pp. 5–12, 1957; Harold T. Friis, "Karl Jansky: His Career at Bell Telephone Laboratories," *Science*, vol. 149, pp. 841–842, 1965.

DIRECTIONAL STUDIES OF ATMOSPHERICS AT HIGH FREQUENCIES*

By

Karl G. Jansky

(Bell Telephone Laboratories, New York City)

Summary—*A system for recording the direction of arrival and intensity of static on short waves is described. The system consists of a rotating directional antenna array, a double detection receiver and an energy operated automatic recorder. The operation of the system is such that the output of the receiver is kept constant regardless of the intensity of the static.*

Data obtained with this system show the presence of three separate groups of static: Group 1, static from local thunderstorms; Group 2, static from distant thunderstorms, and Group 3, a steady hiss type static of unknown origin.

Curves are given showing the direction of arrival and intensity of static of the first group plotted against time of day and for several different thunderstorms.

Static of the second group was found to correspond to that on long waves in the direction of arrival and is heard only when the long wave static is very strong. The static of this group comes most of the time from directions lying between southeast and southwest as does the long wave static.

Curves are given showing the direction of arrival of static of group three plotted against time of day. The direction varies gradually throughout the day going almost completely around the compass in 24 hours. The evidence indicates that the source of this static is somehow associated with the sun.

Introduction

FOR some time various investigators have made records of one type or another of the direction of arrival of static on the long wavelengths. Watson Watt has made a comprehensive study of the direction of arrival of static in England. Others working under him have used apparatus similar to his in Australia and Africa. Captain Bureau has done considerable work on the study of static in France. In this country, L. W. Austin with E. B. Judson working with him has worked on the long-wave static problem. Harper and Dean, also of this country, have made a study of the direction of arrival of long-wave static in Maine. Very little work, however, has been done on the direction of arrival of short and very short-wave static with the exception of the series of observations made by Mr. Potter as described in his paper on short-wave noise.[1]

* Decimal classification: R114. Original manuscript received by the Institute, May 26, 1932. Presented at the meeting of the American Section of the U.R.S.I. at Washington, D. C., April 29, 1932.

[1] R. K. Potter, "High-frequency atmospheric noise," Proc. I.R.E., vol. 19, p. 1731; October, (1931).

Reprinted from *Proc. IRE*, vol. 20, pp. 1920–1921, 1930–1932 (extract of a longer paper), Dec. 1932.

Description of Apparatus

Since the middle of August, 1931, records have been taken at Holmdel, N. J., of the direction of arrival and the intensity of static on 14.6 meters. Fig. 1 shows a schematic diagram of the recording system. It consists of a rotating antenna array, a short-wave measuring set, and a Leeds and Northrup temperature recorder revamped to record field strengths.[2]

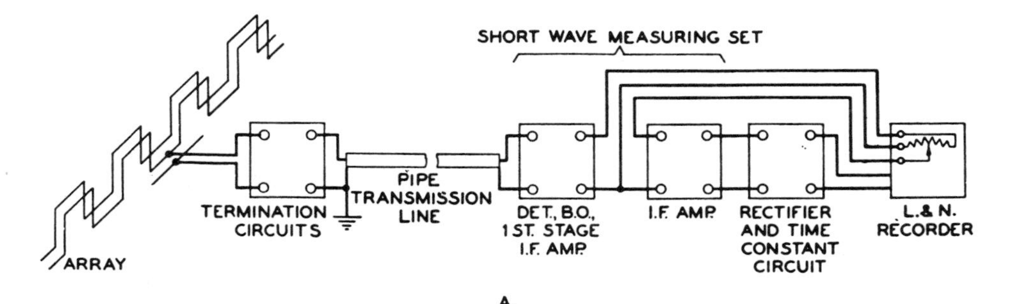

A.
SCHEMATIC DIAGRAM OF SHORT WAVE STATIC RECORDING SYSTEM

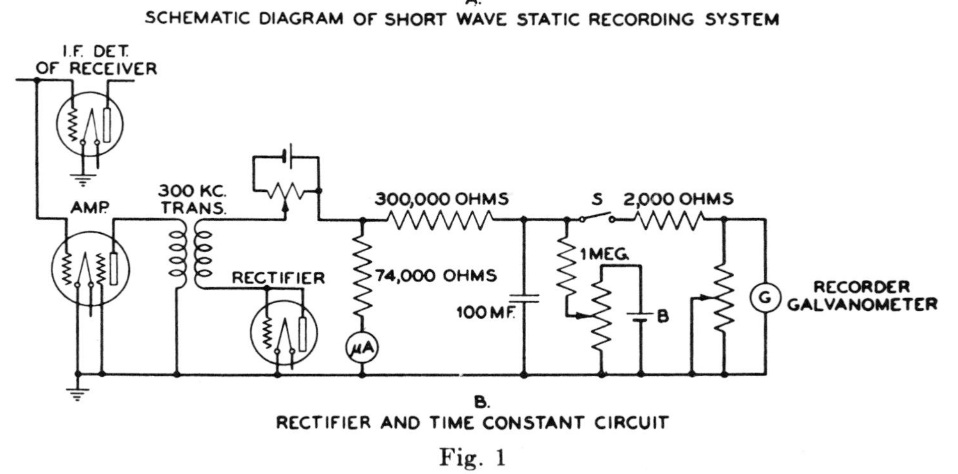

B.
RECTIFIER AND TIME CONSTANT CIRCUIT

Fig. 1

The rotating antenna, a photograph of which is shown in Fig. 2, is a Bruce type broadside receiving array[3] two wavelengths long made of 3/4-inch brass pipe. The array was designed to operate on a wavelength of 14.5 meters. As shown in the photograph it is mounted on a wooden framework which in turn is mounted on a set of four wheels and a central pivot. The structure is connected by a chain drive to a small synchronous motor geared down so that the array makes a complete rotation once every twenty minutes.

[2] A detailed description of the measuring set and recorder is given in a paper by W. W. Mutch, Proceedings, this issue, pp. 1914–1919.

[3] A. A. Oswald, "Transoceanic telephone service, short wave equipment," *Jour. A.I.E.E.*, vol. 49, p. 267; April, (1930).

Since this static is so weak that it cannot be recorded much of the time, the crash method[1] of measuring static as used by Potter could probably be used to great advantage to measure it.

The static of the third group is also very weak. It is, however, very steady, causing a hiss in the phones that can hardly be distinguished from the hiss caused by set noise. It is readily distinguished from ordinary static and probably does not originate in thunderstorm areas. The direction of arrival of this static changes gradually throughout the day going almost completely around the compass in twenty-four hours. It does not quite complete the cricuit, but in the middle of the night when it reaches the northwest, it begins to die out and at the same time static from the northeast begins to appear on the record. This new static then gradually shifts in direction throughout the day and dies out in the northwest also and the process is repeated day after day. Fig. 13 shows the direction of arrival of this static for three different days plotted against time of day. Curve 1 is for January 2, 1932, curve 2 is for January 26, 1932, and curve 3 is for February 24, 1932. Fig. 14 is a photograph of a section of one of the records.

This type of static was first definitely recognized only this last January. Previous to this time it had been considered merely as interference from some unmodulated carrier. Now, however, that it has been detected it is possible to go back to the old records and trace its position on them.

During the latter part of December and the first part of January the direction of arrival of this static coincided, for most of the daylight hours, with the direction of the sun from the receiver. (See curve 1, Fig. 13.) However, during January and February the direction has gradually shifted so that now (March 1) it precedes in time the direction of the sun by as much as an hour. It will be noticed that the curves 2 and 3 of Fig. 13 have shifted to the left.[9] Since December 21, the sun's rays have been getting more and more perpendicular at the receiving location causing sunrise to occur at the receiver earlier and earlier each day. It would appear that the change in the latitude of the sun is connected with the changing position of the curves. However, the data as yet only cover observations taken over a few months and more observations are necessary before any hard and fast deductions can be drawn.

The fact that the direction of arrival changes almost 360 degrees

[9] Since this paper was written the curve has shifted much further to the left. Now (May 25) it crosses south at 4:30 A.M.

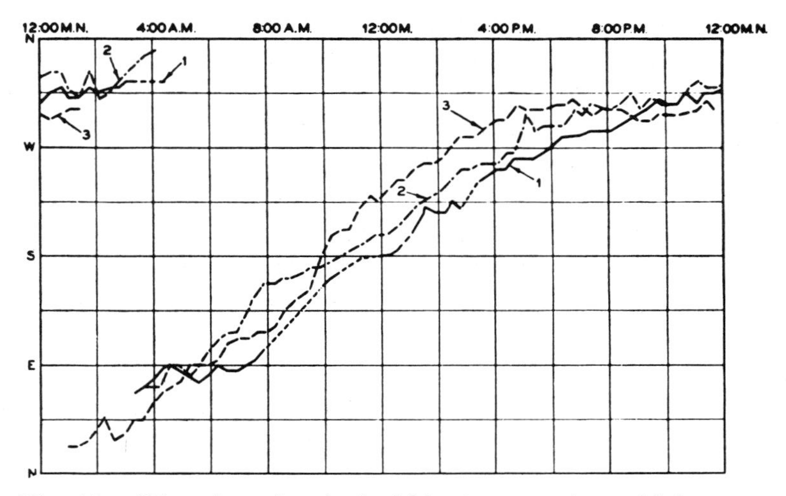

Fig. 13.—Direction of arrival of hiss type static on 14.6 meters.

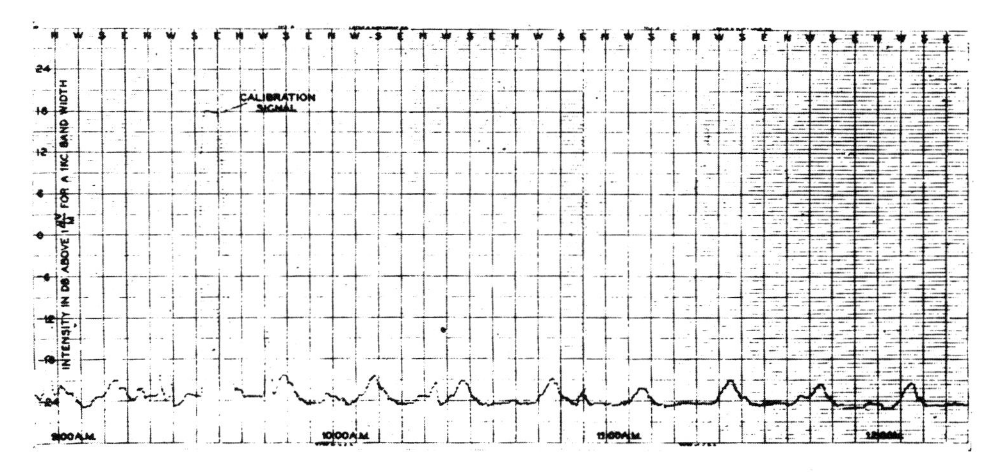

Fig. 14—Sample record of short-wave hiss type static. February 24, 1932.

during twenty-four hours and that the shift in the position of the curve observed during the three months over which data has been taken corresponds to the change in latitude of the sun affords definite indication that the source of this static is somehow associated with the position of the sun. It may be that the static comes directly from the sun or, more likely, it may come from the subsolar point on the earth.

The intensity of this static is never very high. At no time during the period that records have been taken has it exceeded 0.39 microvolts per meter for 1-kilocycle band width. As will be noticed from the record (Fig. 14), however, its presence during otherwise quiet periods is unmistakable.

The experiments which have been described in this paper were carried out at Holmdel, New Jersey. The writer wishes to acknowledge his indebtedness to Mr. Friis for his many helpful suggestions.

◆◆◆ ◆◆◆ ◆◆◆

Zworykin on the Kinescope

IN this paper a leading pioneer in television reported on a technological breakthrough in the design of television receivers. Through the use of an intensity-modulated cathode ray tube (the kenescope), Zworykin and associates had been able to eliminate the mechanical scanning feature of early television receivers, thus making possible an all-electronic receiver. Zworykin's iconoscope achieved a similar improvement in the television camera, although the system described here still employed a scanning disk in the transmitter. Other pioneers in television included Charles F. Jenkin, Philo T. Farnsworth, and E. F. W. Alexanderson. See George Shiers, "Early Schemes for Television," *IEEE Spectrum*, pp. 24–34, May 1970. Also see a paper by D. G. Fink on the NTSC Committees and a paper by E. W. Herold on the history of color television displays in the special issue of the PROCEEDINGS OF THE IEEE for September 1976.

Vladimir K. Zworykin (1889–) was born in Russia and graduated from the Institute of Technology in Petrograd in 1912. He continued his education at the College of France in Paris until 1914 and served as an Officer in the Signal Corps of the Russian Army during World War I. He came to the U.S. in 1919 and was hired by the Westinghouse Electric Company to do research on radio tubes and photocells. He applied for patents on the iconscope in 1923 and the kinescope in 1924. Zworykin demonstrated an all-electronic television system in 1929. He received a Ph.D. degree from the University of Pittsburgh in 1926. He joined RCA in 1929 and became Director of Electronic Research at RCA in 1946. He was awarded the IRE Medal of Honor in 1951 and the Edison Medal of the AIEE in 1952. See *Engineers of Distinction* (Engineers Joint Council, 1970), p. 457 and *Who's Who in Engineering*, p. 2105, 1964.

DESCRIPTION OF AN EXPERIMENTAL TELEVISION SYSTEM AND THE KINESCOPE*

By

V. K. Zworykin

(RCA Victor Company, Inc., Camden, New Jersey)

Summary—*A general description is given of an experimental television system using a cathode ray tube (kinescope) as the image reproducing element in the receiver. The fundamental considerations underlying the design and use of the kinescope for television are outlined. A description of the circuits associated with the kinescope and an explanation of the application to an experimental receiver are included.*

Introduction

THE experimental television system placed in operation by RCA Victor in New York late in 1931, and on which practical tests were made during the first half of 1932, was based on the use of a cathode ray tube as the image reproducing element in the receiver. This allowed the use of a system with 120 scanning lines and a frame

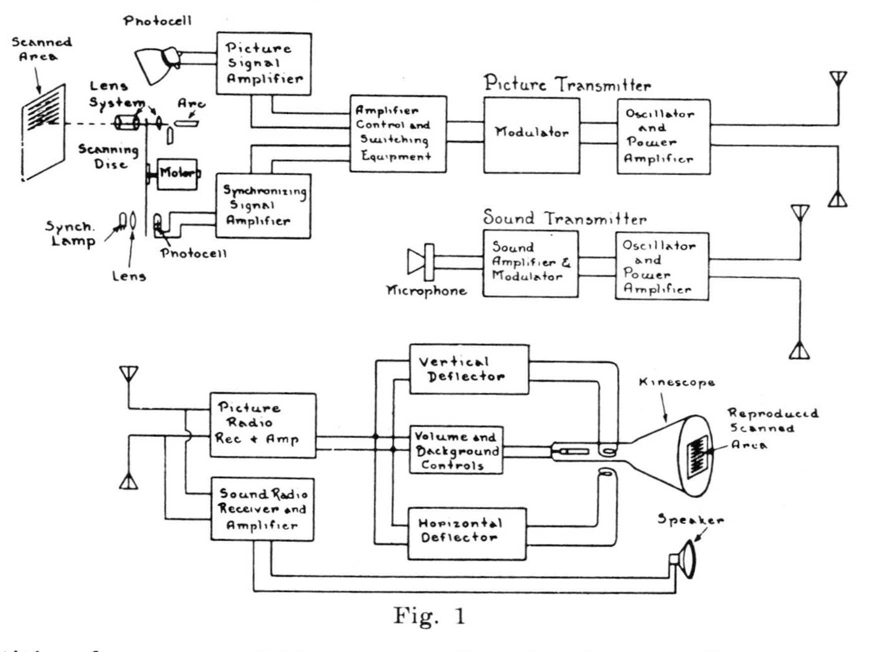

Fig. 1

repetition frequency of 24 per second with adequate illumination for the reproduced image.

A block diagram of the system is shown in Fig. 1, where the components and their location in the system are indicated. Naming the

* Decimal classification: R583. Original manuscript received by the Institute, July 10, 1933.

Reprinted from *Proc. IRE*, vol. 21, pp. 1655–1657, Dec. 1933.

units in order, we have for television from the studio: The photoelectric tubes, the flying spot scanning equipment, the picture signal and synchronizing signal amplifiers, the control and switching equipment, and the modulating and radio transmitter equipment. The units comprising the television receiver are: An antenna system feeding two radio receivers, one for sight, including the cathode ray unit with its associated horizontal and vertical deflecting equipment, and the other for sound, including the usual loud speaker.

The Kinescope

The name "kinescope" has been applied to the cathode ray tube used in the television receiver to distinguish it from ordinary cathode

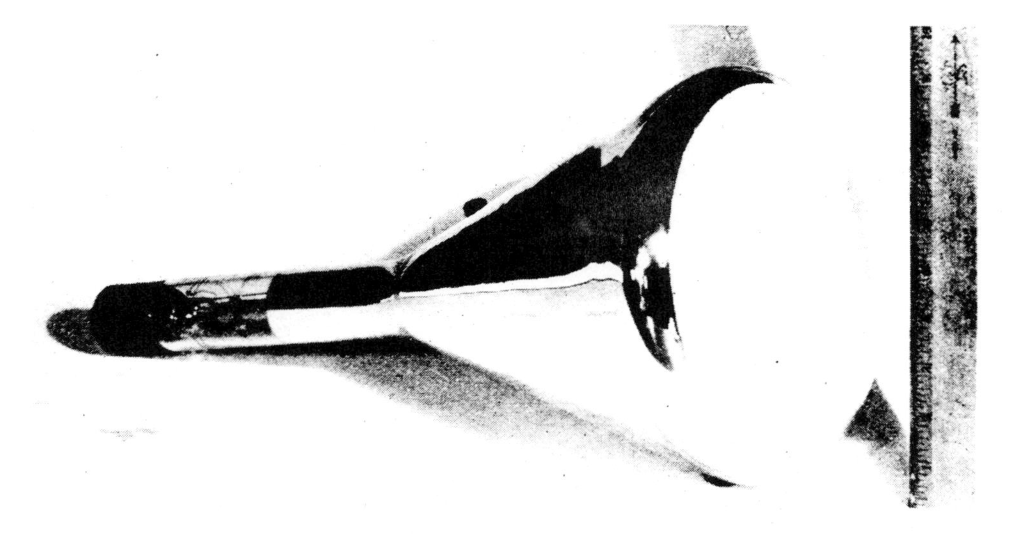

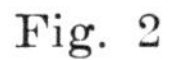

Fig. 2

ray oscilloscopes because it has several important points of difference; for instance, an added element to control the intensity of the beam. Fig. 2 gives the general appearance of the tube which has a diameter of 9 inches, permitting a reproduced image of approximately $5\frac{1}{2} \times 6\frac{1}{2}$ inches. Fig. 3 is a cross-section view of one of these tubes, showing the relative position of the electrodes, especially the cathode and its surrounding assembly, which is usually referred to as the "electron gun." The indirectly heated cathode, *C*, operates on alternating current. Its emitting area is located at the tip of the cathode sleeve and is formed by coating with the usual barium and strontium oxides. The control electrode, corresponding to the grid in the ordinary triode, is shown at *G*. It has an aperture, *O*, directly in front of the cathode emitting surface, and besides functioning as the control element it also serves as a shield for the cathode.

The first anode, A_1, has suitable apertures which limit the angle of the emerging electron beam. The electron gun is situated in the long, narrow neck attached to the large cone-shaped end of the kinescope, the inner surface of the cone being silvered or otherwise metallized, and serves as the second anode. The purpose of the second anode, A_2, is to accelerate the electrons emerging from the electron gun and to form the electrostatic field to focus them into a very small, thread-like beam. The first anode usually operates at a fraction of the second anode voltage.

The focusing is accomplished by an electrostatic field set up by potential differences applied between elements of the electron gun and the gun itself and the metallized portion of the neck of the kinescope.

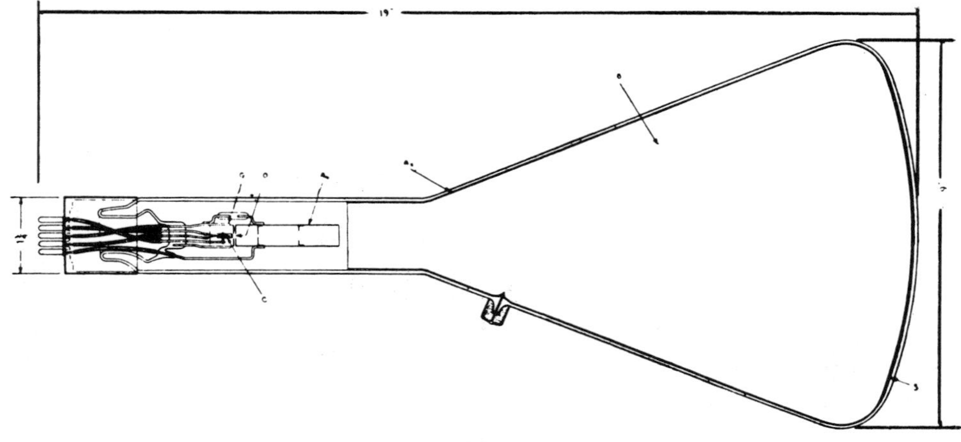

Fig. 3

The theory of the electrostatic focusing is described in detail in a recent paper by the writer.[1] The lines of force of the electrostatic field, between properly shaped electrodes, force the electrons of the beam to move toward the axis, overcoming the natural tendency of electrons to repel each other. This action is analogous to the focusing of light rays by means of optical lenses. The electrostatic lenses, however, have a peculiarity in that their index of refraction for electrons is not confined to the boundary between the optical media, as in optics, but varies throughout all the length of the electrostatic field. Also, it is almost impossible to produce a simple single electron lens; the field always forms a combination of positive and negative lenses. However, by proper arrangement of electrodes and potentials, it is always possible to produce a complex electrostatic lens which will be equivalent to either positive or negative optical lenses.

[1] V. K. Zworykin, *Jour. Frank. Inst.*, pp. 535–555, May, (1933).

Armstrong on Frequency Modulation

IN this paper a famous and controversial radio engineer challenged the long-held conviction that frequency modulation was inherently inferior to amplitude modulation in radio communications. As Armstrong points out in his historical introduction, John Carson and other theoreticians had reached conclusions "uniformly adverse to frequency modulation," yet he had now demonstrated that FM broadcast service could be "very greatly superior" to that provided by AM stations. In the deleted section of this paper, Armstrong noted that, although most authorities believed that signal-to-noise ratio was inversely proportional to bandwidth, "that principle is not of general application. In the present method an opposite rule applies."

Edwin H. Armstrong (1890–1954) was born in New York City and graduated in electrical engineering from Columbia University in 1913. The first of his four important inventions was the regenerative feedback triode oscillator in 1913. This invention became the focal point of a long dispute over priority with Lee de Forest that was finally resolved by the Supreme Court in 1934 (in de Forest's favor). (See the exchange between Armstrong and deForest in a discussion of a paper by Armstrong in the PROCEEDINGS OF THE IRE of 1915.) Armstrong was apparently the first to publish characteristic curves and oscillograms of vacuum tube amplifiers and oscillators in 1914. He served as a Signal Corps Officer during World War I and conceived his second major invention, the superheterodyne receiver, while serving in France. He became quite wealthy from the sale of patent rights to this and his third invention, the superregenerative circuit. After the War he returned to the Hartley Laboratory at Columbia and later became a Professor of Electrical Engineering at Columbia. His 4th innovation, FM, was patented in 1933. After failing to interest RCA in his FM system, Armstrong became its leading promoter and later rejected an offer from RCA of a million dollars for the rights to develop FM. See the *DSB*, vol. 1, pp. 287–288 and an obituary in *Electrical Engineering*, vol. 73, p. 283, 1954. Also see Bernard S. Finn, "Controversy and the Growth of the Electrical Art," *IEEE Spectrum*, pp. 52–56, January 1966.

A METHOD OF REDUCING DISTURBANCES IN RADIO SIGNALING BY A SYSTEM OF FREQUENCY MODULATION*

BY

EDWIN H. ARMSTRONG

(Department of Electrical Engineering, Columbia University, New York City)

Summary—*A new method of reducing the effects of all kinds of disturbances is described. The transmitting and receiving arrangements of the system, which makes use of frequency modulation, are shown in detail. The theory of the process by which noise reduction is obtained is discussed and an account is given of the practical realization of it in transmissions during the past year from the National Broadcasting Company's experimental station on the Empire State Building in New York City to Westhampton, Long Island, and Haddonfield, New Jersey. Finally, methods of multiplexing and the results obtained in these tests are reported.*

PART I

IT IS the purpose of this paper to describe some recent developments in the art of transmitting and receiving intelligence by the modulation of the frequency of the transmitted wave. It is the further purpose of the paper to describe a new method of reducing interference in radio signaling and to show how these developments may be utilized to produce a very great reduction in the effects of the various disturbances to which radio signaling is subject.

HISTORICAL

The subject of frequency modulation is a very old one. While there are some vague suggestions of an earlier date, it appears to have had its origin shortly after the invention of the Poulsen arc, when the inability to key the arc in accordance with the practice of the spark transmitter forced a new method of modulation into existence. The expedient of signaling (telegraphically) by altering the frequency of the transmitter and utilizing the selectivity of the receiver to separate the signaling wave from the idle wave led to the proposal to apply the principle to telephony. It was proposed to effect this at the transmitter by varying the wave length in accordance with the modulations of the voice, and the proposals ranged from the use of an electrostatic micro-

* Decimal classification: R400×R430. Original manuscript received by the Institute, January 15, 1936. Presented before New York meeting, November 6, 1935.

Reprinted from *Proc. IRE*, vol. 24, pp. 689–691, 739 (extract of a longer paper), May 1936.

phone associated with the oscillating circuit to the use of an inductance therein whose value could be controlled by some electromagnetic means. At the receiver it was proposed to cause the variations in frequency of the received wave to create amplitude variations by the use of mistuned receiving circuits so that as the incoming variable frequency current came closer into or receded farther from the resonant frequency of the receiver circuits, the amplitude of the currents therein would be correspondingly varied and so could be detected by the usual rectifying means. No practical success came from these proposals and amplitude modulation remained the accepted method of modulating the arc. The various arrangements which were tried will be found in the patent records of the times and subsequently in some of the leading textbooks.[1] The textbooks testify unanimously to the superiority of amplitude modulation.

Some time after the introduction of the vacuum tube oscillator attempts were again made to modulate the frequency and again the verdict of the art was rendered against the method. A new element however, had entered into the objective of the experiments. The quantitative relation between the width of the band of frequencies required in amplitude modulation and the frequency of the modulating current being now well understood, it was proposed to narrow this band by the use of frequency modulation in which the deviation of the frequency was to be held below some low limit; for example, a fraction of the highest frequency of the modulating current. By this means an economy in the use of the frequency spectrum was to be obtained. The fallacy of this was exposed by Carson[2] in 1922 in the first mathematical treatment of the problem, wherein it was shown that the width of the band required was at least double the value of the highest modulating frequency. The subject of frequency modulation seemed forever closed with Carson's final judgment, rendered after a thorough consideration of the matter, that "Consequently this method of modulation inherently distorts without any compensating advantages whatsoever."

Following Carson a number of years later the subject was again examined in a number of mathematical treatments by writers whose results concerning the width of the band which was required confirmed those arrived at by Carson, and whose conclusions, when any were expressed, were uniformly adverse to frequency modulation.

[1] Zenneck, "Lehrbuch der drahtlosen Telegraphy," (1912).
Eccles, "Wireless Telegraphy and Telephony," (1916).
Goldsmith, "Radio Telephony," (1918).

[2] "Notes on the theory of modulation," Proc. I.R.E., vol. 10, pp. 57–82; February, (1922).

In 1929 Roder[3] confirmed the results of Carson and commented adversely on the use of frequency modulation.

In 1930 van der Pol[4] treated the subject and reduced his results to an excellent form for use by the engineer. He drew no conclusions regarding the utility of the method.

In 1931, in a mathematical treatment of amplitude, phase, and frequency modulation, taking into account the practical aspect of the increase of efficiency at the transmitter which is possible when the frequency is modulated, Roder[5] concluded that the advantages gained over amplitude modulation at that point were lost in the receiver.

In 1932 Andrew[6] compared the effectiveness of receivers for frequency modulated signals with amplitude modulated ones and arrived at the conclusion that with the tuned circuit method of translating the variations in frequency into amplitude variations, the frequency modulated signal produced less than one tenth the power of one which was amplitude modulated.

While the consensus based on academic treatment of the problem is thus heavily against the use of frequency modulation it is to the field of practical application that one must go to realize the full extent of the difficulties peculiar to this type of signaling.

Problems Involved

The conditions which must be fulfilled to place a frequency modulation system upon a comparative basis with an amplitude modulated one are the following:

1. It is essential that the frequency deviation shall be about a fixed point. That is, during modulation there shall be a symmetrical change in frequency with respect to this point and over periods of time there shall be no drift from it.

2. The frequency deviation of the transmitted wave should be independent of the frequency of the modulating current and directly proportional to the amplitude of that current.

3. The receiving system must have such characteristics that it responds only to changes in frequency and that for the maximum change of frequency at the transmitter (full modulation) the selective characteristic of the system responsive to frequency changes shall be such that substantially complete modulation of the current therein will be produced.

[3] "Ueber Frequenzmodulation," *Telefunken-Zeitung* no. 53, p. 48, (1929).

[4] "Frequency modulation," Proc. I.R.E., vol. 18, pp. 1194–1205; July, (1930).

[5] "Amplitude, phase, and frequency modulation," Proc. I.R.E., vol. 19, pp. 2145–2176; December, (1931).

[6] "The reception of frequency modulated radio signals," Proc. I.R.E., vol. 20, pp. 835–840; May, (1932).

••• ••• •••

Acknowledgment

On account of the ramifications into which this development entered with the commencement of the field tests many men assisted in this work. To some reference has already been made.

I want to make further acknowledgment and express my indebtedness as follows:

To the staff of the National Broadcasting Company's station W2XDG for their help in the long series of field tests and the conducting of a large number of demonstrations, many of great complexity, without the occurrence of a single failure;

To Mr. Harry Sadenwater of the RCA Manufacturing Company for the facilities which made possible the Haddonfield tests and for his help with the signal-to-noise ratio measurements herein recorded;

To Mr. Wendell Carlson for the design of many of the transformers used in the modulating equipment;

To Mr. M. C. Batsel and Mr. O. B. Gunby of the RCA Manufacturing Company for the sound film records showing the comparison, at Haddonfield, of the Empire State transmission with that of the regular broadcast service furnished by the New York stations;

To Mr. C. R. Runyon for his development of the two-and-one-half-meter transmitters and for the solution of the many difficult problems involved in the application of these principles of modulation thereto;

To Mr. T. J. Styles and particularly to Mr. J. F. Shaughnessy, my assistants, whose help during the many years devoted to this research has been invaluable.

Conclusion

The conclusion is inescapable that it is technically possible to furnish a broadcast service over the primary areas of the stations of the present-day broadcast system which is very greatly superior to that now rendered by these stations. This superiority will increase as methods of dealing with ignition noise, either at its source or at the receiver, are improved.

Appendix

Since the work which has been reported in this paper on forty-one megacycles was completed attention has been paid to higher frequencies. On the occasion of the delivery of the paper a demonstration of transmission on 110 megacycles from Yonkers to the Engineering Societies Building in New York City was given by C. R. Runyon, who described over the circuit the transmitting apparatus which was used.

••• ••• •••

Espenschied and Strieby Discuss a Wide-Band Coaxial Communication System

THE authors of this extract from a longer paper described what would become a major innovation in the technology of wire communications. As they point out, it was not the coaxial structure that was new, but rather the combination with repeaters and other apparatus in such a way as to permit use of bandwidths of the order of millions of cycles per second. The new system would effectively have an isolated "frequency spectrum of its own and shuts it off from its surroundings so that it may be used again and again in different systems without interference." They foresaw that such a system would accommodate the wide bands required in television or 200 or more telephone channels. It was in fact to be used in the first trans-Atlantic telephone cable that was completed in 1956. (See Paper 47.)

Lloyd Espenschied (1889–) was born in St. Louis. The family moved to Brooklyn, N.Y. in 1901 where he became an avid radio enthusiast and operated an amateur station as early as 1904. He worked as a wireless telegraph operator for the United Wireless Telegraph Company aboard ships beginning in 1907 and also took courses at Pratt Institute. He became an assistant engineer with the Telefunken Wireless Telegraph Company in 1909 and joined A.T.&T. the following year, remaining with the Bell System until his retirement in 1954. He was a member of the Wireless Institute and a charter member of the IRE (see Paper 15). Espenschied received about 130 patents, including a 1924 patent on a railroad warning system that led in 1926 to his conception of a radio altimeter for airplanes. He was co-inventor of the wide-band coaxial system (with H. A. Affel) described in this paper. Espenschied had a strong avocational interest in electrical history and has published a number of papers on that subject. See the announcement of his retirement in the *Reporter*, vol. 3, p. 22, 1954 and *Who's Who in Engineering*, p. 735, 1959.

Maurice E. Strieby (1893–1975) was born in Colorado Springs and graduated from Colorado College in 1914. He received B.S. degrees from Harvard and M.I.T., both in 1916, and joined the N.Y. Telephone Company the same year. After serving in the Army Signal Corps, Strieby returned to the Department of Development and Research at A.T.&T. in 1919 and remained with the Bell System in various capacities until his retirement in 1956. After retirement from Bell, he served as a consultant on cable television systems. See *Who's Who in Engineering*, p. 2386, 1959 and an obituary in *IEEE Spectrum*, p. 109, April 1975.

Wide Band Transmission Over Coaxial Lines

By
LLOYD ESPENSCHIED
FELLOW A.I.E.E.

M. E. STRIEBY
MEMBER A.I.E.E.

Both of Bell Tel. Labs., Inc., New York, N. Y.

In this paper systems are described whereby frequency band widths of the order of 1,000 kc or more may be transmitted for long distances over coaxial lines and utilized for purposes of multiplex telephony or television. A coaxial line is a metal tube surrounding a central conductor and separated from it by insulating supports.

IT appears from recent development work that under some conditions it will be economically advantageous to make use of considerably wider frequency ranges for telephone and telegraph transmission than are now in use[1,2] or than are covered in the recent paper on carrier in cable.[3] Furthermore, the possibilities of television have come into active consideration and it is realized that a band of the order of 1,000 kc or more in width would be essential for television of reasonably high definition if that art were to come into practical use.[4,5]

This paper describes certain apparatus and structures which have been developed to employ such wide frequency ranges The future commercial application of these systems will depend upon a great many factors, including the demand for additional large groups of communication facilities or of facilities for television. Their practical introduction is, therefore, not immediately contemplated and, in any event, will necessarily be a very gradual process.

Types of High Frequency Circuits

The existing types of wire circuits can be worked to frequencies of tens of thousands of cycles, as is evidenced by the widespread application of carrier systems to the open wire telephone plant and by the development of carrier systems for telephone cable circuits.[2,3] Further development may lead to the operation of still higher frequencies over the existing types of plant. However, for protection against external interference these circuits rely upon balance, and as the frequency band is widened, it becomes more and more difficult to maintain a sufficiently high degree of balance. The balance requirements may be made less severe by using an individual shield around each circuit, and with sufficient shielding balance may be entirely dispensed with.

A form of circuit which differs from existing types in that it is unbalanced (one of the conductors being grounded), is the coaxial or concentric circuit. This consists essentially of an outer conducting tube which envelops a centrally disposed conductor. The high-frequency transmission circuit is formed between the inner surface of the outer conductor and the outer surface of the inner conductor. Unduly large losses at the higher frequencies are prevented by the nature of the construction, the inner conductor being so supported within the tube that the intervening dielectric is largely gaseous, the separation between the conductors being substantial, and the outer conductor presenting a relatively large surface. By virtue of skin effect, the outer tube serves both as a conductor and a shield, the desired currents concentrating on its inner surface and the undesired interfering currents on the outer surface. Thus, the same skin effect which increases the losses within the conductors provides the shielding which protects the transmission path from outside influences, this protection being more effective the higher the frequency.

The system which this paper outlines has been based primarily upon the use of the coaxial line. The repeater and terminal apparatus described, however, are generally applicable to any type of line, either balanced or unbalanced, which is capable of transmitting the frequency range desired.

The Coaxial System

A general picture of the type of wide band transmission system which is to be discussed is briefly as follows: A coaxial line about 0.5 in. in outside diameter may be used to transmit a frequency band of about 1,000 kc, with repeaters capable of handling the entire band placed at intervals of about 10 miles. Terminal apparatus may be provided which will enable this band either to be subdivided into more than 200 telephone circuits or to be used *en bloc* for television.

Such a wide band system is illustrated in Fig. 1. It is shown to comprise several portions, namely, the line sections, the repeaters, and the terminal apparatus, the latter being indicated in this case as for multiplex telephony. Two-way operation is secured by using 2 lines, one for either direction. It would be possible, however, to divide the frequency band and use different parts for transmission in opposite directions.

A form of flexible line which has been found con-

Full text of a paper recommended for publication by the A.I.E.E. committee on communication, and scheduled for discussion at the A.I.E.E. winter convention, New York, N. Y., Jan. 22–25, 1935. Manuscript submitted Aug. 21, 1934; released for publication Sept. 4, 1934. ***Not published in pamphlet form.***

1. For all numbered references, see list at end of paper.

Reprinted from *Elec. Eng.*, vol. 53, pp. 1371–1374 (extract of a longer paper), Oct. 1934.

venient in the experimental work is illustrated in Fig. 2 and will be described more fully subsequently. Such a coaxial line can be constructed to have the same degree of mechanical flexibility as the familiar telephone cable. While this line has a relatively high loss at high frequencies, the transmission path is particularly well adapted to the frequent application of repeaters, since the shielding permits the transmission currents to fall to low power levels at the high frequencies.

Of no little importance also is the fact that the attenuation-frequency characteristic is smooth throughout the entire band and obeys a simple law of change with temperature. (This is due to the fact that the dielectric is largely gaseous and that insulation material of good dielectric properties is employed.) This smooth relation is extremely helpful in the provision of means in the repeaters for automatically compensating for the variations which occur in the line attenuation with changes of temperature. This type of system is featured by large transmission losses which are offset by large amplification, and it is necessary that the 2 effects match each other accurately at all times throughout the frequency range.

It will be evident that the repeater is of outstanding importance in this type of system, for it must not only transmit the wide band of frequencies with a transmission characteristic inverse to that of the line, with automatic regulation to care for temperature changes, but must also have sufficient freedom from inter-modulation effects to permit the use of large numbers of repeaters in tandem without objectionable interference. Fortunately, recent advances in repeater technique have made this result possible, as will be appreciated from the subsequent description.

An interesting characteristic of this type of system is the way in which the width of the transmitted band is controlled by the repeater spacing and line size, as follows:

1. For a given size of conductor and given length of line, the band width increases nearly as the square of the number of the repeater points. Thus, for a coaxial circuit with about 0.3-in. inner diameter of outer conductor, a 20-mile repeater spacing will enable a band up to about 250 kc to be transmitted, a 10-mile spacing will increase the band to about 1,000 kc, and a 5-mile spacing to about 4,000 kc.

2. For a given repeater spacing, the band width increases approximately as the square of the conductor diameter. Thus, whereas a tube of 0.3-in. inner diameter will transmit a band of about 1,000 kc, 0.6-in. diameter will transmit about 4,000 kc, while a diameter corresponding to a full sized telephone cable might transmit something of the order of 50,000 kc, depending upon the dielectric employed and upon the ability to provide suitable repeaters.

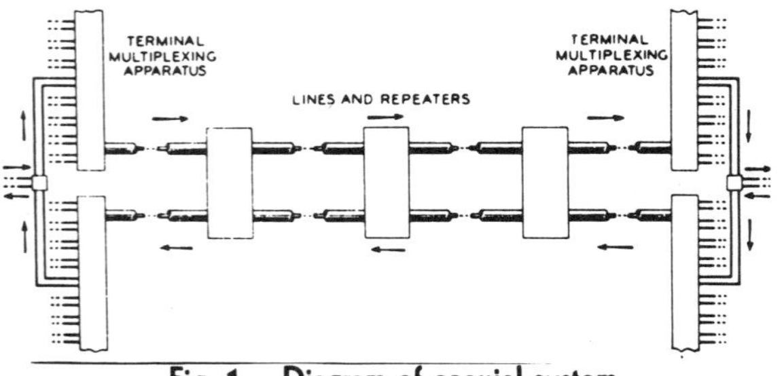

Fig. 1. Diagram of coaxial system

Earlier Work

It may be of interest to note that as a structure, the coaxial form of line is old—in fact, classical. During the latter half of the last century it was the object of theoretical study, in respect to skin effect and other problems, by some of the most prominent mathematical physicists of the time. Reference to some of this work is made in a paper by Schelkunoff, dealing with the theory of the coaxial circuit.[6]

On the practical side, it is found on looking back over the art that the coaxial form of line structure has been used in 2 rather widely different applications: First, as a long line for the transmission of low frequencies, examples of which are usage for submarine cables[7,8] and for power distribution purposes, and second as a short-distance high-frequency line serving as an antenna lead-in.[9,10]

The coaxial conductor system herein described may be regarded as an extension of these earlier applications to the long distance transmission of a very wide range of frequencies suitable for multiplex telephony or television.[11] Although dealing with radio frequencies, this system represents an extreme departure from radio systems in that a relatively broad band of waves is transmitted, this band being confined to a small physical channel which is shielded from outside disturbances. The system, in effect, comprehends a frequency spectrum of its own and

Fig. 2. Small flexible coaxial structure

shuts it off from its surroundings so that it may be used again and again in different systems without interference.

This new type of facility has not yet been commercially applied, and is, in fact, still in the development stage. Sufficient progress has already been made, however, to give reasonable assurance of a satisfactory solution of the technical problems involved. This progress is outlined below under 3 general headings: (1) the coaxial line and its transmission properties, (2) the wide band repeaters, and (3) the terminal apparatus.

The Coaxial Line

An Experimental Verification

One of the first steps taken in the present development was in the nature of an experimental check of the coaxial conductor line, designed primarily to determine whether the desirable transmission properties which had been disclosed by a theoretical study could be fully realized under practical conditions. For this purpose a length of coaxial structure capable of accurate computation was installed near Phoenixville, Pa. The diagram of Fig. 3 shows a sketch of the structure used and gives its dimensions. It comprised a copper tube of 2.5-in. outside diameter, within which was mounted a smaller tube which, in turn, contained a small copper wire. Two coaxial circuits of different sizes were thus made available, one between the outer and the inner tubes, and the other between the inner tube and the central wire. The installation comprised 2 2,600-ft lengths of this structure.

The diameters of these coaxial conductors were so chosen as to obtain for each of the 2 transmission paths a diameter ratio which approximates the optimum value, as discussed later. The conductors were separated by small insulators of isolantite. The rigid construction and the substantial clearances between conductors made it possible to space the insulators at fairly wide intervals, so that the dielectric between conductors was almost entirely air.

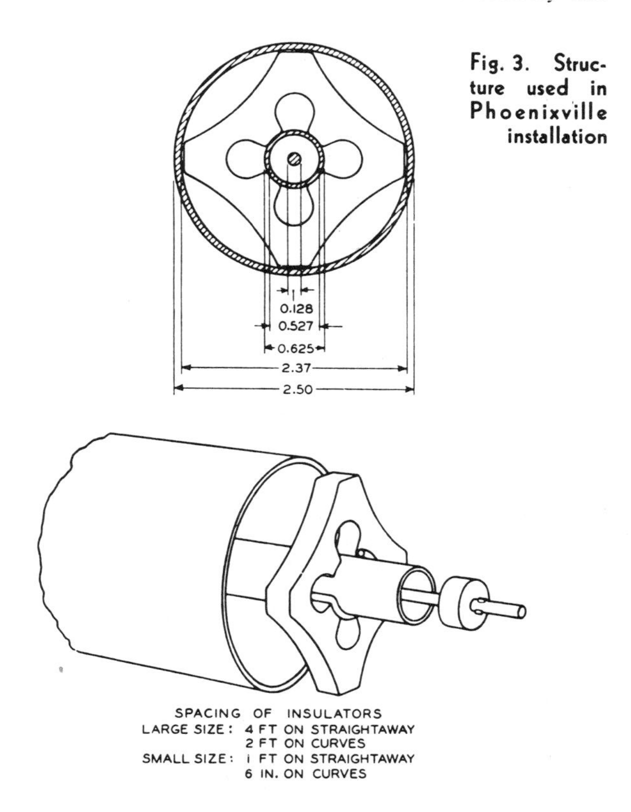

Fig. 3. Structure used in Phoenixville installation

The outer conductor was made gastight, and the structure was dried out by circulating dry nitrogen gas through it. The 2 triple conductor lines were suspended on wooden fixtures and the ends brought into a test house, as shown in Fig. 4.

The attenuation was measured by different methods over the frequency range from about 100 kc to 10,000 kc. Investigation showed that the departures from ideal construction occasioned by the joints, the lack of perfect concentricity, etc., had remarkably little effect upon the attenuation. In order to study the effect of eccentricity upon the

Fig. 4. Phoenixville installation showing conductors entering test house

attenuation, tests were made in which this effect was much exaggerated, and the results substantiated theoretical predictions. The impedance of the circuits was measured over the same range as the attenuation. A few measurements on a short length were made at frequencies as high as 20,000 kc.

Measurements were secured of the shielding effect of the outer conductor of the coaxial circuit up to frequencies in the order of 100 to 150 kc, the results agreeing closely with the theoretical values. Above these frequencies, even with interfering sources much more powerful than would be encountered in practice, the induced currents dropped below the level of the noise due to thermal agitation of electricity in the conductors (resistance noise) and could not be measured.

The preliminary tests at Phoenixville, therefore, demonstrated that a practical coaxial circuit, with its inevitable mechanical departures from the ideal, showed transmission properties substantially in agreement with the theoretical predictions.

Small Flexible Structures

Development work on wide band amplifiers, as discussed later, indicated the practicability of employing repeaters at fairly close intervals. This pointed toward the desirability of using sizes of coaxial circuit somewhat smaller than the smaller of those used in the preliminary experiments, and having correspondingly greater attenuation. Furthermore, it was desired to secure flexible structures which could be handled on reels after the fashion of ordinary cable. Accordingly, several types of flexible construction, ranging in outer diameter from about 0.3 in. to 0.6 in., have been experimented with. Structures were desired which would be mechanically and electrically satisfactory, and which could be manufactured economically, preferably with a continuous process of fabrication.

One type of small flexible structure which has been developed is shown in Fig. 2. The outer conductor is formed of overlapping copper strips held in place with a binding of iron or brass tape. The insulation consists of a cotton string wound spirally around the inner conductor, which is a solid copper wire. This structure has been made in several sizes of the order of 0.5-in. diameter or less. When it is to be used as an individual cable, the outer conductor is surrounded by a lead sheath, as shown, to prevent the

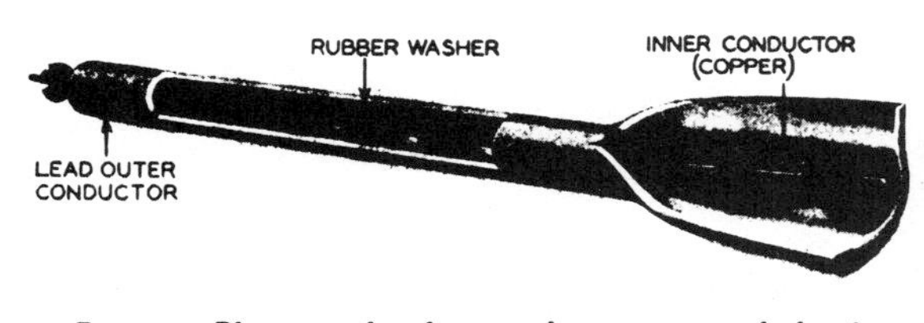

Fig. 5. Photograph of coaxial structure with lead outer conductor and rubber disk insulators

entrance of moisture. One or more of the copper tape structures without individual lead sheath may be placed with balanced pairs inside a common cable sheath.

Another flexible structure is shown in Fig. 5. The outer conductor in this case is a lead sheath which directly surrounds the inner conductor with its insulation. Since lead is a poorer conductor than copper, it is necessary to use a somewhat larger diameter with this construction in order to obtain the same transmission efficiency. Lead is also inferior to copper in its shielding properties and to obtain the same degree of shielding the lead tube of Fig. 5 must be made correspondingly thicker than is necessary for a copper tube.

The insulation used in the structure shown in Fig. 5 consists of hard rubber disks spaced at intervals along the inner wire. Cotton string or rubber disk insulation may be used with either form of outer tube. The hard rubber gives somewhat lower attenuation, particularly at the higher frequencies.

Another simple form of structure employs commercial copper tubing into which the inner wire with its insulation is pulled. Although this form does not lend itself readily to a continuous manufacturing process, it may be advantageous in some cases.

Transmission Characteristics—Attenuation

At high frequencies the attenuation of the coaxial circuit is given closely by the well-known formula:

$$\alpha = \frac{R}{2}\sqrt{\frac{C}{L}} + \frac{G}{2}\sqrt{\frac{L}{C}} \quad (1)$$

where R, L, C and G are the 4 so-called "primary constants" of the line, namely, the resistance, inductance, capacitance, and conductance per unit of length. The first term of eq 1 represents the losses in the conductors, while the second term represents those in the dielectric.

When the dielectric losses are small, the attenuation of a coaxial circuit increases, due to skin effect in the conductors, about in accordance with the square root of the frequency. With a fixed diameter ratio, the attenuation varies inversely with the diameter of the circuit. By combining these relations there are obtained the laws of variation of band width in accordance with the repeater spacing and the size of circuit, as stated previously.

The attenuation-frequency characteristic of the flexible structure illustrated in Fig. 2, with about 0.3-in. diameter, is given in Fig. 6. The figure shows also that the conductance loss due to the insulation is a small part of the total.

It is interesting to compare the curves of the transmission characteristics of the coaxial circuit with those of other types of circuits. The diagram of Fig. 7 shows the high-frequency attenuation of 2 sizes of coaxial circuit using copper tube outer conductors, of 0.3-in. and 2 5-in. inner diameter, and that of cable and open wire pairs in the same frequency range.

Effect of Eccentricity

The small effect of lack of perfect coaxiality upon the attenuation of a coaxial circuit is illustrated by the curve of Fig. 8, which shows attenuation ratios plotted as a function of eccentricity, assuming a fixed ratio of conductor diameters and substantially air insulation.

Fig 6. Attenuation of small flexible coaxial structure shown in Fig. 2

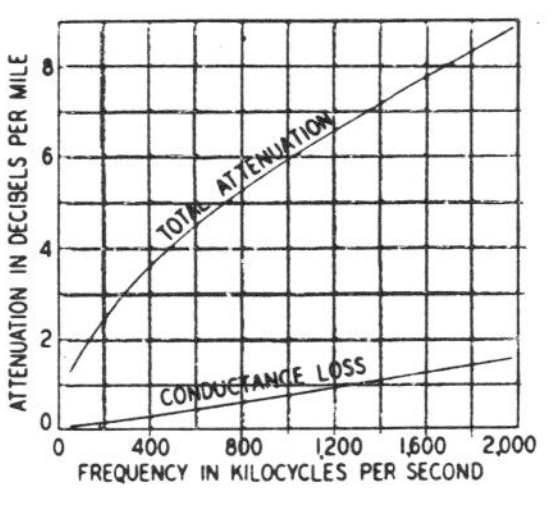

Temperature Coefficient

With a coaxial circuit, as with other types of circuits, the temperature coefficient of resistance de creases as the frequency is increased, due to the action of skin effect, and approaches a value of $1/2$ the d-c temperature coefficient.[12] Thus, for conductors of copper the a-c coefficient at high frequencies is approximately 0.002 per degree Centigrade. When the dielectric losses are small, the temperature coefficient of attenuation at high frequencies is the same as the temperature coefficient of resistance.

Diameter Ratio

An interesting condition exists with regard to the relative sizes of the 2 conductors. For a given size of outer conductor there is a unique ratio of inner diameter of outer conductor to outer diameter of inner conductor which gives a minimum attenuation. At high frequencies, this optimum ratio of diameters (or radii) is practically independent of frequency. When the conductivity is the same for both conductors, and either the dielectric losses are small or the insulation is distributed so that the dielectric flux follows radial lines, the value of the optimum diameter ratio is approximately 3.6. When the outer and inner conductors do not have the same conductivity, the optimum diameter ratio differs from this value. For a lead outer conductor and copper inner conductor, for example, the ratio should be about 5.3.

◆◆◆ ◆◆◆ ◆◆◆

Southworth on Waveguides

THE broad-band coaxial cable system described in Paper 43 was complemented and extended to far higher frequencies by the waveguide system that also dates from the 1930's. In this paper, George Southworth, largely responsible for stimulating interest in waveguides within the Bell System, reported on some of his experiments. The work of Southworth, W. L. Barrow, and others laid the groundwork for the microwave radar systems of World War II. (See Paper 45.) For an interesting summary of microwave history, see Southworth's paper "Survey and History of the Progress of the Microwave Arts," PROCEEDINGS OF THE IRE, vol. 50, pp. 1199–1206, 1962.

George C. Southworth (1890–1972) was born in Pennsylvania and graduated from Grove City College in 1914. He later received a master's degree from Grove City in 1916 and a Ph.D. degree from Yale in 1923. Southworth worked at the National Bureau of Standards during World War I and then taught at Yale from 1918 until 1923 when he joined the Bell System as an editor of the *Bell System Technical Journal*. He was a charter member of the Bell Telephone Laboratories and remained with the Labs until his retirement in 1955. Although Southworth's interest in high-frequency transmission dated from his work at Yale, he did not pursue waveguide development actively until 1931. By 1933 he demonstrated transmission through a 5-inch hollow metal pipe that was 875 feet in length. He reported his work in several lecture demonstrations at IRE meetings until the approach of World War II brought on a policy of secrecy. See Southworth's autobiography *Forty Years of Radio Research* (N.Y., 1962). Also see Orrin E. Dunlap, Jr., *Radio's 100 Men of Science* (N.Y., 1944), pp. 247–250 and an obituary in *IEEE Spectrum*, p. 113, December 1972.

SOME FUNDAMENTAL EXPERIMENTS WITH WAVE GUIDES*

By

G. C. Southworth

(Bell Telephone Laboratories, Inc., New York City)

Summary—*This paper describes in considerable detail the early apparatus and methods used to verify some of the fundamental properties of wave guides. Cylinders of water about ten inches in diameter and four feet long were used as the experimental guides. At one end of these guides were launched waves having frequencies of roughly 150 megacycles. The lengths of the standing waves so produced gave the velocity of propagation. Other experiments utilizing a probe made up of short pickup wires attached to a crystal detector and meter enabled the configuration of the lines of force in the wave front to be determined. This was done for each of four types of waves. For certain types the properties had already been predicted mathematically. For others the properties were determined experimentally in advance of analysis. In both cases analysis and experiment proved to be in good agreement.*

PREVIOUS papers† have set forth the mathematical theory and a few of the experimental results concerning the transmission of electromagnetic waves through hollow metal tubes, through pipes filled with insulation, and also through cylinders of dielectric material. The purpose of this paper is to describe in considerable detail the methods by which the verifying experiments were made and to add data which it was not feasible to present previously. The methods described are believed to be of interest not only for the way in which they are able to verify the theory but also because they seem to be pointing toward a new technique of electrical measurements.

The previous papers have described four of the many forms of waves that may be propagated through guides. These waves may be distinguished by such propagating characteristics as velocity and characteristic impedance and also by the various configurations of the lines of force which go to make up their wave fronts. Also there is a critical or limiting frequency below which power is not transmitted. For convenience of reference certain of the standard configurations have been reproduced as Fig. 1 below for the particular case of a hollow metal pipe of circular cross section. These four configurations have been designated rather arbitrarily as E_0, E_1, H_0, and H_1 waves, re-

* Decimal classification: R110. Original manuscript received by the Institute, March 9, 1937. Several illustrations in this paper are published through the courtesy of the Engineering Institute of Canada and appear in its *Engineering Journal*, vol. 20, pp. 186–190; April, (1937).

† See references (6) and (7) of Bibliography.

Reprinted from *Proc. IRE*, vol. 25, pp. 807–814, July 1937.

spectively. Somewhat similar waves are also possible in wires of dielectric material where no metal shield is present but in that case the lines of force which are shown above as terminating on the conducting boundary extend into the surrounding space and close as loops.

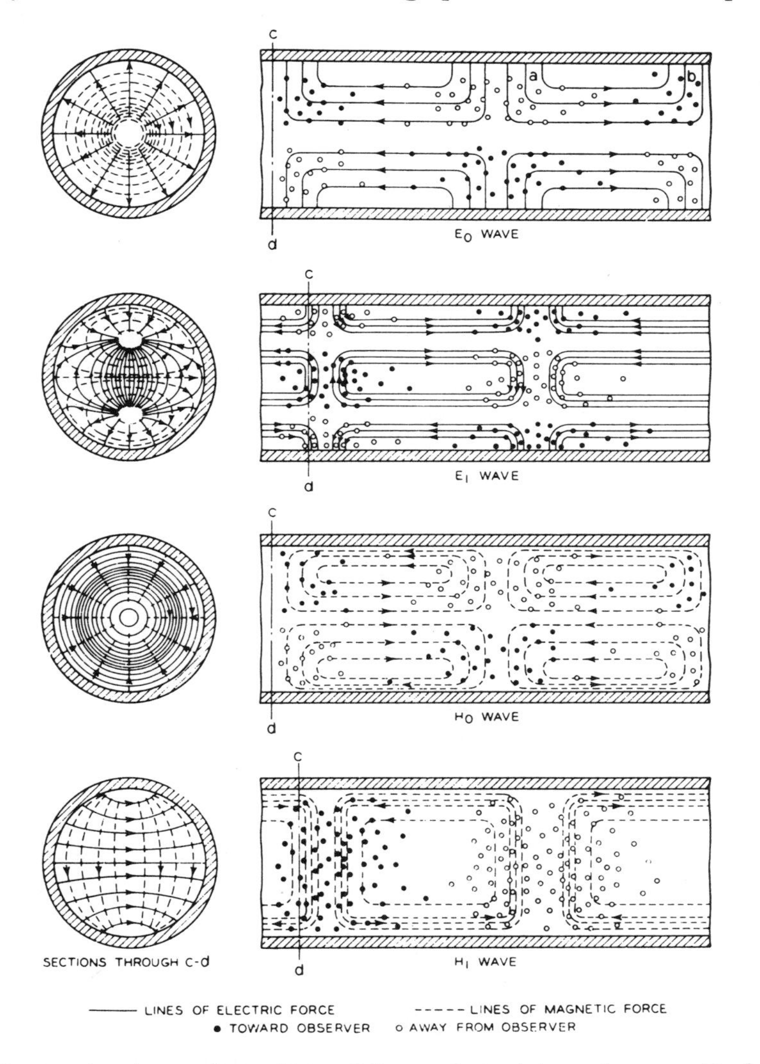

Fig. 1—Approximate configuration of lines of electric and magnetic force in a typical wave guide. Small solid circles represent lines of force directed toward the observer. Propagation is assumed to be directed to the right and away from the observer.

Obvious questions that might be asked at this time are: (1) Do the nice geometrical configurations shown in Fig. 1 actually exist in practice? (2) How are these waves launched and received? (3) How can their velocities be measured? The following paragraphs are directed at these answers.

Experimental Verification

The verification of the above-mentioned properties of waves and their guides has involved two rather different ranges of frequencies. In the earlier work which is the particular object of this paper, frequencies between 100 and 400 megacycles were used. In this case the waves were propagated through cylinders of water which served as the experimental guides. More recently it has been feasible to use air-core guides. However, this has called for much higher frequencies.

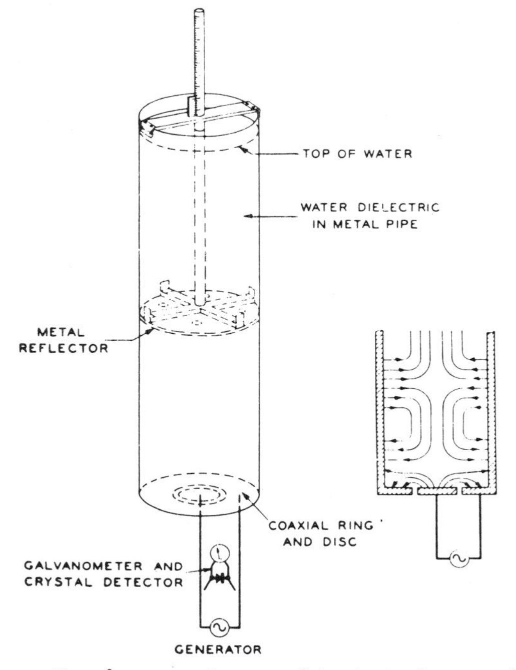

Fig. 2—Schematic of apparatus used to test theory of propagation through wave guides.

The latter are of the order of 1000 megacycles ($\lambda = 30$ centimeters) and 4000 megacycles ($\lambda = 7.5$ centimeters). The results obtained by the two methods are, however, very similar so that the more recent work can be said to be largely confirmatory.

Apparatus

The apparatus used in this preliminary study is shown in very simple schematic form in Fig. 2 and in greater detail in Fig. 3. As already explained the dielectric guide under observation consisted of a cylinder of water about four feet long. In the course of the experiments four such columns were used. Two were enclosed in copper tubes ten inches and six inches in diameter, respectively, and two were within bakelite tubes of these same diameters. The bakelite was considered sufficiently thin and its dielectric constant so low compared with water that the whole could be regarded as a cylinder of water only.

Waves generated by the source (1) (Fig. 3) were set up in the water column (2) through the intermediary of tuned parallel wires

Fig. 3—Arrangement of oscillator, Lecher wires, and water column used in testing the theory of waves in dielectric guides.

(3) to which the water column was only loosely coupled. The kind and the degree of coupling could be varied through the use of different arrangements at the base of the column. These were sometimes referred to as "pads." In some cases this coupling was analogous to a mutual capacitance and in others it might be regarded as conductive or

inductive. The shape of the conductors on the pad determined the type of wave produced. When the E_0 type of wave was desired the pad took the form of metal disk surrounded by a metal ring. The two Lecher wires were then connected to the center and outside rings of the pad, respectively. If other types were wanted pads were used made up of other electrode arrangements as illustrated in Fig. 4 below.

The essential components of the assembled apparatus are shown in detail in Fig. 3. A shielded platform about six feet square separated in a general way the water column from the rest of the apparatus. The removable pads which afforded the coupling made close contact with the shield, thereby minimizing the wave power that might reach the dielectric from spurious sources. This isolation was acceptable but by no means perfect as the whole room was sometimes found to contain standing waves.

The parallel wires were capable of adjustment for length by means of a 4-×4-inch brass bridge (5) which was movable along the wooden supporting frame. A permanently mounted scale enabled the position of the bridge to be read. A 0- to 200-microampere meter (6), to which was attached a sensitive crystal detector, was loosely coupled to the parallel wire system. This showed conditions of resonance in the parallel wires. This parallel wire system proved to be a moderately accurate wavemeter but it was not relied upon for this purpose. Its main function was that of impressing on the end of the dielectric guide waves of a definite configuration. This was considered to be more feasible than to couple the oscillator direct to the column.

The source (1) was mounted on the end of a wooden arm (7) hinged at two points. This permitted a wide range of couplings to the parallel wires. In some instances the separation was but a few inches but in other cases it was two or more feet. This arm was also capable of vertical displacement in order to meet the wide range of conditions due to wave length changes. The power leads to the oscillator were cabled and laced to the arm. There was little trouble from standing waves on these leads.

An independent wavemeter system (8) also made up of parallel wires was arranged vertically on a second hinged arm (9). This would be brought in from the right when it was desired to measure wave length. In this case the wires were much larger and the mechanical arrangement was such that more accurate bridge settings could be made than for the case of the parallel wires described above. This arrangement permitted measurements when the oscillator was under normal load. Its sensitivity was such that it led to no appreciable reaction on the source.

Standing waves were detected in the water column by three rather different methods. Substantially the same results were obtained by all. In one case the parallel wire system was adjusted for resonance, as shown by the meter (6), with no water in the column. Water was then admitted slowly through a hole in the bottom of the column. As the level of the water rose the deflection of the meter varied periodically through rather wide limits, indicating points of resonance in the column with a corresponding absorption of power from the parallel wire system. When the coupling between the column and the parallel wire circuit could be made small, the reaction between the two was also small except at points close to where absorption took place. This

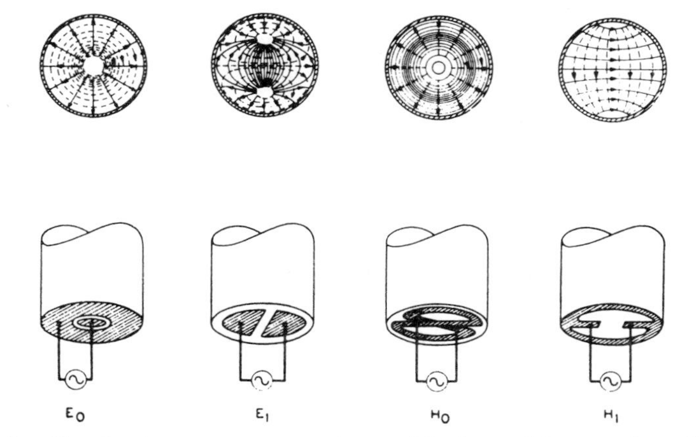

Fig. 4—Conductor arrangements used in launching various forms of guided waves.

method is very similar, of course, to the scheme that was at one time generally used in radio measurements where resonance in a secondary circuit was indicated by absorption in a primary. In this experiment the distance between successive absorption points was taken as one half of the wave length as measured in the medium.

In another method of detecting standing waves, the column was nearly filled with water and the parallel wires again adjusted for resonance. A circular disk of sheet copper which contacted the sides of the column was lowered into the water. This showed the same reactionary effects on the primary circuit as that of the changing water level. In the third method which was used in connection with E_0 waves, the copper reflector was divided into three coaxial rings between which "waterproofed" crystal detectors were connected. This is shown in Fig. 5(b) below. Connecting wires from the detectors were brought out to an external microammeter which indicated maximum conditions. The second of these three methods was most generally used.

Fig. 4 shows various coupling pads that were used. Each of the four pads was designed respectively for one of the types of waves shown in Fig. 1.

Fig. 5 shows some of the reflectors. That shown in (a) was most generally used. It would of course reflect any of the principal types

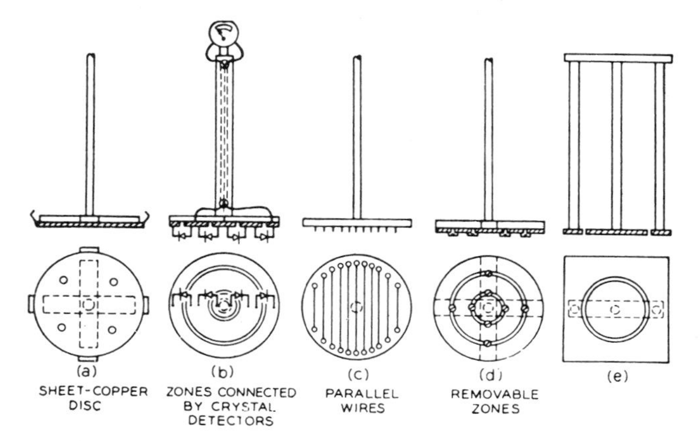

Fig. 5—Various arrangements used to reflect guided waves.

of waves. That shown in (c) was used to investigate the state of polarization in H_1 waves. That shown in (d) had removable sections and indicated what zones of the cross section of the guide contained the greatest E_0 wave power. That shown in (e) was used to determine the magnitude of fields outside of the guide itself. Two other reflectors not shown made up respectively of radial wires and wires arranged as coaxial circles were used in investigating the orientations in E_0 and H_0 waves.

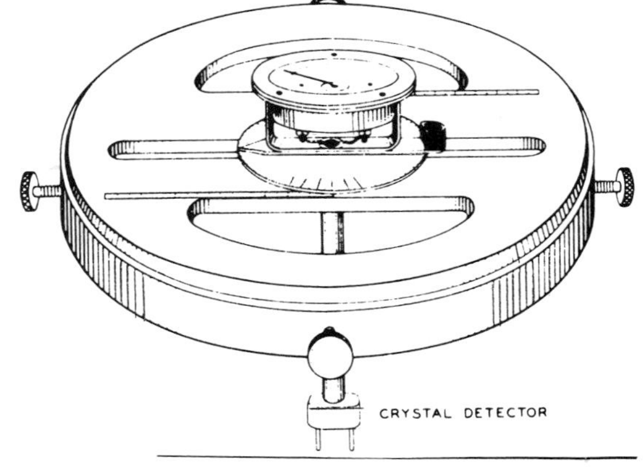

Fig. 6—Crystal detector and meter used as a probe in verifying the arrangement of the lines of electric force in a wave guide.

A meter and detector combination similar in principle to that used on the wavemeter served as a probe in determining the direction and intensity of the field in and about the dielectric wire. Fig. 6 shows the details of both the probe and its bakelite mounting. The latter was clamped to the top of the water column and adjusted so that the probe wires just reached the water. The probe could rotate about its axis as well as be moved laterally along the slot shown. The part in which the slot is located was also free to rotate with respect to the column so that the probe could be carried over the entire surface of the water. The two rotating parts were laid off in degrees and the slot carried a centimeter scale. Typical data obtained by this method are shown in Figs. 11, 12, and 13 below.

Experimental Results

The primary purpose of these experiments was to establish the existence of guided electric waves, particularly for the case of a guide surrounded by a conductor. This simple objective was readily accomplished as standing waves of a kind were produced with little or no difficulty. However, the waves first found turned out to be rather complicated usually indicating a mixture of two or more types each traveling with different velocities. It then became necessary to disentangle the component parts.

It is not feasible to discuss here all of the difficulties that were encountered except to say that, unless certain precautions are used in the design of the launching mechanism, many spurious waves may be set up in a guide. In most cases, however, these spurious waves may now be identified as a mixture of two or more of the standard types. The various coupling pads shown in Fig. 4 were evolved from some of these early experiments with a view to a fairly pure wave of the desired kind. It was in connection with this problem that the simple probe meter shown as Fig. 6 was first used. This disentangling process led to forms of waves not at first appreciated. However, mathematical analysis was soon able to account for these and other waves as well.

The method for determining velocity of propagation using the principle of standing waves was not essentially different than that commonly involved for the velocity of sound waves in air columns or for electric waves on Lecher wires or in free space. It is based on the simple fact that two oppositely directed waves of the same length and approximately the same amplitude give rise to a series of successive maxima and minima that appear to be at rest. Measurements of the distances between alternate maxima (or minima) give wave length λ_g as observed in the guide. This together with a knowledge of the frequency f permits velocity in the guide to be calculated by the simple relation $V_g = f\lambda_g$.

Schneider on Radar

THIS is an extract from one of the first papers published by an engineering journal after World War II that undertook to give a comprehensive summary of one of the major technical achievements made during the War. According to an authoritative account published the same year, more than three billion dollars had been expended on radar by July of 1945 and about 150 different radar systems had been developed. (See James Phinney Baxter, *Scientists Against Time*, p. 142.) As Schneider notes, a substantial portion of the technical manpower had been assigned to the crash program to develop radar and ECM systems during the war, resulting in an acceleration of "possibly ten times the normal peacetime rate." The deleted section of the paper outlined the features of the basic radar system and also discussed the design features of such devices as the cavity resonator, magnetron, klystron, and microwave antenna. He predicted that the major peace-time use of radar would be for navigation and control of aircraft. See Henry C. Guerlac, "The Radio Background of Radar," *Journal of the Franklin Institute*, vol. 250, pp. 285–308, 1950. Also see Gordon D. Friedlander, "World War II: Electronics and the U.S. Navy," *IEEE Spectrum*, pp. 56–70, November 1967.

Edwin G. Schneider received a Ph.D. degree from Harvard in 1934 and was involved in radar development work with the Radiation Laboratory at M.I.T. during World War II. See the biographical note in the PROCEEDINGS OF THE IRE, vol. 34, p. 578, 1946.

Radar*

EDWIN G. SCHNEIDER†

I. Introduction

SINCE a fairly large percentage of the effort of physicists and electrical engineers in this country and in England was expended on radar development during the war, techniques in electronics have advanced at possibly ten times the normal peacetime rate. The purpose of this paper is to give a brief survey of the wartime developments in electronics and to show how these were used in a few radar sets. Because the applications of radar to civilian activities will probably be of minor importance when compared with the sum total of other electronic applications, this survey makes no attempt to give a complete description of any particular radar set. Those readers who are interested in a detailed description of a radar installation will find complete information on the SCR-268 in the literature.[1,2] On the other hand, for those readers who are interested in further details of the electronic devices mentioned here, a set of treatises is being prepared. These are to be published by McGraw-Hill during the next year, the royalties to go to the Federal Government.

II. Basic Principles of Radar Electronic Techniques of Radar

1. General

Basically, radar determines the existence of an object by observation of reflected radio energy. In the case of tail-warning radar and one type of air-search equipment mounted on submarines, the prime use is to give warning of the presence of aircraft in the neighborhood, the exact location being a secondary consideration. However, most radar sets are designed to give reasonably accurate information on the position of the objects which are reflecting energy. Bearing or azimuth with respect to north is normally determined by rotating a directional antenna to the position of maximum echo, a process similar to locating an object by use of a searchlight beam. Similarly, one method of measuring angle of elevation is to tilt the antenna to center the beam on the target. In contrast with the optical case, in which range must be measured by difficult triangulation methods, radar simply measures the time for a short pulse of radio energy to travel to the target and return. Although a great deal of work has gone into development of timing circuits, range measurement is operationally an extremely simple process compared with optical measurements. Furthermore, the range accuracy with properly designed equipment is such that secondary bench marks may be located for topographical surveys by measuring the range to primary points. An interesting example of the discovery of a map error by radar occurred in northern Italy. During the operations, a blind-bombing system called shoran was very successfully used for bombing pin-point targets such as bridges. This system consists of an airborne radar which triggers two coded radio beacons at known points on the ground. The signals from the two beacons, which merely serve to give signals which cannot be mistaken for other objects, are displayed on a cathode-ray tube in the aircraft. By suitable timing circuits the range to each of these beacons is measured with an accuracy of a few yards, and the position of the plane is located by triangulation, using the beacons as reference points. With one beacon set up in Corsica and the other in Italy, the positions being accurately located from maps, a bombing mission was run. Strike photographs showed a miss of almost a thousand yards. Since this was about 50 to 100 times the error expected, a recheck of the calculation of target position was made but no error was discovered. The suggestion was then made that the position of Corsica as shown on the map might be wrong. The problem was therefore worked backward to correct the position of Corsica. A correction of very nearly 1000 yards in the map position of Corsica was used on the next bombing run with strike results indistinguishable from optical bombing. That this result was not fortuitous was borne out by several months of successful bombing using the corrected position for Corsica.

Since the pulse energy travels the distance to the target twice, once on the way out and again on the return trip, the time required to receive an echo from an object a mile away is that necessary for a radio wave to travel two miles. With radio waves traveling at a speed of 186,000 miles a second, 10.7 microseconds elapse between the time the pulse is sent out and the time it is received from a target one mile away. It is apparent, therefore, that time measurements must be made in terms of millionths of a second. Furthermore, the transmitted pulse may be a large fraction of a mile or even several miles in length. For this reason the range must be measured by determining the time interval between the start of the transmitted pulse and the beginning of the received pulse. Methods for measuring range and displaying target positions will be discussed in detail later.

2. Maximum Range of a Radar Set

Before discussing details of actual equipments, let us examine some of the factors which determine whether

* Decimal classification: R537. Original manuscript received by the Institute, February 20, 1946; revised manuscript received, May 7, 1946.

† Formerly, Radiation Laboratory, Massachusetts Institute of Technology, Cambridge, Mass.; now, Stevens Institute of Technology, Hoboken, N. J. This paper is based on work done for the Office of Scientific Research and Development under Contract OEMsr-262 with the Massachusetts Institute of Technology.

[1] "The SCR-268 radar," *Electronics*, vol. 18, pp. 100–109; September, 1945.

[2] "The SCR-584 radar," *Electronics*, vol. 18, pp. 104–109; November, 1945.

Reprinted from *Proc. IRE*, vol. 34, pp. 528–530 (extract of a longer paper), Aug. 1946.

or not a measurable reflection of energy can be obtained from an object. To be detectable, the received power P_r must obviously be greater than the minimum power sensitivity of the receiver. The problem of determing the maximum range of the set is, therefore, one of calculating the range at which the received power is just measurable. The relationship between the major factors which affect the received power may be derived as follows.

The power per unit area at the "target" will be proportional to the instantaneous peak power transmitted by the radar P_T, and will be proportional to the gain G of the transmitting antenna, where the gain in a given direction represents the increase in power resulting from focusing of the radio energy by the antenna as compared with that which would have been present if the energy had been radiated equally in all directions. For example, placing a dipole at the focus of a searchlight type of reflector may result in an increase of a factor of 1000 in the amount of power sent in one direction. The reflector has, therefore, resulted in a gain of 1000 in the direction of the focused beam as compared with the dipole, while in other directions the amount of power has been correspondingly greatly reduced. Unless otherwise specified, the gain is understood to mean the ratio of power increase in the strongest part of the beam. A so-called "isotropic radiator," one which radiates equally in all directions, is usually considered as the source for measuring gain. On this basis a dipole in its strongest directions of radiation has a gain of 3/2; hence, the antenna in the above example has a gain $G=1500$. Using the above facts, we may write for a target a distance R from an isotropic radiator

$$\text{power per unit area} = P_T/4\pi R^2 \tag{1}$$

and for an actual antenna with a gain G,

$$\text{power per unit area} = P_T G/4\pi R^2. \tag{2}$$

This energy is intercepted by the target and scattered in many directions, a portion being reflected to the receiving antenna which may or may not be the same as the transmitting antenna. Since most targets are of a very complicated form, it is customary to use an effective scattering cross section for the target which is defined as the cross section of a perfectly reflecting sphere which would give the same strength of reflection in the direction of the radar as does the actual object. (The scattering by such a sphere can be shown to be isotropic.) If this cross section is denoted by S, the total power received by the equivalent sphere is

$$\frac{P_T G S}{4\pi R^2}.$$

This power will be reradiated equally in all directions, and the amount intercepted by the receiving antenna and thence transmitted to the receiver will be

$$P_r = \frac{P_T G S}{4\pi R^2}\,\frac{A}{4\pi R^2} \tag{3}$$

where A is the area of the receiving antenna.

A rather complex calculation shows that the gain of an antenna is given by

$$G = K\frac{A_T}{\lambda^2} \tag{4}$$

where A_T is the area of the antenna, λ is the wavelength, and K is a constant which varies with the type and efficiency of the antenna but is, in general, between 3 and 10 in magnitude.

Since most radars use the same antenna for transmitting and receiving, we may take the antenna areas in (3) and (4) as being equal for simplicity of discussion. Substituting (4) in (3) and rearranging we obtain

$$R = \sqrt[4]{\frac{P_T S K A^2}{16\pi^2 P_r \lambda^2}} = \sqrt[4]{\frac{C P_T S A^2}{P_r \lambda^2}} \tag{5}$$

where

$$C = K/16\pi^2.$$

If P_r is considered to be the minimum energy detectable by the receiver, (5) shows that the maximum range at which an object can be detected is proportional to the fourth root of the power, the square root of the antenna area, and the square root of the frequency since frequency bears a reciprocal relationship to the wavelength. We see, therefore, that changing the power output or the receiver sensitivity does not alter the maximum range as rapidly as a change in antenna area or wavelength. Before making a sample substitution in (5), let us see what values may be considered reasonable for the above factors.

If a more accurate value for K is not known from experimentally measured gains on antennas of the type to be used, a value of 5 will serve as a fairly representative number for range calculations.

The effective cross-section area of the target is usually not under the control of the radar set designer. For detection of aircraft, for example, the set must be built with due regard to the effective reflection from standard planes. In a few cases, such as the use of radar to locate a rubber life raft carrying a metal reflector, the size of the reflecting surface can be varied within reasonable limits to obtain the desired performance. For objects such as aircraft or ships, the effective cross section S will obviously vary with the size, being larger for the larger craft. On the other hand, because of the complicated shape of such objects the reflection will vary considerably with the orientation. This is especially true where there are flat surfaces which may give an intense directional reflection for a very specific orientation. The flash of the sun from the windshield of an approaching car when it is in just the right position is a common example of such a highly directional reflection. This variation in reflection results in a radar signal which varies considerably in strength as a plane is deflected slightly from its course by rough air or by small changes in steering by the pilot. In the case of a ship, the rolling caused by the waves may easily be seen in the changing signal strength. The net result is that the maximum range of a set is not a definite quantity. It not only varies with the size of the object, but will vary from one trial to another with the same object. For example, a plane flying away from a radar set will show a continuous but fluctuating signal when near the set. As it gets farther away the signal will at times drop below the minimum detectable value, causing the tracking to become intermittent. As the distance increases still more, the intervals during which the signal is lost become longer because only the peaks are seen. Eventually even the strongest peaks are too weak to see. The range at which the last signal is seen will, therefore, depend on just how the plane happened to be bounced around by rough air. For radar-set design, it is customary to consider the range of a set for a particular target as that distance for which the signal is visible half of the time. It has also been customary to use the effective cross section of a medium-sized plane such as an A-20 or B-25 for computing ranges on planes. Fighters will then give ranges about 25 per cent less, small planes such as "Cubs" will be about 35 per cent less, while large bombers and transports may be followed 20 to 30 per cent farther, depending on the size. Experimentally, it has been determined that a value of $S=2\pi^2$ will give a reasonable prediction of the range on a medium bomber when used in (5) for frequencies between 100 and 9000 megacycles. This value is only approximate, and the π^2 is used not for theoretical reasons but in order to simplify the numerical calculation.

In general, the value of S for a given object is very small when the dimensions are small compared with the wavelength. As the wavelength is decreased, S rises rapidly as half-wave resonance is approached, and then falls very gradually with further decrease in wavelength. The exact shape of the curve is dependent on the shape of the object, the maximum at resonance being much more pronounced for a properly oriented wire than for a wide object.

The maximum transmitter power available will vary with the frequency chosen for the set. For example, at extremely short wavelengths, say in the neighborhood of 1 centimeter, the antenna dipoles and transmission lines may impose the eventual limit to the power which can be handled, whereas at 1 meter the transmitter tube may always remain the limiting factor. At present, in all frequency regions used for radar, the power output of existing transmitters is the limiting factor. For wavelengths longer than about 7 centimeters, present techniques will permit a peak pulse-power output greater than 1,000,000 watts, provided the spacing between pulses is great enough to keep the average power within the ratings of the tubes. In the neighborhood of 3 centimeters, peak powers of about 250 kilowatts are available, while near 1 centimeter only about 50-kilowatt output may be realized with present tubes. Where weight and size of the transmitter are not considerations and it is desired to obtain as great a range as possible, a radar system should be designed to use the maximum power available with existing techniques at the chosen frequency. Frequently weight and size are important and may set the limit on the practical power for the application. For example, in an aircraft it may not be feasible to carry a sufficiently large transmitter unit nor to supply the power to operate a set with an output of 1 megawatt. In such cases, consideration of the proper allotment of weight between the transmitter and antenna must be made to obtain the optimum-range performance.

In computing the maximum range to be expected from a radar, P_r is the minimum power which the receiver can detect. For the frequency range commonly used in radar, receivers can detect signals as small as 1 to 0.1 micromicrowatt. Since the cost in weight, size, and complexity of a good receiver is at most only slightly greater than that of a poor one, the receiver is always designed to have the greatest possible sensitivity.

Antenna sizes are usually determined by considerations of mechanical design, space, or weight. It is at present considered impractical to build antennas much over 30 feet wide for transportable field equipment, although larger sizes might be quite practical for permanent fixed installations. For ship or airborne sets and for many ground purposes even this size of antenna is impractical. Furthermore, in many applications the maximum possible range is unnecessary; hence, a small antenna is adequate. The aspects of antenna design will be more fully covered in later sections on antennas and radar systems.

The factors controlling the choice of frequency are wave-propagation effects and antenna beamwidth. These will be discussed in detail in later chapters. Let us now use (5) to compute the range of a large microwave set which might have the following values:

$$A = 200 \text{ square feet}$$
$$\lambda = 4 \text{ inches} = 1/3 \text{ foot}$$
$$P_T = 750 \text{ kilowatts} = 750{,}000 \text{ watts}$$
$$P_r = 5\times10^{-13} \text{ watt}$$

$$R = \sqrt[4]{\frac{5}{16\pi^2}\,\frac{7.5\times10^5\times2\pi^2}{5\times10^{-13}}\times\frac{(200)^2}{(1/3)^2}}$$
$$= 760{,}000 \text{ feet} = 144 \text{ miles}$$

or roughly 150 miles on a medium bomber.

◆◆◆ ◆◆◆ ◆◆◆

Shannon's Revolutionary Theory of Communication

THE year 1948 was marked by two major events in electrical history as Claude Shannon published this classic paper and the first public announcement of the transistor was made. Shannon's paper launched the modern era in information theory by defining "the fundamental problem of communication" and giving a generalized analysis of a communication system. In a discussion of the impact of Shannon's theory, J. R. Pierce compared it to Carnot's theory of thermodynamics in that it "divided the world into two parts—that which was possible and that which was not. . . ingenious people no longer invented coding or modulation schemes that were analogous to perpetual motion. But they were offered the possibility of efficient error-free transmission over noisy channels." For an informed discussion of information theory in the "pre-Shannon world" and the background and impact of Shannon's paper, see John R. Pierce, "The Early Days of Information Theory," IEEE TRANSACTIONS ON INFORMATION THEORY, pp. 3–8, January 1973. Subsequent issues of this TRANSACTIONS contain papers by David Slepian (March 1973) and Andrew Viterbi (May 1973) on the history of information theory since publication of Shannon's paper.

Claude E. Shannon (1916–) was born in Michigan and graduated from the University of Michigan in 1936. He received a master's degree in electrical engineering from M.I.T. and a Ph.D. degree in mathematics in 1940, also from M.I.T. He joined the staff of the Bell Telephone Laboratories in 1941 and remained affiliated with the Labs until his retirement in 1972. He became Visiting Professor of Communications Science at M.I.T. in 1956 and has continued at M.I.T. since leaving the Bell System. See *Engineers of Distinction*, p. 277, 1973 and *Who's Who in Engineering*, p. 1684, 1964. The announcement of his retirement from BTL appeared in *IEEE Spectrum*, p. 110, November 1972.

A Mathematical Theory of Communication

By C. E. SHANNON

Introduction

THE recent development of various methods of modulation such as PCM and PPM which exchange bandwidth for signal-to-noise ratio has intensified the interest in a general theory of communication. A basis for such a theory is contained in the important papers of Nyquist[1] and Hartley[2] on this subject. In the present paper we will extend the theory to include a number of new factors, in particular the effect of noise in the channel, and the savings possible due to the statistical structure of the original message and due to the nature of the final destination of the information.

The fundamental problem of communication is that of reproducing at one point either exactly or approximately a message selected at another point. Frequently the messages have *meaning*; that is they refer to or are correlated according to some system with certain physical or conceptual entities. These semantic aspects of communication are irrelevant to the engineering problem. The significant aspect is that the actual message is one *selected from a set* of possible messages. The system must be designed to operate for each possible selection, not just the one which will actually be chosen since this is unknown at the time of design.

If the number of messages in the set is finite then this number or any monotonic function of this number can be regarded as a measure of the information produced when one message is chosen from the set, all choices being equally likely. As was pointed out by Hartley the most natural choice is the logarithmic function. Although this definition must be generalized considerably when we consider the influence of the statistics of the message and when we have a continuous range of messages, we will in all cases use an essentially logarithmic measure.

The logarithmic measure is more convenient for various reasons:

1. It is practically more useful. Parameters of engineering importance

[1] Nyquist, H., "Certain Factors Affecting Telegraph Speed," *Bell System Technical Journal*, April 1924, p. 324; "Certain Topics in Telegraph Transmission Theory," *A. I. E. E. Trans.*, v. 47, April 1928, p. 617.

[2] Hartley, R. V. L., "Transmission of Information," *Bell System Technical Journal*, July 1928, p. 535.

such as time, bandwidth, number of relays, etc., tend to vary linearly with the logarithm of the number of possibilities. For example, adding one relay to a group doubles the number of possible states of the relays. It adds 1 to the base 2 logarithm of this number. Doubling the time roughly squares the number of possible messages, or doubles the logarithm, etc.

2. It is nearer to our intuitive feeling as to the proper measure. This is closely related to (1) since we intuitively measure entities by linear comparison with common standards. One feels, for example, that two punched cards should have twice the capacity of one for information storage, and two identical channels twice the capacity of one for transmitting information.

3. It is mathematically more suitable. Many of the limiting operations are simple in terms of the logarithm but would require clumsy restatement in terms of the number of possibilities.

The choice of a logarithmic base corresponds to the choice of a unit for measuring information. If the base 2 is used the resulting units may be called binary digits, or more briefly *bits*, a word suggested by J. W. Tukey. A device with two stable positions, such as a relay or a flip-flop circuit, can store one bit of information. N such devices can store N bits, since the total number of possible states is 2^N and $\log_2 2^N = N$. If the base 10 is used the units may be called decimal digits. Since

$$\begin{aligned} \log_2 M &= \log_{10} M/\log_{10} 2 \\ &= 3.32 \log_{10} M, \end{aligned}$$

a decimal digit is about $3\frac{1}{3}$ bits. A digit wheel on a desk computing machine has ten stable positions and therefore has a storage capacity of one decimal digit. In analytical work where integration and differentiation are involved the base e is sometimes useful. The resulting units of information will be called natural units. Change from the base a to base b merely requires multiplication by $\log_b a$.

By a communication system we will mean a system of the type indicated schematically in Fig. 1. It consists of essentially five parts:

1. An *information source* which produces a message or sequence of messages to be communicated to the receiving terminal. The message may be of various types: e.g. (a) A sequence of letters as in a telegraph or teletype system; (b) A single function of time $f(t)$ as in radio or telephony; (c) A function of time and other variables as in black and white television—here the message may be thought of as a function $f(x, y, t)$ of two space coordinates and time, the light intensity at point (x, y) and time t on a pickup tube plate; (d) Two or more functions of time, say $f(t)$, $g(t)$, $h(t)$—this is the case in "three dimensional" sound transmission or if the system is intended to service several individual channels in multiplex; (e) Several functions of

several variables—in color television the message consists of three functions $f(x, y, t)$, $g(x, y, t)$, $h(x, y, t)$ defined in a three-dimensional continuum—we may also think of these three functions as components of a vector field defined in the region—similarly, several black and white television sources would produce "messages" consisting of a number of functions of three variables; (f) Various combinations also occur, for example in television with an associated audio channel.

2. A *transmitter* which operates on the message in some way to produce a signal suitable for transmission over the channel. In telephony this operation consists merely of changing sound pressure into a proportional electrical current. In telegraphy we have an encoding operation which produces a sequence of dots, dashes and spaces on the channel corresponding to the message. In a multiplex PCM system the different speech functions must be sampled, compressed, quantized and encoded, and finally interleaved

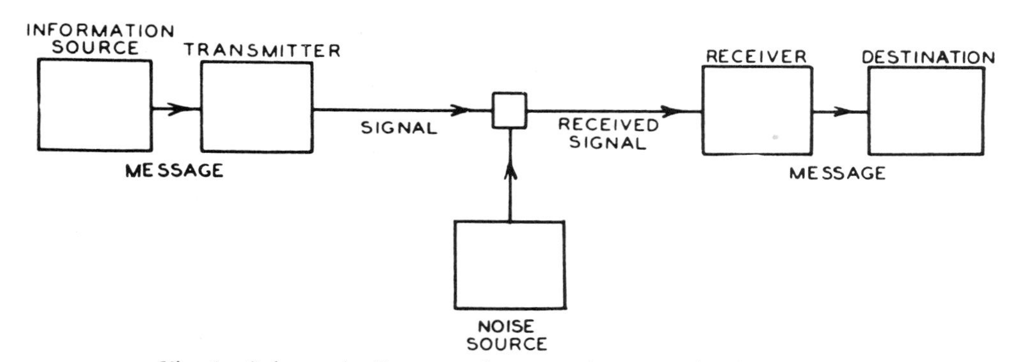

Fig. 1—Schematic diagram of a general communication system.

properly to construct the signal. Vocoder systems, television, and frequency modulation are other examples of complex operations applied to the message to obtain the signal.

3. The *channel* is merely the medium used to transmit the signal from transmitter to receiver. It may be a pair of wires, a coaxial cable, a band of radio frequencies, a beam of light, etc.

4. The *receiver* ordinarily performs the inverse operation of that done by the transmitter, reconstructing the message from the signal.

5. The *destination* is the person (or thing) for whom the message is intended.

We wish to consider certain general problems involving communication systems. To do this it is first necessary to represent the various elements involved as mathematical entities, suitably idealized from their physical counterparts. We may roughly classify communication systems into three main categories: discrete, continuous and mixed. By a discrete system we will mean one in which both the message and the signal are a sequence of discrete symbols. A typical case is telegraphy where the message is a sequence of letters and the signal a sequence of dots, dashes and spaces. A continuous system is one in which the message and signal are both treated as continuous functions, e.g. radio or television. A mixed system is one in which both discrete and continuous variables appear, e.g., PCM transmission of speech.

We first consider the discrete case. This case has applications not only in communication theory, but also in the theory of computing machines, the design of telephone exchanges and other fields. In addition the discrete case forms a foundation for the continuous and mixed cases which will be treated in the second half of the paper.

PART I: DISCRETE NOISELESS SYSTEMS

1. The Discrete Noiseless Channel

Teletype and telegraphy are two simple examples of a discrete channel for transmitting information. Generally, a discrete channel will mean a system whereby a sequence of choices from a finite set of elementary symbols $S_1 \cdots S_n$ can be transmitted from one point to another. Each of the symbols S_i is assumed to have a certain duration in time t_i seconds (not necessarily the same for different S_i, for example the dots and dashes in telegraphy). It is not required that all possible sequences of the S_i be capable of transmission on the system; certain sequences only may be allowed. These will be possible signals for the channel. Thus in telegraphy suppose the symbols are: (1) A dot, consisting of line closure for a unit of time and then line open for a unit of time; (2) A dash, consisting of three time units of closure and one unit open; (3) A letter space consisting of, say, three units of line open; (4) A word space of six units of line open. We might place the restriction on allowable sequences that no spaces follow each other (for if two letter spaces are adjacent, it is identical with a word space). The question we now consider is how one can measure the capacity of such a channel to transmit information.

In the teletype case where all symbols are of the same duration, and any sequence of the 32 symbols is allowed the answer is easy. Each symbol represents five bits of information. If the system transmits n symbols per second it is natural to say that the channel has a capacity of $5n$ bits per second. This does not mean that the teletype channel will always be transmitting information at this rate—this is the maximum possible rate and whether or not the actual rate reaches this maximum depends on the source of information which feeds the channel, as will appear later.

◆◆◆ ◆◆◆ ◆◆◆

Kelly and Radley Reflect on the First Trans-Atlantic Telephone Cable

THIS paper was written as an introduction to an issue of the *Bell System Technical Journal* devoted to the technical features of the newly completed trans-Atlantic telephone cable. This new mode of communication between the U.S. and Europe became operational almost 100 years after the first attempt to lay a trans-Atlantic telegraph cable in 1857. (See Paper 9.) For the electrical historian, both the design and timing of the new telephone link are of particular interest. It was completed near the watershed between the age of thermionic amplifiers and the age of solid-state amplifiers. It employed devices and techniques that had been developed in organized industrial laboratories (especially at the Bell Laboratories) since World War I. (See Papers 37, 38, 39, 43, and 58.) It was completed on the eve of a dramatic new technological system for communication, the communication satellite. (The first Sputnik was launched in 1957, a few months after this paper was published.)

Mervin J. Kelly (1894–1971) was born in Missouri and graduated from the Missouri School of Mines at Rolla in 1914. He later received a Ph.D. degree from the University of Chicago in 1918 and then joined the Engineering Department of the Western Electric Company under F. B. Jewett. He worked on the improvement of repeater amplifiers and radio transmitting tubes for the next several years at Western Electric and at the Bell Telephone Laboratories. He was named Research Director of BTL in 1936 and later became Executive Vice President in 1944 and President in 1951. He retired in 1959, although he continued in consulting work, especially with IBM. See the *NAS*, vol. 46, pp. 191–219 and an obituary in *IEEE Spectrum*, p. 119, May 1971.

William Gordon Radley (1898–   ) received a degree in engineering from the University of London and spent most of his career with the British Post Office (the Post Office had been in charge of wire communications in Britain since the 19th century). Radley became Engineer-in-Chief of the Post Office in 1951 and shared in the planning of the cooperative trans-Atlantic telephone cable project. He later was Director-General of the Post Office from 1955 to 1960. See *Who's Who of British Engineers*, p. 397, 1966.

Transatlantic Communications— An Historical Resume

By DR. MERVIN J. KELLY* and SIR GORDON RADLEY†

(Manuscript received July 30, 1956)

The papers that follow describe the design, manufacture and installation of the first transatlantic telephone cable system with all its component parts, including the connecting microwave radio-relay system in Nova Scotia. The purpose of this introduction is to set the scene in which this project was undertaken, and to discuss the technical contribution it has made to the development of world communications.

Electrical communication between the two sides of the North Atlantic started in 1866. In that year the laying of a telegraph cable between the British Isles and Newfoundland was successfully completed. Three previous attempts to establish transatlantic telegraph communication by submarine cable had failed. These failures are today seen to be the result of insufficient appreciation of the relation between the mechanical design of the cable and the stresses to which it is subjected as it is laid in the deep waters of the Atlantic. The making and laying of deep sea cables was a new art and designers had few experiments to guide them.

During the succeeding ninety years, submarine telegraph communication cables have been laid all over the world. Cable design has evolved from the simple structure of the first transatlantic telegraph cable — a

* Bell Telephone Laboratories. † British Post Office.

stranded copper conductor, insulated with gutta-percha and finished off with servings of jute yarn and soft armoring wires — to the relatively complex structure of the modern coaxial cable, strengthened by high tensile steel armoring for deep sea operation. The coaxial structure of the conducting path is necessary for the transmission of the wide frequency band width required for many telephone channels of communication. The optimum mechanical design of the structure for this first transoceanic telephone cable has been determined by many experiments in the laboratory and at sea. As a result, the cable engineer is confident that the risk of damage is exceedingly small even when the cable has to be laid and recovered under conditions which impose tensile loads approaching the breaking strength of the structure.

The great difference between the transatlantic telephone cable and all earlier transoceanic telegraph cables is, however, the inclusion of submerged repeaters as an integral part of the cable at equally spaced intervals and the use of two separate cables in the long intercontinental section to provide a separate transmission path for each direction. The repeaters make possible a very large increase in the frequency band width that can be transmitted. There are fifty-one of these submerged repeaters in each of the two cables connecting Clarenville in Newfoundland with Oban in Scotland. Each repeater provides 65 db of amplification at 164 kc, the highest transmitted frequency. The working frequency range of 144 kc will provide thirty-five telephone channels in each cable and one channel to be used for telegraph traffic between the United Kingdom and Canada. Each cable is a one-way traffic lane, all the "go" channels being in one cable and all the "return" in the other.

The design of the repeaters used in the North Atlantic is based on the use of electron tubes and other components, initially constructed or selected for reliability in service, supported by many years of research at Bell Telephone Laboratories. Nevertheless, the use of so many repeaters in one cable at the bottom of the ocean has been a bold step forward, well beyond anything that has been attempted hitherto. There are some 300 electron tubes and 6,000 other components in the submerged repeaters of the system. Many of the repeaters are at depths exceeding 2,000 fathoms ($2\frac{1}{4}$ miles) and recovery of the cable and replacement of a faulty repeater might well be a protracted and expensive operation. This has provided the incentive for a design that provides a new order of reliability and long life.

On the North Atlantic section of the route, the repeater elements are housed in flexible containers that can pass around the normal cable laying gear without requiring the ship to be stopped each time a repeater is laid. The advantages of this flexible housing have been apparent during the laying operations of 1955 and 1956. They have made it possible to continue laying cable and repeaters under weather conditions which would have made it extremely difficult to handle rigid repeater housings with the methods at present available.

A single connecting cable has been used across Cabot Strait between Newfoundland and Nova Scotia. The sixteen repeaters in this section have been arranged electronically to give both-way amplification and the single cable provides "go" and "return" channels for sixty circuits. "Go" and "return" channels are disposed in separate frequency bands. The design is based closely on that used by the British Post Office in the North Sea. Use of a single cable for both-way transmission has many attractions, including that of flexibility in providing repeatered cable systems, but no means has yet been perfected of laying as part of a continuous operation the rigid repeater housings that are required because of the additional circuit elements. This is unimportant in relatively shallow water, but any operation that necessitates stopping the ship adds appreciably to the hazards of cable laying in very deep water.

The electron tubes used in the repeaters between Newfoundland and Scotland are relatively inefficient judged by present day standards. They have a mutual conductance of 1,000 micromhos. Proven reliability, lower mechanical failure probability and long life were the criteria that determined their choice. Electron tubes of much higher performance with a mutual conductance of 6,000 micromhos are used in the Newfoundland-Nova Scotia cable, and it is to be expected that long repeatered cable systems of the future will use electron tubes of similar performance. This will increase the amplification and enable a wider frequency band to be transmitted; thus assisting provision of a greater number of circuits. If every advantage is to be taken of the higher performance tubes, it will be necessary to duplicate (or parallel) the amplifier elements of each repeater, in the manner described in a later paper, in order to assure adquately long trouble-free performance. This has the disadvantage of requiring the use of a larger repeater housing.

During the three years that have elapsed since the announcement in December, 1953 by the American Telephone and Telegraph Company, the British Post Office, and the Canadian Overseas Telecommunication Corporation, of their intention to construct the first transatlantic telephone cable system, considerable progress has been made in the development and use of transistors. The low power drain and operating voltage required will make practicable a cable with many more submerged repeaters than at present. This will make possible a further widening of the transmission band which could provide for more telephone circuits with accompanying decrease in cost per speech channel or the widened band could be utilized for television transmission. Much work, however, is yet to be done to mature the transistor art to the level of that of the thermionic electron tube and thus insure the constancy of characteristics and long trouble-free life that this transatlantic service demands.

The present transatlantic telephone cable whose technical properties are presented in the accompanying papers, however, gives promise of large reduction in costs of transoceanic communications on routes where the traffic justifies the provision of large traffic capacity repeatered cables. The thirty-six, four-kilocycle channels which each cable of the two-way system provides, are the equivalent of at least 864 telegraph channels. A modern telegraph cable of the same length without repeaters would provide only one channel of the same speed. The first transatlantic telegraph cable operated at a much slower speed, and transmitted only three words per minute. The greater capacity of future cables will reduce still further the cost of each communication circuit provided in them. Such considerations point to the economic attractiveness, where traffic potentials justify it, of providing broad band repeatered cables for all telephone, telegraph and teletypewriter service across ocean barriers.

The new transatlantic telephone cable supplements the service now provided by radio telephone between the European and North American Continents. It adds greatly to the present traffic handling capacity of this service. The first of these radio circuits was brought into operation between London and New York in 1927. As demands for service have grown, the number of circuits has been increased. We are, however, fast approaching a limit on further additions, as almost all possible frequency space has now been occupied. The submarine telephone cable has come therefore at an opportune time; further growth in traffic is not limited by traffic capacity.

Technical developments over the years by the British Post Office and Bell Telephone Laboratories have brought continuing improvement in the quality, continuity and reliability of the radio circuits. The use of high frequency transmission on a single side band with suppressed carrier and steerable receiving antenna are typical of these developments. Even so, the route, because of its location on the earth's surface, is particularly susceptible to ionospheric disturbances which produce quality deterioration and at times interrupt the service completely.

Cable transmission will be free of all such quality and continuity limitations. In fact, service of the quality and reliability of the long distance service in America and Western Europe is possible. This quality and continuity improvement may well accelerate the growth in transatlantic traffic.

The British Post Office and Bell Telephone Laboratories are continuing vigorous programs of research and development on submarine cable systems. Continuing technical advance can be anticipated. Broader transmission bands, lower cost systems and greater insurance of continuous, reliable and high quality services surely follow.

Pierce and Kompfner Discuss Satellite Communications Systems

IN this paper two engineers from Bell Laboratories outlined some of the technical requirements for establishing reliable transoceanic communications by means of passive or active earth satellites. They also called attention to gaps in knowledge that would require further research before such systems were implemented commercially. On the basis of their theoretical analysis, they concluded that they had "a great deal of confidence in the over-all feasibility of satellite communications." Pierce had been among the first proponents of satellite communications and had been stimulated by the announcement in 1953 of the proposed submarine telephone cable (see Paper 47) to publish an analysis of "orbital radio relays" before the first artificial satellite was successfully launched. Even earlier, Arthur C. Clarke had published a paper on "extra-terrestrial relays" in the *Wireless World* in 1945. Reprints of the two early papers by Clarke and Pierce are included in J. R. Pierce, *The Beginnings of Satellite Communications* (San Francisco, 1968).

John R. Pierce (1910–) was born in Des Moines, Iowa and received B.S., M.S., and Ph.D. degrees from the California Institute of Technology in 1933, 1934, and 1936, respectively. He was a science fiction enthusiast and has written several science fiction stories himself. He joined the Bell Telephone Laboratories in 1936 and remained until his retirement in 1971. He was a specialist in high-frequency electronic amplifiers and worked with Kompfner on the development of the traveling-wave tube. He became Director of Electronics Research at the Laboratories in 1952. He and Kompfner were largely responsible for the Echo passive satellite experiment conducted by NASA in 1960. See *Engineers of Distinction*, p. 239, 1973 and the announcement of his retirement from Bell in the *IEEE Spectrum*, p. 109, November 1971.

Rudolf Kompfner (1909–) was born in Vienna and educated as an architect. He moved to Great Britain in 1934 where he became interested in electronics and began to publish articles in radio journals. In 1939 he was interned briefly but was released to participate in high-frequency tube research at the University of Birmingham. His conception of the traveling-wave tube was made in 1942 and reduced to practice the following year. He continued his research at Oxford University and was visited by J. R. Pierce in 1944. Kompfner joined Pierce's group at Bell Labs in 1951. See *Engineers of Distinction*, p. 171, 1973. Also see Rudolf Kompfner, *The Invention of the Traveling-Wave Tube* (San Francisco, 1964). Also see Paper 56.

Transoceanic Communication by Means of Satellites*

J. R. PIERCE†, FELLOW, IRE AND R. KOMPFNER†, FELLOW, IRE

Summary—The existence of artificial earth satellites and of very low-noise maser amplifiers makes microwave links using spherical satellites as passive reflectors seem an interesting alternative to cable or tropospheric scatter for broad-band transatlantic communication.

A satellite in a polar orbit at a height of 3000 miles would be mutually visible from Newfoundland and the Hebrides for 22.0 per cent of the time and would be over 7.25° above the horizon at each point for 17.7 per cent of the time. Out of 24 such satellites, some would be mutually visible over 7.25° above the horizon 99 per cent of the time. With 100-foot diameter spheres, 150-foot diameter antennas, and a noise temperature of 20°K, 85 kw at 2000 mc or 9.5 kw at 6000 mc, could provide a 5-mc base band with a 40-db signal-to-noise ratio.

The same system of satellites could be used to provide further communication at other frequencies or over other paths

I. INTRODUCTION

THE time will certainly come when we shall need a great increase in transoceanic electronic communications. For example, the United States and Western Europe have a wide community of interests and are bound to demand more and more communication facilities across the Atlantic. If we are to be ready to fill these growing needs, we shall have to investigate all promising possibilities.

In doing so, we shall certainly want to keep in mind a rule founded on experience. This rule is that telephone circuits become cheaper the more of them we can handle in one bundle. Then, too, there is the possibility of requirements for television. In either case, there is a premium on availability of wide bands of frequency.

The submarine cable art is presently distinctly limited in bandwidth. No doubt its capability in this respect will improve as the years go by, but we may well run into economic or technical restrictions not suffered by other techniques.

A chain of UHF scatter links over a northern route might provide channels across the Atlantic Ocean but the quality is dubious, the available bandwidth is limited, and the cost is great. Indeed, we cannot now imagine how one might improve quality of bandwidth while at the same time reducing the costs of such a system. Moreover, such links would not serve for some transoceanic routes.

An undersea millimeter wave system using a round waveguide excited in the TE_{01} mode is a possibility for the remote future, but such a system is far beyond present technology.

A microwave system using satellite repeaters may have many advantages over the foregoing alternatives. Present rocket technology is at least close to the point of putting in orbit some structure which could act as a reflector or passive repeater. The maser amplifier, which introduces only around a hundredth the noise of earlier amplifiers, cuts down the transmitting power required to a hundredth of that arrived at in an earlier study.[1] This means that a satellite link with attractive properties could be attained within existing microwave art. The cost of a pair of microwave installations at the terminals would probably be less than the cost of a cable of far less bandwidth.

When highly reliable long-life microwave components and power supplies suited to a space environment are available, active repeaters may provide useful communication. When, in addition, accurate enough guidance is available, together with long-life means for adjusting attitude and position in orbit, a "fixed" repeater in a 24-hour orbit could be used.

Obviously, the present state of knowledge is insufficient for the design of a transatlantic satellite communication system of assured performance and cost. Much remains to be learned. For this very reason, and because they appear to be serious contenders for the future, it is important that research on satellite systems be given serious attention.

II. ALTERNATIVE SATELLITE REPEATER SCHEMES

A number of alternative types of orbital radio relays was discussed by Pierce.[1] These can be divided into classes in two essentially different ways: 1) active and passive repeaters, and 2) fast-revolving (relatively near) repeaters and repeaters in 24-hour orbits (stationary with respect to earth). The most important characteristics of these four possible types of repeater systems are summarized in Table I.

It should be noted that in the case of the passive repeater the bandwidth available is almost unlimited. The passive repeater is a truly linear device, and it can be used simultaneously in many ways, at many frequencies, and with many power levels, without crosstalk. Thus, the only cost in adding new channels is that of adding terminals, and these may be of many sorts.

In contrast, active repeaters have a limited dynamic range, as well as a limited bandwidth and, hence, can be used for a limited number of separate simultaneous signals only, and the levels and natures of several signals passing simultaneously through an active repeater must be carefully controlled if they are not to jam one another.

* Original manuscript received by the IRE, November 20, 1958.
† Bell Telephone Laboratories, Inc., Murray Hill, N. J.

[1] J. R. Pierce, "Orbital radio relays," *Jet Propulsion*, vol. 25, pp. 153 157; April, 1955.

Reprinted from *Proc. IRE*, vol. 47, pp. 372–373, 377–380 (extract of a longer paper), Mar. 1959.

TABLE I

Orbit	Repeater	
	Passive	Active
Near (1- to 3-hour period)	Simplest embodiment. Metallized plastic sphere, 100-foot diameter. (On the ground: large-size steerable antenna.)	Carries lightweight microwave repeater and power supplies. Low-directivity antennas (On the ground: medium-size steerable antennas.)
Far (24-hour period)	Plane reflector. Attitude stabilized. (On the ground: extra-large fixed antennas.)	Carries heavy microwave repeaters and power supplies. High-directivity antennas. Attitude stabilized. (On the ground: fixed medium-size and small antennas.)

On the other hand, smaller ground antennas and transmitter powers can be used with active repeaters, although for broad-band use large antennas and large transmitter powers have certain advantages.

Certain other considerations concerning these various possible systems will be brought out in the following sections.

III. Orbits, Mutual Visibility, and Distances

A. *The 24-hour Orbit*

The 24-hour orbit has received considerable attention and does not need extensive discussion here. If fixed antennas, perhaps of very large size, are to be used, a satellite will have to revolve in the equatorial plane at a distance of roughly 26,000 miles from the earth's center. It is hoped that perturbations from its mean position due to the attraction of the moon and sun will be small enough so that only relatively small motions of the antenna feeds and corrections of orbital position would be required. One repeater would suffice to span the Atlantic Ocean, and one to span the Pacific. Areas of mutual visibility could be controlled by means of the size and orientation of the reflecting surfaces, or antennas, on the satellite. With the advanced technology required for an active 24-hour repeater, it should be easy and desirable to provide switching or readjustment of antennas to provide communication over various paths, and so, to some degree, to overcome the limitations imposed by the fact that active repeaters can handle only a limited number of signals lying in a limited range of power levels.

Many variations on this theme are possible and will be realized in due course. The transmission provided by such a system would of course be uninterrupted.

B. *Near Orbits*

The near orbits, that is, orbits between 1000 and 3000 miles above the earth, can be classified into equatorial, polar, and inclined orbits. It is intuitively clear that the utility of a satellite orbit depends on the distances from the satellite to the terminal points for the portion of the orbit for which the satellite is visible from both terminal points. It is also obvious that this distance should be as small as possible, consistent with the requirement that a substantial portion of the whole orbit should be simultaneously visible from the terminal points.

Transatlantic communications appear to be of the greatest immediate importance, for the North Atlantic routes are already carrying the heaviest traffic in every available medium. Thus, we shall choose a transatlantic route as an example. Equatorial orbits, which are not suited to this route, will not be examined here. In order to determine whether inclined or polar orbits are more advantageous for transatlantic use, computations of mutual visibility have been made for two comparable cases; the result is that polar orbits are more efficient for terminations which might be chosen for a transatlantic link. Therefore, all the subsequent calculations have been made on the basis of polar orbits only.

C. *Visibility Considerations*

A computation has been made of the durations of simultaneous visibility of a satellite at various heights from points in North America and Europe, selected, somewhat arbitrarily, in the following locations:

Terminal A—Newfoundland	48° north, 55° west
Terminal B—Island of Lewis in the Hebrides, Scotland	58° north, 7° west

They are about as close to each other as can be found on a map, keeping in mind that there will have to be microwave links between the terminals and the respective continental communication networks.

Assuming the radius of the earth to be 3950 miles, we find that the distance between terminals A and B is 2060 miles. For the sake of simplicity we have assumed that the satellite becomes visible as soon as it crosses the horizon and that it moves in perfectly circular orbits.

A map showing the regions of mutual visibility of satellites at heights of 500, 1000, 1500, 2000, 2500, and 3000 miles will be found in Fig. 1. These regions are projected onto the earth's surface, with the earth's center as the center of projection, and this in turn is shown in orthogonal projection, with the North Pole at the center. At the higher altitudes these regions extend close to the equator and are very much foreshortened, and another projection (not shown) has been used in computing visibility involving these regions. So far these visibility regions do not depend on the type of orbit. However, it is a simple matter to compute visibility durations for polar orbits, and this is what has been done.

♦♦♦ ♦♦♦ ♦♦♦

Let f be the average fraction of the satellite period when it is visible from both terminals. This is also the probability of mutual visibility, and $(1-f)$ is then the probability of the satellite *not* being seen. Hence, the probability of not seeing any of n satellites:

$$(1 - f)^n.$$

If this is set equal to i, the average fraction of service interrupted, the required number of satellites is obtained:

$$n = \frac{\log i}{\log (1 - f)}.$$

Table IV shows the minimum numbers of satellites required to give service interrupted, on the average, by the amounts indicated. This has been done for various satellite heights, assuming visibility right down to the horizon. In Table V there is a similar set of numbers of satellites calculated for various minimum elevations, assuming all satellites to travel at a height of 3000 miles. Note that doubling the number of satellites reduces the amount of interruption by one tenth.

TABLE IV

NUMBER OF RANDOMLY SPACED SATELLITES TO PROVIDE SERVICE INTERRUPTED BY NO MORE THAN 100. PER CENT

Satellite Height	Number of satellites for indicated percentage of service interruption			
miles	10 per cent	1 per cent	0.1 per cent	0.01 per cent
500	62.5	125.0	187.5	2500.0
1000	32.3	64.6	96.9	129.2
1500	16.7	33.3	50.0	66.6
2000	11.8	23.6	35.4	47.2
2500	10.5	21.0	31.5	42.0
3000	9.3	18.5	27.8	37.0

Assumptions: Terminals in Newfoundland and Hebrides.
Polar orbits.
Visibility from horizon to horizon.

TABLE V

NUMBER OF RANDOMLY SPACED SATELLITES REQUIRED TO GIVE SERVICE INTERRUPTED NO MORE THAN 100. PER CENT, AS A FUNCTION OF MINIMUM ELEVATION FOR ORBITS AT 3000-MILE HEIGHT

Minimum elevation angle	Number of satellites for indicated percentage of service interruption		
degrees	10 per cent	5 per cent	1 per cent
0	9	12	19
3.25	11	14	21
7.25	12	15	24
12.60	17	22	33

Assumptions: Terminals in Newfoundland and Hebrides Polar orbits.

It should be noted at this point that the interruptions, though happening at irregular intervals, should be predictable well in advance, and the communication services can be organized accordingly.

V. PATH-LOSS CALCULATIONS

Many factors enter into a calculation of transmission performance of a system involving satellite repeaters between terminals on earth. Rather than carry out calculations on a variety of schemes, we shall concentrate on one in particular, namely frequency modulation with feedback,[3] which is particularly applicable to satellite communications, as has been pointed out by Ruthroff and Goodall.[4] The proposed system compares favorably with other known systems, such as PPM, PCM, or SSB. The results will be obtained in a fashion so that extrapolations to other systems can be easily performed.

Starting with the case of the passive repeater in the form of a reflecting sphere with a diameter D, the path loss can be written, as for the well-known radar case:

$$L = \frac{P_T}{P_R} = \frac{16\lambda^2 p^4}{A^2 \eta^2 D^2},$$

where

P_T = transmitter power,
P_R = receiver power,
λ = wavelength,
p = geometric mean of the distances between the satellite and the terminals,
A = antenna area (assumed to be the same for both transmitter and receiver),
η = antenna efficiency.

The noise power at the receiver input can be written

$$N = kTB,$$

where

k = Boltzmann's constant (1.38×10^{-23} watt per degree per cycle per second, or -228.9 dbw for 1°K and 1 cps),
T = effective noise temperature of receiver (including sky noise, antenna loss, tube noise, etc.) in degrees Kelvin,
B = the RF bandwidth in cycles per second,
b = base-band modulation bandwidth.

In an FM receiver, noise problems are minimized if feedback is applied to reduce the deviation in the IF stages and the limiter. The operation of such a receiver is described by Chaffee[3] and Carson[5] and has been digested by Goodall and Ruthroff.[1] More recently, a

[3] J. G. Chaffee, "Application of negative feedback to frequency modulation systems," *Bell Sys. Tech. J.*, vol. 18, pp. 404–437; July, 1939.

[4] C. L. Ruthroff and W. M. Goodall, private communication.

[5] J. R. Carson, "Frequency-modulation: theory of the receiving feedback circuit," *Bell Sys. Tech. J.*, vol. 18, pp. 395–403; July, 1939.

closely related system has been described by Jaffe and Rechtin.[6]

An optimized receiver of this kind will use sufficient feedback to reduce the deviation in the IF circuits to near zero. Under these conditions

$$(S/N) \cong 3\left(\frac{C}{N}\right) M^2$$

and

$$B \cong 2(1 + M)b,$$

where

M = modulation index,
(S/N) = signal-to-noise ratio at the receiver output,
(C/N) = signal-to-noise ratio at the IF frequency.

To give reliable operation

$$\left(\frac{C}{N}\right) \geqq 16,$$

which is the usual 12-db threshold. Now, since the IF bandwidth required is twice the modulation bandwidth

$$N_{IF} = 2kTb,$$

and since

$$C = \frac{P_T}{L},$$

then

$$\frac{P_T}{L} \geqq 32\,kTb, \text{ or 15.1 db above a power } kTb.$$

It follows from the above that if

$$\left(\frac{S}{N}\right) = 10^4 \text{ (40 db)},$$

then $M = 14.5$, and a 5-mc signal bandwidth requires a 155-mc RF bandwidth. To accomplish this exactly requires infinite feedback, but for practical purposes a feedback factor of 145 (43 db), reducing the modulation index in the IF to 0.1, should suffice.

The transmitter power can be kept constant and the output signal-to-noise ratio increased (up to the limit given by information theory) by further increasing in equal amounts the transmitter index, receiver feedback, and bandwidth in the medium.

Consider what this implies in terms of an actual system.

A. FM Feedback—Passive Repeater

Assuming a maser amplifier is in use, the receiver noise becomes negligible. The sky temperature is then the dominant factor and, referring to Fig. 3, choose 2000 mc for the operating frequency; assuming that the antennas point at elevations no lower than 7.25°, the effective receiver noise temperature is 20°K, so that the noise is 13 db above that at 1°K. With no more than one per cent interruption of service, this requires 24 satellites at orbits 3000 miles high. The maximum distance (RMS) is 5240 miles (Table III). The shortest distance, that is, when the satellite is half-way between the terminals, is 3300 miles (Table II).

Suppose the antenna diameters are 150 feet, and efficiencies 60 per cent. Further, assume that the passive repeater is a metallic sphere of 100-foot diameter. Thus, the maximum path loss:

$$L = \frac{16 \times (\frac{1}{2})^2 \times (5240)^4 \times (5280)^4}{(\pi \times 75^2)^2 \times (0.60)^2 \times (100)^2}$$

$$L = 2.04 \times 10^{18} \text{ or } 183.1 \text{ db}.$$

The minimum loss turns out to be 8 db less.

For one-cps base-band bandwidth and 1°K, the received power must be 15.1 db above the thermal level of −228.9 dbw. A temperature of 20°K increases the required power by 13.0 db, and a bandwidth b of 5 mc increases the power by 67.0 db.

In db-watt

$$P_T = +183.1 + 15.1 - 228.9 + 13 + 67 \text{ dbw}$$

$$P_T = 49.3 \text{ dbw or 85 kw}.$$

This amount of CW power can, in principle, be supplied with existing tubes (Varian Associates VA-800, 1.7 to 2.4 kmc, 10-kw klystrons), eight of them driven in parallel.

Suppose a higher frequency is chosen, for example, 8000 mc. This reduces the power requirement by a factor

$$\left(\frac{8000}{2000}\right)^2 = 16.$$

On the other hand, the antenna noise temperature will be increased by a few degrees, *e.g.*, from 20 to 25°K, giving a factor 1.25. We assume that antenna size and efficiency are unchanged, which actually implies a considerable increase in the cost of the antennas, and arrive at a transmitter power of:

$$P_T = \frac{85 \times 1.25}{16} = 6.65 \text{ kw}.$$

This amount of power can be supplied by four klystrons in parallel, such as the Varian Associates VA-806. Further, it appears that single tubes could be developed for either of the cases discussed above.

B. FM Feedback—Active Repeater

In the active repeater it is assumed that there is a wide-band low-noise RF amplifier with a power output limited to P_a by weight, life, and bulk considerations.

[6] R. Jaffe and E. Rechtin, "Design and performance of phase-lock circuits capable of near-optimum performance over a wide range of input signal and noise levels," IRE TRANS. ON INFORMATION THEORY, vol. IT-1, pp. 66–76; March, 1955.

It is connected to the earth terminals by means of non-directional antennas, namely dipoles with effective areas $3\lambda^2/8\pi$.

It can be shown that the crucial factors are the size and noise temperature of the antenna at the receiving terminal and the satellite output power. The transmitter power on the ground and the noise figure of the satellite amplifier are of minor importance.

The path loss of importance now is given by

$$\frac{P_R}{P_a} = \frac{A_{sat}A_R \cdot \eta}{\lambda^2 p^2} = \frac{1}{L},$$

where

$A_{sat} = 3\lambda^2/8\pi$,

$A_R = \pi a^2/4$, a being the antenna diameter,

η = antenna efficiency,

p = path length.

Hence

$$L = \frac{32}{3}\left(\frac{p^2}{a^2\eta}\right).$$

Employing FM with feedback as before, use the formula

$$\frac{P_a}{L} = 32\ kTb:$$

Thus, the necessary RF power at the satellite

$$P_a = 326\ \frac{p^2}{a^2\eta}\ kTb.$$

Take the same antenna, frequency, noise temperature, bandwidth, and distance as used before in the passive repeater case, namely:

p = 5240 miles,

a = 150 foot,

η = 0.6,

T = 20°K at 2000 mc,

b = 5 mc.

This requires a satellite transmitter power of P_a = 25.4 mw.

This does not seem to be much in the way of RF power; an increase of one order of magnitude is perhaps not out of the question. Over-all efficiencies of something like a few per cent may perhaps be achieved, leading to a continuous power requirement of a few watts. This can be satisfied by a combination of solar and storage batteries, but owing to the limited life of existing storage batteries, the life of such a repeater would at present be restricted.

An active repeater in a 24-hour orbit with the ground installations much as described above will have to put out approximately $(25{,}000/5000)^2 = 25$ times the transmitter power calculated above, that is, about 625 mw. This is more than enough, since a 24-hour satellite will, in all probability, be attitude stabilized and therefore could carry a high-gain antenna to great advantage.

The power available on this kind of repeater has to include provision for the maintenance of accurate position, velocity, and attitude; how much, it is difficult to estimate at this time. Life will again be a serious problem.

VI. Modulation Systems

Microwave tubes such as klystrons, traveling-wave tubes, and backward-wave oscillators lend themselves very conveniently to frequency modulation. Magnetrons and amplitrons are operated advantageously under pulse conditions.

Ruthroff and Goodall have shown[3] that FM with feedback and PPM give practically the same performance when the same mean transmitter power is employed. It would not be surprising if other modulation schemes were to be found to give similar performance when all requisite conditions are optimized.

All modulation systems will have to operate in the presence of large and continuously varying Doppler-frequency shifts. With approximately spherical reflectors at an elevation higher than a few degrees above the horizon, no fading, scintillating, or glinting is expected to occur; thus, no fading margin was deemed to be necessary in the path-loss calculations. The noise temperature, antenna size, power, etc, are the most reasonable estimate that can be made at present of what is necessary to provide the performance specified. In an actual system, a greater power might be used in order to provide a margin of safety.

VII. Unknowns in Satellite Communications

The authors have a great deal of confidence in the over-all feasibility of satellite communications. Nevertheless, quite a few unknowns exist. Experiments, development, and experience are needed before all problems can be considered solved. Some problems are of a fundamental nature, such as the influence of the earth's atmosphere on the propagation of radiation; others are instrumental, such as the limits on receiver noise temperature due to losses and mismatches.

A. Satellite Construction

The passive repeater envisioned in Section V-A is a metallized plastic sphere of 100-foot diameter. A considerable amount of work is being done by the National Aeronautics and Space Agency (NASA), who have announced that they will place one or more balloons of this type into orbit sometime in 1959. The major unknown at present is the life of such balloons. Also in question is the ultimate shape.

Other satellite constructions, such as metallic wire-mesh spheres, the mesh being small compared with a wavelength, have been suggested, and methods of placing them in orbits deserve to be studied.

A spherical satellite scatters the radiation which it intercepts isotropically. Satellites of shapes other than spherical could be used to reflect a greater fraction of the radiation striking them from the transmitter to the receiver.

A great deal of work needs to be done on active repeaters, particularly on components such as microwave tubes, storage batteries, capacitors, etc., before a sufficiently long life can be assured. Many components may have to be constructed on an entirely new basis, taking into account that the environment will be radically different from any encountered so far, namely ultra-high vacuum, intense ultraviolet, X-ray and cosmic ray bombardment and micrometeorites.

B. Propagation Effects

The whole of the earth's atmosphere will have to be traversed twice in every satellite radio link; it is therefore important to know the effect it will have on the beams of microwave radiation. It is fairly certain that nothing harmful will happen to beams pointing at more than 10° above the horizon, except perhaps when traversing auroral regions. Nevertheless, propagation measurements are required.

It would be desirable to know the actual instantaneous angle of arrival of beams of radiation coming from all possible directions over a wide frequency range and covering a wide range of climatic conditions. Data concerning rotation of the plane of polarization are also important. The effective sky temperature as a function of frequency and elevation must be ascertained.

C. Antenna Considerations

Some of the research results of the preceding section will determine the largest size of antenna which can be used with advantage.

Another problem to be solved is that of an antenna with low effective noise temperature; that is, one in which the losses and the side and back lobes have been reduced below a certain tolerable limit.

Problems of very large steerable antennas call for solution; these may be divided into electrical and mechanical problems. It may be that both will be solved eventually by adopting the principles of multielement steerable arrays.

D. Over-all System Noise Figure

This is largely an RF input problem, once the antenna itself has been "cleaned up." With very large antennas it will be necessary, for economical and mechanical reasons, to combine into one unit both transmitter and receiver feed horns and also perhaps the output and input of a tracking radar. The frequencies employed for all these functions may be spaced widely apart so that efficient filtering should not be too difficult.

Nevertheless, to achieve an effective low-noise temperature will require much competent and painstaking experimentation. Considerable development work is already in progress on masers and parametric amplifiers. Not so much has been done on tying them in with a particular communication system.

E. Tracking of Satellites

Satellites move in smooth, regular orbits, predictable with high precision. This makes it attractive to think of using computers, analog or digital, for the purpose of steering antennas on them.

The alternative method employs a tracking radar. For relatively small antennas, or in case only a feed system has to move, the tracking radar may have a separate antenna, and the communication antenna be "slaved" to the radar.

With large antennas, which may distort, sag, or twist as they are slewed about or in the presence of high winds, it might be necessary to make the radar output and input integral with the communication feed system in order to point the antenna accurately despite distortions with respect to the mounting and drive.

Similar considerations also apply to the Doppler-shift of the reflected radiation, which can be computed beforehand, or which can be derived instantaneously from the radar data.

The results of the research on propagation effects will affect solutions to the tracking problems. Any satellite communication system involving very large antennas at microwave frequencies will depend entirely on an accurate and dependable tracking system such as probably has never been built before.

VIII. Acknowledgment

The subject matter of this paper has been discussed with many people, and the authors have greatly benefited from their comments. Where possible, individual acknowledgment has been made.

Jordan and King on Antennas

TWO engineering educators and antenna specialists contributed this paper to the 50th anniversary issue of the PROCEEDINGS OF THE IRE on significant developments in antenna analysis and practice since 1945. They noted that important advances had been achieved in the quantitative analysis of antennas and that such "frequency-independent" antennas as the log-periodic and equiangular spiral antennas had been introduced. Several other papers on the earlier history of antennas and propagation were included in the same issue of the PROCEEDINGS.

Edward C. Jordan (1910–   ) was born in Edmonton, Canada and received B.S. and M.S. degrees from the University of Alberta in 1934 and 1936. He worked as an engineer for a radio broadcast station from 1928 to 1935. Jordan received the Ph.D. degree from Ohio State University in 1940, and after teaching for a year at the Worcester Polytechnic Institute, returned to teach at OSU from 1941 to 1945. He became head of the Electrical Engineering Department at the University of Illinois in 1945 and was author of a popular text on applied electromagnetics. He received the IEEE Education Medal in 1963. See *Engineers of Distinction*, p. 159, 1973. Also see *Who's Who in Engineering*, p. 953, 1964.

Ronald W. P. King (1905–   ) was born in Massachusetts and received the A.B. and M.S. degrees from the University of Rochester in 1927 and 1929. He studied in Munich, Germany for a year and at Cornell for a year before receiving the Ph.D. degree from the University of Wisconsin in 1932. He taught at Wisconsin and Lafayette College and spent another year in Germany on a Guggenheim fellowship before joining the faculty at Harvard in 1938. He remained at Harvard until his retirement in 1972 and is credited with directing the work of more than 80 doctoral candidates. See *Who's Who in Engineering*, p. 1006, 1964 and an announcement of his retirement in *IEEE Spectrum*, p. 99, June 1972.

Advances in the Field of Antennas and Propagation Since World War II: Part I—Antennas*

E. C. JORDAN†, FELLOW, IRE, AND R. W. P. KING‡, FELLOW, IRE

Summary—**Progress in the quantitative understanding of antennas as circuit elements, transmitters, receivers and scatterers of electromagnetic radiation is reviewed briefly for the period 1945–1961. Advances in the design of selected radiating systems with special properties are indicated.**

Specific reference is made to the impedance, current distribution, and pattern characteristic of cylindrical dipoles, singly and in arrays. Particular developments touched upon include slot and surface wave antennas, microwave antennas and microwave lenses, super-gain antennas, and very large antennas and arrays for radio astronomy and satellite communication. Frequency-independent "angle" antennas and log-periodic structures are reviewed briefly.

Introduction

THE WIRELESS transmission of signals depends critically on the radiating and receiving properties of antennas. In the period 1945–1961 significant advances were made in the quantitative understanding of the fundamental properties of antennas and many new and useful types of radiating systems were developed. In this brief review references can be made to only a small cross section of this rapidly growing field.

The Dipole Antenna

The basic radiating element is the dipole in the broad sense of a symmetrical center-driven conductor, half of such a conductor with its image in a metal screen as shown in Fig. 1 (or, by complementarity, a slot in a metal plane). The groundwork in the quantitative study of cylindrical, biconical, and spheroidal dipoles as boundary-value problems in electromagnetic theory instead of as sources with arbitrarily assumed distributions of current was laid in the years just preceding 1945. Research since that time has been extensive and so fruitful that the behavior of these antennas is now well understood in a detailed quantitative sense.

The transmitting and receiving properties of the cylindrical dipole have been determined theoretically in overlapping ranges by iterative, variational, Fourier-series, and Wiener-Hopf methods. The impedance is now known for antennas of any length over a broad range of ratios of radius to wavelength. Typical curves are shown in Fig. 2. A simple and quite accurate formula for the current when $kh = 2\pi h/\lambda \leq 3\pi/2$ is

$$I(z) = \frac{j2\pi V}{\zeta\Psi \cos kh} \cdot \left[\sin k(h - |z|) + T(h)(\cos kh - \cos kz)\right], \quad (1)$$

where $\zeta = 120\pi$ ohms and where Ψ and $T(h)$ are complex coefficients that depend on the length and the radius of the conductor. Excellent correlation between theory and extensive series of measurements was achieved as soon as the interesting problem of end and coupling effects for various types of transmission lines had been solved and the properties of the idealized generators used in the theory were understood. Extensions of both the theoretical and the experimental techniques have been made

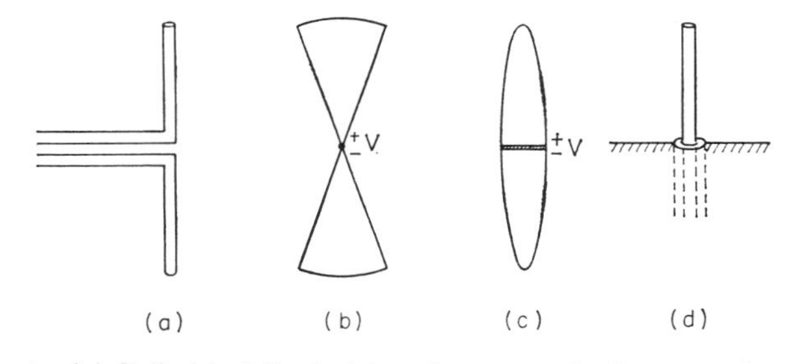

Fig. 1—(a) Cylindrical dipole driven from two-wire line. (b) Biconical antenna driven by idealized point generator. (c) Spheroidal antenna driven by idealized slice generator. (d) Cylindrical half dipole over conducting image plane.

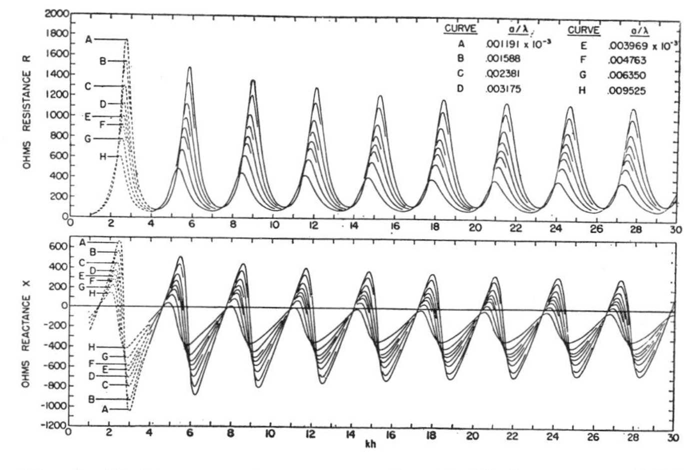

Fig. 2—Resistance and reactance of cylindrical antenna. Solid line = Wu. Broken line = King-Middleton 2nd-order theory.

* Received by the IRE, July 21, 1961; revised manuscript received, November 1, 1961.

† College of Engineering, University of Illinois, Urbana, Ill.

‡ Division of Engineering and Applied Physics, Harvard University, Cambridge, Mass.

Reprinted from *Proc. IRE*, vol. 50, pp. 705–708, May 1962.

to cylindrical antennas in dissipative media including their application as probes to study ionized regions in space. The response of dipoles to voltage pulses has also been studied theoretically and experimentally.

Corresponding progress has been made in the study of asymmetrical, folded and special antennas, as well as of the closely related loop antenna both with and without a ferrite core (which behaves like a magnetic dipole when electrically small).

The power transferred to the load of a cylindrical receiving antenna in an arbitrarily polarized plane-wave field incident from any direction has been obtained from an analysis of the distribution of current (which differs significantly from that in a transmitting antenna). The directional properties and the EMF in an equivalent circuit are expressed in terms of a complex effective length. The scattering of electromagnetic waves from variously loaded cylindrical antennas has been analyzed and measured; in particular, the back-scattering cross section has been determined for a thin dipole of any length.

The complete electromagnetic field of the driven dipole with an assumed sinusoidally distributed current was well known in 1945. More correct fields have since been obtained with the current given by (1) and from the application of the reciprocal theorem to the quite accurately known effective length of the receiving antenna. For very long antennas a Wiener-Hopf solution has yielded good results. Measured field patterns are in satisfactory agreement with theory.

Many of the properties obtained for thin cylindrical antennas have been translated to apply approximately to narrow slot antennas in a conducting plane with the help of the principle of complementarity. Advances similar to those summarized for the cylindrical dipole have also been made for biconical and spheroidal antennas.

Dipole Arrays

An unlimited variety of directional patterns may be achieved with suitably designed arrays of dipole antennas in air or slots in metal surfaces. In 1945 a very substantial knowledge existed about the properties of arrays of elements that are geometrically alike and are *assumed to have identically distributed* currents. Important advances in arrays of this type since 1945 include the proper choice of currents in the elements in order to achieve optimum relationships especially between the beamwidth and the minor lobe level, the use of elements that are of unequal length and unequally spaced in order to obtain broad-band properties, the development of methods for the synthesis and for the scanning of patterns, and new mathematical techniques that involve the application of potential theory and the treatment of the currents in the elements as sampled values of continuous functions.

Significant progress has also been made in the determination of the actual current distributions and the self- and mutual impedances of coupled elements in collinear, circular, and curtain arrays including Yagi, corner-reflector and special types of antennas. Except in arrays of half-wave dipoles and regular phase-sequences in circular arrays, the distributions of current along the several elements are far from alike, so that the actual field patterns with specified input currents may differ greatly from those predicted by conventional theory, especially with reference to the nulls and minor lobes. Distributions of current on a couplet of two full-wave elements with specified input currents, $I_{20}=jI_{10}$, are shown in Fig. 3 and the resulting field pattern in Fig. 4, together with the ideal pattern for identically distributed currents.

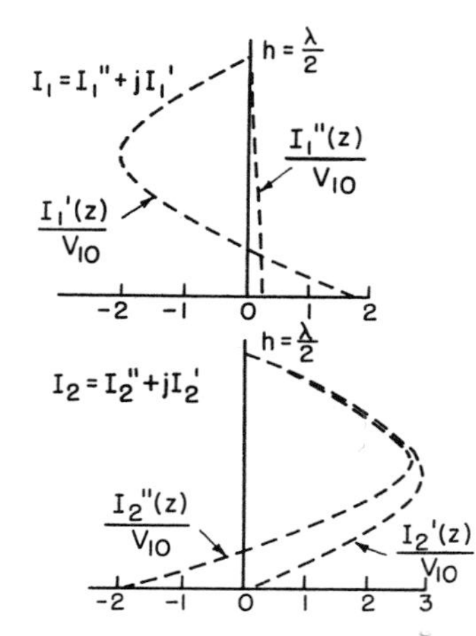

Fig. 3—Currents in the upper halves of the two elements of a full wave couplet; $I_2(0)=jI_1(0)$; $h/a=75$.

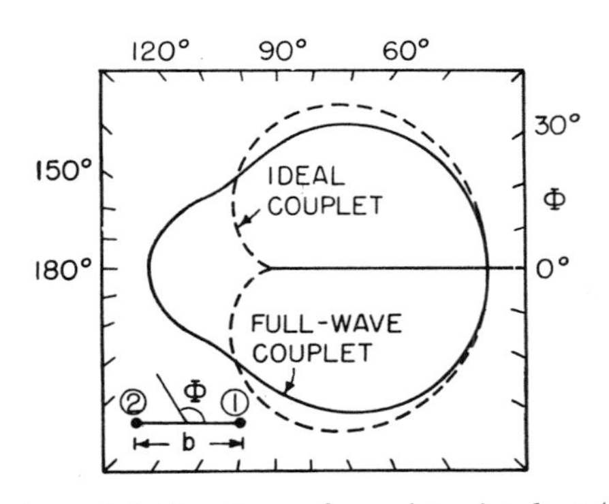

Fig. 4—Horizontal field pattern of couplet wjen $b=\lambda/4$; for ideal couplet with $I_2(z)=jI_1(z)$ and for full-wave couplet with currents shown in Fig. 3.

Surface Wave Antennas

Surface waves are associated with an interface between two media, their most important characteristics being the exponential decay of their field components in an equiphase surface perpendicular to the direction of propagation. The surface wave does not transmit power away from a lossless supporting interface unless the supporting structure is terminated or a discontinuity of some kind is introduced. The radiation pattern may be analyzed in essentially the same way as that of other aperture antennas from the field distribution in the terminal plane. For a nonuniform or modulated supporting structure there are, in addition, the radiation fields generated by the nonuniformities.

Although many of the characteristics of surface waves have been known for some time, not much attention was paid to practical uses in transmission lines and radiating systems until after the development of high-powered microwave oscillators during and after World War II. It was found that suitable interfaces that support surface waves generally consist of plane or cylindrical dielectric materials, such as sheets, rods or tubes as well as plane or cylindrical transversely corrugated metallic interfaces.

The dielectric rod antenna has been extensively used both singly and in arrays (see Fig. 5). It may have a square or circular cross section and may be excited in one of several possible surface-wave modes. The maximum of the radiation pattern is along the axis of the rod.

Corrugated end-fire antennas were developed where a higher degree of rigidity and heat resistance was required (see Fig. 6). In many cases it has been found possible to support the surface wave by means of corrugations in the aerodynamic structure itself. The corrugations are filled with a suitable dielectric material so that streamlining of the aerodynamic body is undisturbed.

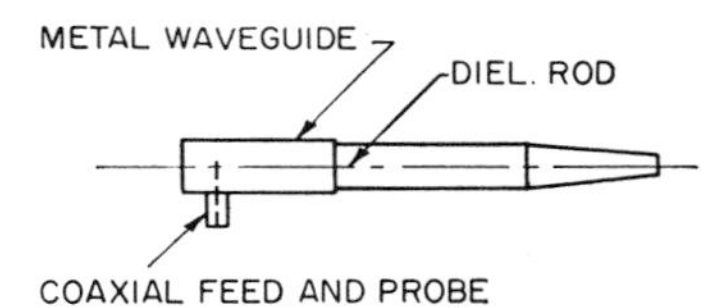

Fig. 5—Dielectric rod antenna.

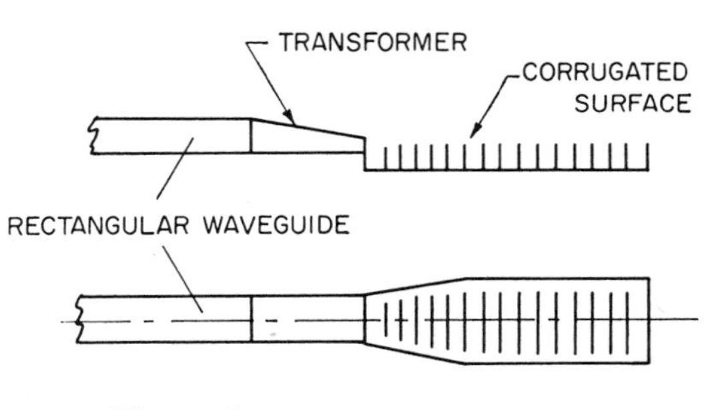

Fig. 6—Corrugated surface antenna.

Other antennas whose behavior may be quite well understood in terms of surface-wave modes are the Yagi and helical antennas which support surface waves with a phase velocity that is less than the velocity of light. Radiation occurs only from the terminal plane or from other discontinuities such as tapers. Antennas of this type are useful where end-fire operation over a wide band of frequencies and low wind resistance are required.

Corrugated or dielectric-coated surfaces operating in the circularly-symmetrical cylindrical surface-wave mode have been designed for use as omnidirectional beacon antennas.

Microwave Antennas

Advances in microwave antennas have been associated with increased size, more precise beam shaping, and improved scanning capabilities. Conventional paraboloidal, spherical, and parabolic cylinder reflectors with apertures of several hundred wavelengths have been constructed. Primarily this development represents a mechanical achievement in maintaining close tolerances although automatic servomechanism control of the surface alignment is also in use. For scanning or multiple-beam operation numerous microwave systems have been developed from their optical counterparts. Microwave lenses have been constructed with artificial dielectrics of many forms, and recent advances in the polymer industry are making available a wide range of natural dielectrics suitable for this purpose. For beam scanning, the Luneberg lens is a particularly interesting type that has been developed in many variations. In its simplest form it is a dielectric sphere with the useful property that rays from a point source, located at any point on its spherical surface, emerge as parallel rays from the other side of the sphere.

Broad-Band Antennas

A highly significant advance in the antenna field since 1945 is the successful development of frequency-independent antennas. Before 1945, a bandwidth of 2 to 1 was considered large, and few antenna engineers would then have predicted the possibility of designing antennas with impedance and pattern characteristics that are *independent* of frequency over as wide a frequency band as the designer may specify (and, of course, is willing to pay for in terms of dimensional requirements).

The concept of frequency-independent antennas has been realized in two separate but related practical developments: equiangular structures and log-periodic structures. The first of these uses the "angle concept" which is based upon the observation that an antenna whose geometry can be specified entirely by angles should have characteristics which are independent of frequency. Of course, all such structures extend to infinity, so the problem is to determine which retain their frequency-independence when truncated to a finite length. (The infinitely-long biconical structure is an example which does *not* remain frequency-independent when truncated to form a practical antenna.) The equiangular spiral antenna of Fig. 7 was the first of a class of antennas which are frequency independent when truncated. The feed voltage is applied between the two arms at the center, and the decay of current is sufficiently rapid that the current is negligible at a distance of a wavelength measured along the arms. Hence, the antenna may be truncated at this length, and its characteristics will be frequency independent for all higher frequencies (up to that for which the diameter of the feed region becomes an appreciable fraction of a wavelength).

In a more practical version of this antenna the spiral arms are developed on the surface of a cone as shown in Fig. 8. This structure is frequency independent, and yields a desirable unidirectional, circularly-polarized beam, having its maximum off the apex of the cone. Antennas of this type have been constructed to cover a frequency band of more than 20 to 1.

Log-periodic structures have radiation characteristics that repeat periodically with the logarithm of the frequency and in addition, that do not vary greatly over a period. Such antennas are essentially frequency independent. One of the simplest of many types is the log-periodic dipole array. In this array adjacent elements have a fixed length-ratio τ, and adjacent spacings bear the same ratio. From the principle of scaling, whatever radiation characteristics result at a frequency f, must also result at a frequency τf, and indeed at all frequencies $\tau^n f$, where n is an integer. When appropriately excited (with phase reversals introduced between adjacent elements) a truncated version of this array radiates a single-lobe linearly-polarized beam directed toward the apex. In operation, the active region of the antenna remains near the element of length $\lambda/2$. As the frequency is increased the active region moves forward through the array.

If the dipole arms are slanted forward, the log-periodic resonant-V array of Fig. 9 results. With care such an array can be designed so that when the active portion runs off the front end with increasing frequency, it returns to the rear end in the $3\lambda/2$ mode, moves through the array and again returns to the rear to advance through the array in a still higher-order mode. By this means very large bandwidths can be achieved from a relatively short array.

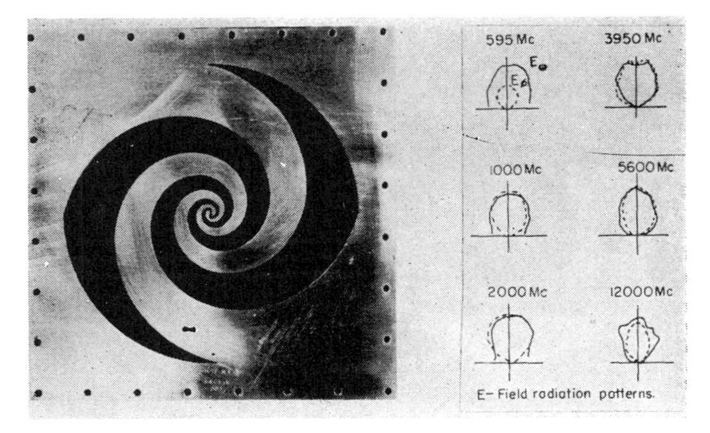

Fig. 7—Planar balanced equiangular spiral slot antenna.

Fig. 8—Conical balanced equiangular spiral antenna.

Fig. 9—Log-periodic resonant-V array.

Large Antennas and Related Problems

Applications in radio astronomy and in satellite tracking and communication have spurred the development of very large antennas and arrays. In radio astronomy where resolution can be more important than gain, large interferometer-type antennas, Mill's cross antennas, and synthetic arrays are in use. The latter, instead of using many elements simultaneously, use a fixed and a movable element to measure the relative phase and amplitude over the aperture of the equivalent array being synthesized. The data are stored and operated upon by a computer to yield information equivalent to that given by an ordinary array. In some large arrays utilizing many elements data processing techniques are being applied to combine the data from individual elements to produce a maximum of information.

During the 1940's it was demonstrated that there is no limit, theoretically, to the narrowness of the beamwidth obtainable from a given-sized aperture, or array. Although several excellent analyses showed conclusively the impracticality of these "super-gain" antennas, considerable effort was expended in some quarters to obtain a small degree of supergaining.

The advent of extremely low-noise receivers, using masers and parametric amplifiers, has introduced the antenna engineer to the notion of antenna temperature as a unit of measurement for the noise picked up by the antenna. The extremely low-noise temperatures seen by a microwave antenna looking at a "cold" sky has required that additional effort be devoted to reducing side and backlobes which "see" the warm earth. By careful design, antenna temperatures of less than 5°K have been achieved for an upward-looking antenna.

Acknowledgment

The authors acknowledge with thanks the assistance given by their colleagues, particularly J. D. Dyson, Y. T. Lo, and Y. Shefer.

Section II-D
Electronics

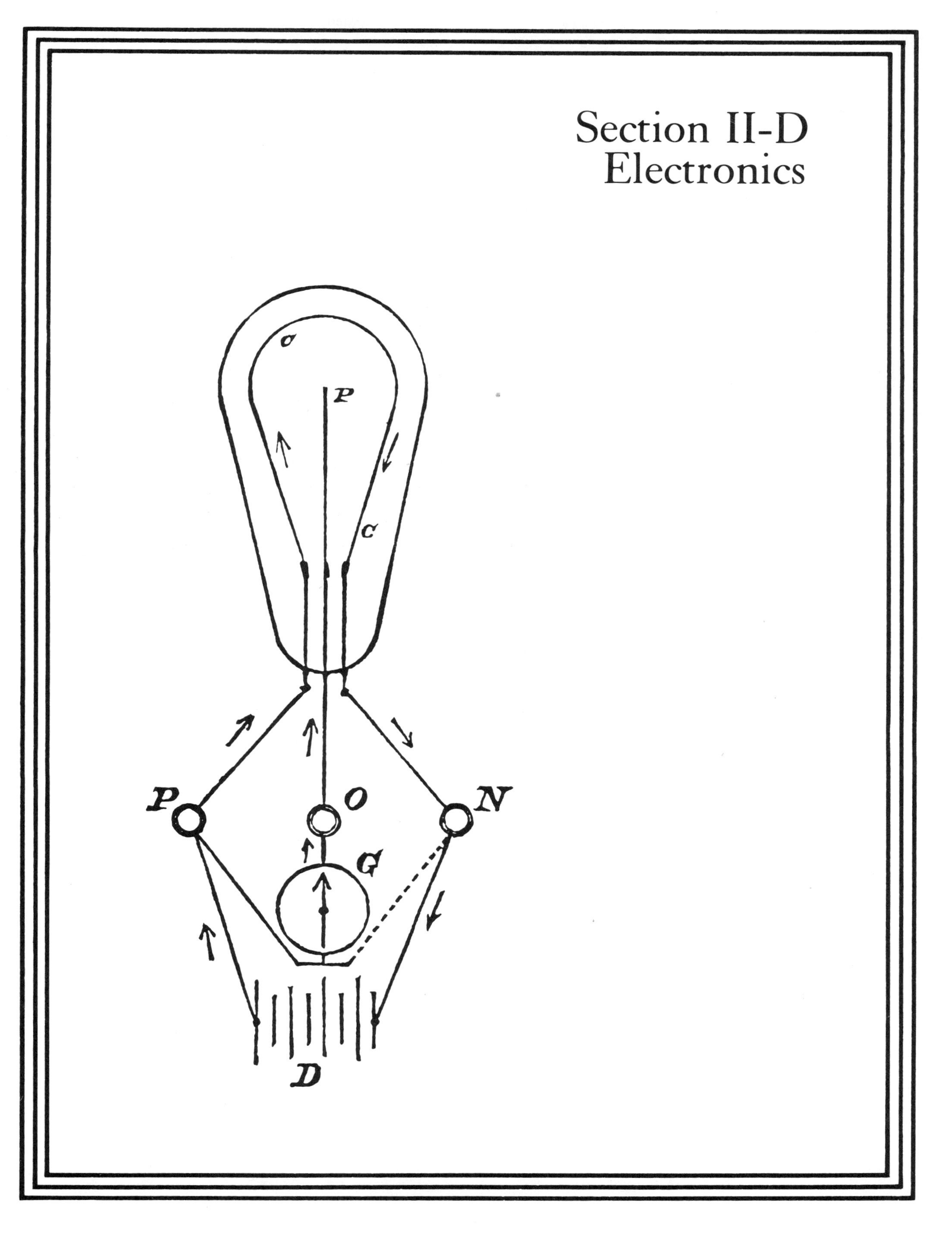

Houston on the Edison Effect

IN this paper presented at the first meeting of the AIEE held in Philadelphia in 1884, Houston reported on a puzzling phenomenon that Edison had encountered in some incandescent lamps. That Houston's puzzlement was shared by others at the meeting is clear from the discussion that followed. Twenty years later, J. A. Fleming converted the Edison effect into the thermionic vacuum diode that marked the arrival of the age of electronics. (See Paper 12.) Also see George Shiers, "The First Electron Tube," *Scientific American*, pp. 104–112, March 1969.

For a biographical note on Houston, see Paper 17.

A paper read before the American Institute of Electrical Engineers, at Philadelphia, October, 1884.

NOTES ON PHENOMENA IN INCANDESCENT LAMPS.

BY PROF. EDWIN J. HOUSTON.

Prof. HOUSTON:—I have not prepared a paper, but merely wish to call your attention to a matter which, I suppose, you have all seen and puzzled over. Indeed, I wish to bring it before the society for the purpose of having you puzzle over it. I refer to the peculiar high vacuum phenomena observed by Mr. Edison in some of his incandescent lamps. I have in my hand an Edison incandescent lamp, having the same vacuum as the ordinary incandescent lamp. This one, however, has, in addition to the carbon filament, a platinum plate, or strip, that is thoroughly insulated from the filament, and supported in the manner seen between the two branches of the filament, as shown in Fig. 1.

FIG. 1.

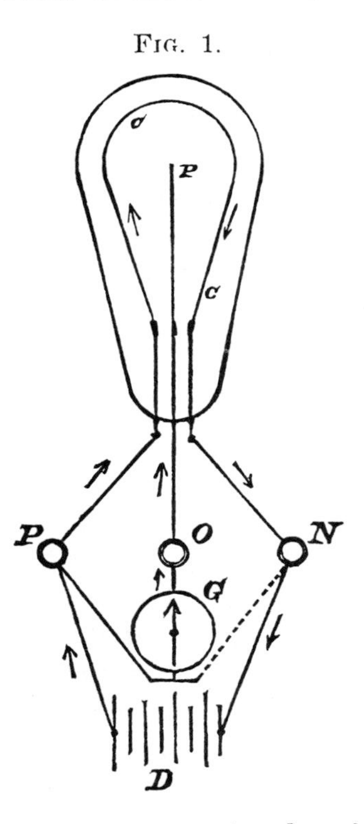

An Edison carbon filament, *c*, *c*, is placed inside an inclosing glass case in the usual manner. A strip of platinum foil, P, is supported as shown inside the loop. The binding posts, P and N, are connected to the ends of the carbon loop, and O to one end

Reprinted from *Trans. AIEE*, vol. I, pp. 1–8, 1884.

of the platinum plate. P is the positive, and N the negative terminal of the electric source.

The lamp being placed in the circuit of any electrical source, as, for example, a battery, D, the current will flow in the direction indicated by the arrows. The galvanometer, G, has one of its terminals connected with the positive terminal, P, of the electrical source, and the other terminal connected with one end of the platinum strip, or plate.

P, representing the positive terminal, the current may be conceived as flowing through the filament in this manner. [Indicating.] If the ordinary current used in producing the light is passing, then no unusual effects are noticed. If, however, the current is increased, so that the incandescence of the filament is raised from its normal, say, eight candle-power, to twenty—, thirty—, forty—, fifty—, or one hundred candle-power, then the needle of the galvanometer is violently deflected by a current passing through its coils.

The question is, what is the origin of this current? How is it produced? Since we have within the globe a nearly complete vacuum, we cannot conceive the current as flowing across the vacuous space, as this is not in accordance with our pre-conceived ideas connected with high vacua. I should mention here, that if the galvanometer terminal, instead of being connected with positive, P, of the electric source, be connected with the negative, N, then we also have a current flowing through the galvanometer, but this time in the opposite direction. But in this case its amount is much less, being but about $\frac{1}{40}$ of the amount of the first current.

I have no theory to propound as to the origin of these phenomena. I make these desultory remarks merely because I wish to call your attention to the phenomena, if you have not already seen them. It is very clear as to the direction in which the current flows, when one end of the galvanometer is connected with the positive terminal, P. This I have traced by observing the direction of deflection of the galvanometer needle. The current flows, not as we might suppose, from the carbon to the platinum; it flows apparently in this direction, as if it came from the platinum. If it flowed as though it went from the carbon to the platinum, then it might be ascribed to various causes. It may be electricity flowing through empty space, which I don't think probable; or it might be the effect of what may be called electrical convection, if there is such a term. The Crookes' discharge from one of the poles might produce an electrical bombardment against the plate, each molecule taking a small charge that might produce the effect of a current.

If we conceive, as I think most probable, a flow of molecules passing from the platinum to the carbon, then the phenomena may be readily explained as a Crookes' effect, since then we can regard a current as flowing in parallel circuit, from P, to N, through the carbon loop, and from P, through G, O, and P, to the carbon loop. But, remembering that the direction of the current is reversed, or apparently so, when the galvanometer is connected with the negative terminal, N, then the difficulty is to understand how the current there produced could possibly overcome the current from the source supplying the lamp. I should say here, however, that I am not entirely convinced from the few experiments I have tried myself, as to the actual existence of such a current. The deflection of the galvanometer needle, when connection was made with the negative terminal, being quite feeble. I am assured, however, that decided deflections have been observed.

I saw another lamp, in which two platinum plates are placed inside the lamp, but unfortunately I have not succeeded in finding the gentleman exhibiting it, so as to obtain from him an explanation of its peculiarities.

FIG. 2.

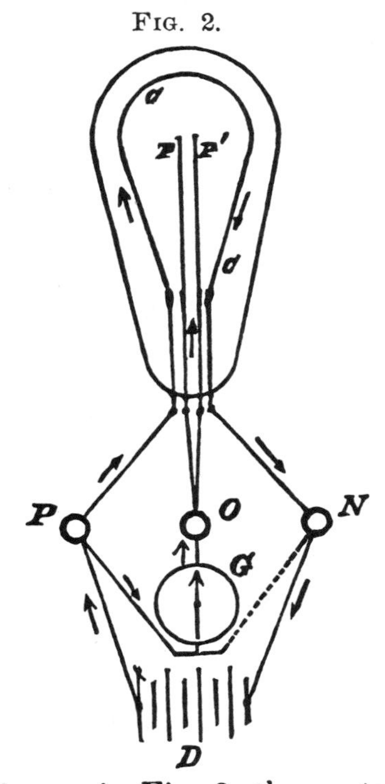

In this lamp, as shown in Fig. 2, the parts are the same as in Fig. 1, only two platinum strips, P, P′, placed parallel to each other, are employed. If a current is produced when the terminals, O, are connected with the galvanometer, then the phenomena is still more difficult of explanation. I believe, however, that one pole of the electrical source is always connected with the galvanometer, the other being connected with either or both of the platinum terminals. If such is the case, then the phenomena are simply modifications of the preceding.

I believe we shall find the preceding phenomena worth looking into. For my own part I am somewhat inclined to believe that we may possibly have here a new source of electrical excitement. That in some way the molecular bombardments against the platinum plate may produce an electrical current. At least, supposing it to be true that when the terminal, N, is connected with the galvanometer, the current through the galvanometer flows in the opposite direction to the current from the electrical source, it would seem that the phenomena could not be ascribed to the Crookes' effect.

Fig. 3.

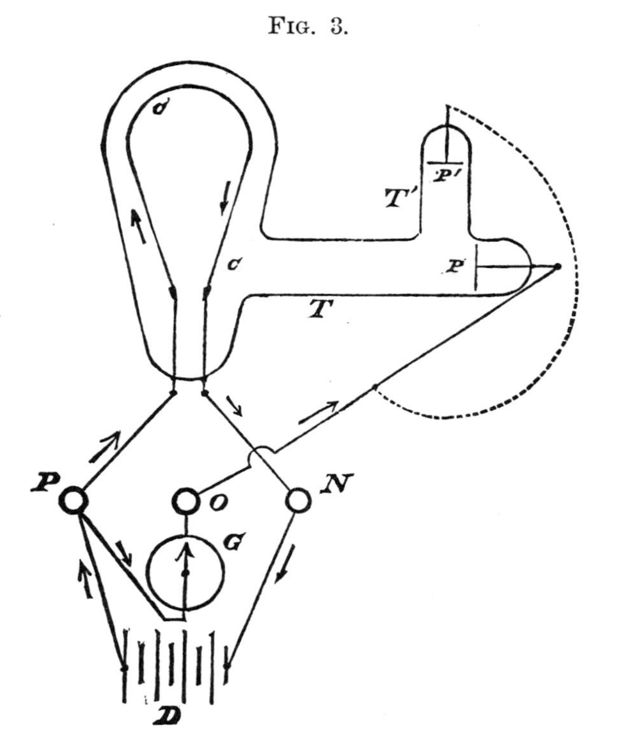

Supposing, however, this opposite current out of the way, then a more careful inquiry may show the sufficiency of the Crookes' effect for the explanation of the phenomena. Indeed, an experiment Mr. Edison has made, would appear to throw no little light on the matter. Instead of placing the platinum pole, P, inside the carbon loop, it was placed in the end of a long tube, T, T′, forming a part of the lamp case. When the connections were made, as shown, with the platinum at P, so as to place it in line with the carbon, and, therefore, expose the filament to the bombardment of the molecules shot out from the platinum, the needle of the galvanometer was deflected, even though the tube, T, was surrounded by a freezing mixture. When, however, the platinum was placed at P′, in the branch tube T′, out of the direct line of the carbon, no effects were observed in the galvanometer.

I have brought this matter before you in this crude way, thinking it may be interesting, and that it might possibly elicit a discussion which would aid us in ascertaining its cause.

Mr. KEITH:—May I ask if the current always flows from the platinum plate towards the carbon?

Prof. HOUSTON:—I am not sure; I think in one case it flows from the carbon to the platinum.

Mr. KEITH:—That mystifies the matter still further.

Prof. HOUSTON:—Certainly.

Mr. KEITH:—It seems to me that there is a galvanic couple. It will be, provided that at all the times the current flows from the platinum, reaction may take place; or even if it flows from the carbon to the platinum, no matter whether constantly or not. The remarks that I was going to make are spoiled by lack of data on the subject.

Mr. PREECE:—Prof. Houston, may I ask you on what grounds you assert, or anybody asserts, that electricity flows in one direction rather than another?

Prof. HOUSTON:—Simply on the ground of its being a definite idea. I don't know that any one could prove that electricity flows more readily in one direction than another; but we have a convenient way of speaking of it as flowing from a higher to a lower potential. I think the facts would appear to warrant us in saying so. Of course, I do not know whether it is so or not. For my own opinion I would say that I think electricity is the transmission of chemical or atomic action, and I speak of it as flowing, or being transmitted, in a definite direction.

Mr. PREECE:—For that reason I asked the question, because there is no difference as to the use of the term or the definition or value of the current; it is a pure convention. I do not think that we have any facts of any sort or kind to justify us in assuming that electricity flows in one direction rather than the other; but I would mention this, that in all recent inquiries that Crookes has made, and also Spottiswoode, before his death, they all had the tendency to show that if electricity does flow at all in any direction, that direction is rather from the lower to the higher potential, than, as we are all accustomed to admit and believe, from the higher potential to the lower potential. In other words, the tendency of the inquiry of Spottiswoode, De la Rue, and Crookes, and one or two German inquirers, whose names at present I forget, all rather tend to confirm your suspicion that electricity flows in the reverse direction to that which we are accustomed to admit.

Every other electrician present at the Exhibition, I think, has watched this experiment with very great interest. I feel puzzled in reference to it, and I feel that it is one of those things that wants to be very carefully and cautiously examined. It is also one upon which it is very dangerous to express an opinion. I was foolish enough the other night, when the experiment was shown to me at Mr. Edison's exhibit, to express an opinion, but I regret that I did so. However, I intend to exercise my persuasive eloquence upon Mr. Edison when I see him next week to induce him to give me one of those lamps, and when I go back to England I shall certainly make an illustration before our society there, and then make a careful inquiry into it. My own idea of electricity and the nature of this experiment is this: That for the production of the current you must have two things; you must have two points at different potentials, separated from each other by matter, and in this whether the direction be from the positive to the negative, or vice versa, is a matter of very little consequence; but to produce a current, as we have it here, we must have two points separated from each other by matter. The principal point here is what is that matter, and where is that matter.

I saw at Montreal an excessively beautiful experiment, tending to show that you had a Crookes' effect from platinum and carbon, just as much as we know that we have an effect from carbon and anything else. We know, for instance, in every incandescent lamp, that when you raise the filament to a certain condition of incandescence, there you have the Crookes' effect, because we see the carbon deposited upon the glass. But when we have platinum, it is a very difficult thing to prove that there exists the Crookes' effect. Dr. Oliver Lodge showed an extremely pretty experiment, to illustrate the fact that fogs resulted from the formation of moisture around a nucleus or molecule of matter. He took a large glass globe, apparently chemically clean, exhausted the air, but with moisture in it. It remained perfectly transparent as long as no matter was admitted, but the moment a spark from an induction coil was passed between the two platinum electrodes in this apparently chemically clean tube, at once a cloud was formed throughout the whole of the tube, showing that the passage of the spark between those two platinum points produces a Crookes' effect. That is, there was a bombardment of the platinum molecules, filling the whole of the tube, and the result was, the fog was formed. Now, here, suppose that that shall form, it may be, as has been suggested, that there is a new source of electricity. It may be it is due to something that we don't know of at the present. I am quite certain that we shall find the cause of this remarkable phenomena to be due to the Crookes' effect.

Prof. Houston :—I would like to say, in answer to the gentleman who has just spoken, that I am aware of the experiments that he has mentioned, some of which have been urged as proving the existence of a negative current. It seems to me that all they show is the transference of matter from the negative to the positive pole. Now, whether that constitutes an electric current or not, I do not think we can quite determine. The peculiar thing about the experiments I have mentioned is in the reverse direction of the current when the galvanometer is connected to the negative terminal. The difficulty exists, however, in finding a satisfactory explanation of the phenomena observed ; whether the current is flowing from the positive or from the negative.

Mr. Keith :—Mr. Chairman, I think the remark, or word, the convention, of assuming a direction for the current, may be used in electrolysis. The current may be considered as carrying metal from the anode to the cathode, and thereby producing a deflection of the galvanometer needle, or a positive-direction wave constantly to the right hand, or to the left hand, as indicated by the deflection of the needle. However, I have often noticed in arc lamps this fact: when we have a very long arc there oftentimes comes a sweeping movement around the ends of the positive and the negative carbon, and there is a little flame proceeding from the negative carbon towards the positive, which seems to rise, and have its point, or beginning, as we may say, upon the negative, running up like a brush towards the positive. You have current enough to spread around this surface, or portion of the surface of the upper carbon, seeming to emanate from the negative and proceeding to the positive. It probably has been noticed by everyone who has worked the arc lights, that same phenomena, as though something was passing from the negative and going to the positive. This cannot be quoted as an illustration of the transference of carbon from the positive to the negative, as is manifested upon the negative carbon when the short arc is used.

Mr. PREECE:—I think that some similar phenomena are noticed with the long platinum wire when it is raised to the state of incandescence. You find the heat commences at the negative end and flows in the other direction from some cause.

Mr. KELLY:—Some explanation ought to be given in reference to incandescent lamps at a high temperature. As to the current, I would say there is a discharge from one side of the carbon to the other. That is noticed to some extent in the same plate. Perhaps the induction is strong between the two sides of the carbon. Now, I don't see any special reason for supposing that when the platinum circuit is closed the discharge is interrupted; instead of passing from one carbon to the other, the discharge takes place between the negative side of the carbon and the platinum plate.

de Forest on the Audion

IN this paper delivered at an IRE meeting in 1913, one of the most colorful and controversial pioneers in radio and electronics discussed his most famous invention, the grid triode amplifier. It was one of the seminal inventions of the 20th century and gave rise to a major new industry. Although de Forest had patented the device in 1908, its potential usefulness as a telephone amplifier and radio amplifier was not generally recognized until 1912. It was converted into a reliable device by Arnold, Langmuir, and others in time to play a significant role in communications during World War I. See Robert A. Chipman, "De Forest and the Triode Detector," *Scientific American*, pp. 93–100, March 1965. Also see the papers by Susskind, Norberg, and Clark in the special issue of the PROCEEDINGS OF THE IEEE for September 1976. See also the discussion of the long controversy between de Forest and Armstrong in the PROCEEDINGS OF THE IEEE, p. 1082, August 1963.

Lee de Forest (1873–1961) was born in Council Bluffs, Iowa and graduated from Yale in 1896. He received a Ph.D. degree from Yale in 1899, and after working a few months for the Western Electric Company, he resigned to develop a radio system. The de Forest Wireless Telegraph Company was organized in 1902. His experiments with the Fleming diode that began in 1905 culminated in his invention of the triode that he patented in 1908. de Forest later sold the patent rights to A.T.&T. after a demonstration of the amplifier by de Forest and John Stone in October 1912. See the *DSB*, vol. 4, pp. 6–7 and the *NCAB*, vol. A, pp. 18–19.

THE AUDION—DETECTOR AND AMPLIFIER *

BY DR. LEE DE FOREST

Past-President, Society of Wireless Telegraph Engineers

Notwithstanding the now wide-spread use of the Audion as a detector in radio telegraph and telephone service few accounts of independent research into the nature of this instrument, or even cursory descriptions of its operation, seem to have been recorded, since the original paper presented before the American Institute of Electrical Engineers in 1906.

At that time, pending patent applications prevented the presentation of a detailed description of certain forms and improvements which are to-day common knowledge among radio engineers. It is my purpose herein to outline briefly the subsequent development of the Audion, both as a radio detector and as an amplifier of minute electric impulses.

For the benefit of those not familiar with radio detectors a brief description of the Audion follows:

A small incandescent lamp bulb is provided with a tantalum filament F (Figure 1), operated at from 4 to 15 volts. There is mounted close to one side of this filament, and parallel to its plane, a small plate of nickel, W. This plate is connected thru a telephone receiver, or other indicating device, R, with the positive terminal of a number of dry cells, B, arranged to give from 15 to 40 volts by the use of a multi-point switch. The negative terminal of this battery is connected to one side or terminal of the filament, which latter is lighted from a suitable storage battery, A.

Between the filament and the plate is mounted a third electrode, G, usually in the form of a grid-shaped wire, or a perforated plate. Approximately 1/16th of an inch (1.5 mm.) separates the grid from the plate on one side, and from the filament on the other.

In Figure 4, a regular commercial Audion receiver set, with its bulbs and adjusting switches, is shown. The two bulbs are mounted in an inverted position in order that the heated filaments do not sag into contact with the grid and their supporting wires after prolonged use. The plates of these bulbs are clearly visible.

* Delivered before The Institute of Radio Engineers, December 3, 1913.

Reprinted from *Proc. IRE*, vol. II, pp. 15–17, 19–21 (extract of a longer paper), 1914.

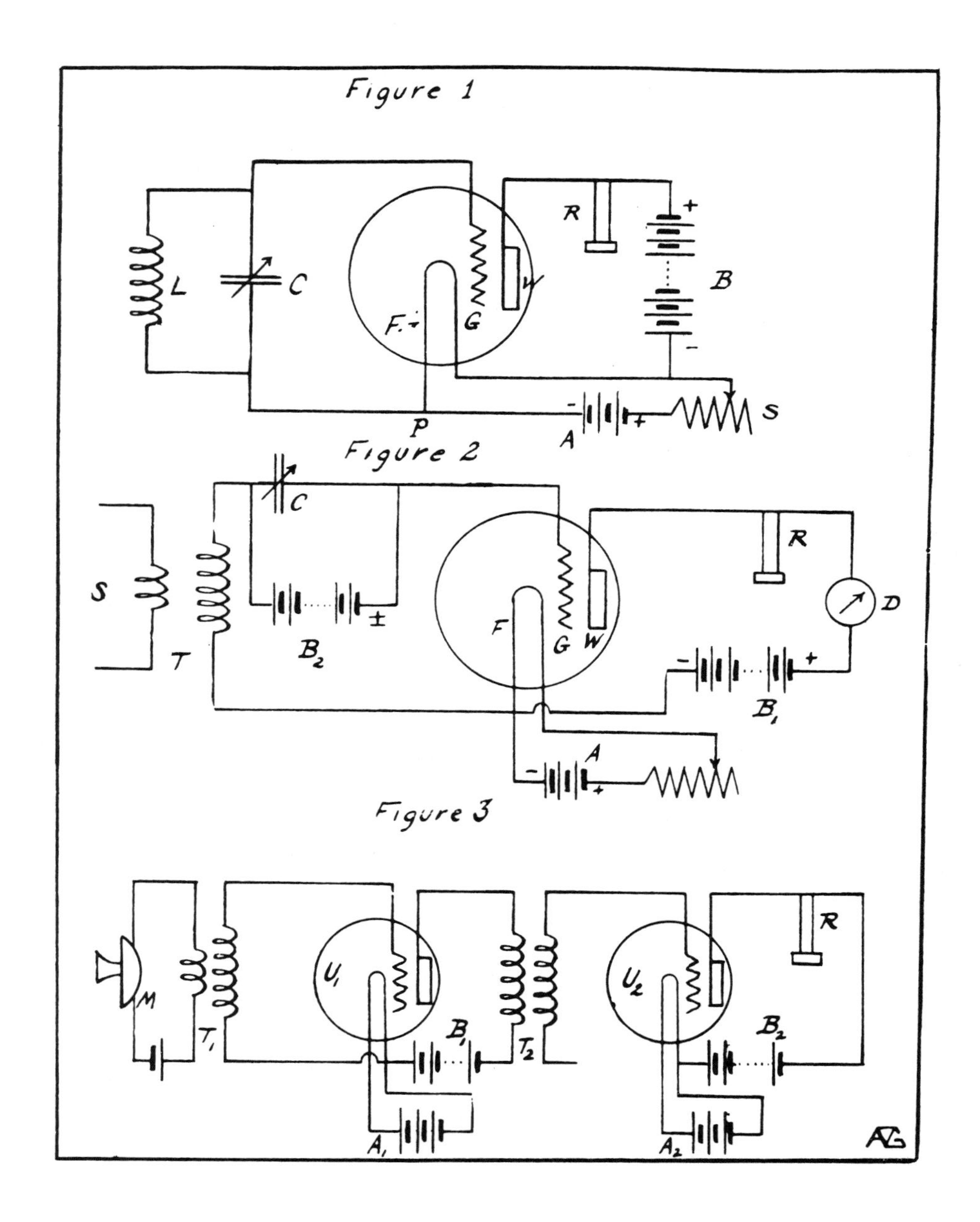
Figure 1
L
C
R
B
+
−
F
G
W
P
A
S
Figure 2
C
S
T
B₂
±
F
G
W
R
D
−
+
B₁
A
Figure 3
M
T₁
U₁
B₁
T₂
U₂
B₂
R
A₁
A₂
AG

From the protected lower ends of these bulbs the terminal wires of the grid and plate pass out. The top three-point switch places either of the Audions in circuit. The center rotary switch inserts more or less resistance in the "A" battery circuit. The lower nine-point switch enables the insertion of extra batteries in the "B" battery circuit; these batteries being inserted in sets of three at a time.

When used as a detector, this grid electrode is connected to one terminal of the source of received radio frequency oscillations, the other terminal being connected to the Audion filament at P.

In the unlighted bulb, the resistance between the electrodes is practically infinite. When the filament is at white heat, a resistance of from 10,000 to 30,000 ohms is found between it and the cold anode. The value of this resistance depends upon a variety of conditions; such as filament temperature; size, shape, condition and location of the two outside electrodes; amount of electromotive force impressed on the gas; degree of vacuum; nature and degree of purity of the gas; magnetic forces to which the ions are subjected; *and the instantaneous value and sign of the electric potential impressed, or residing, on the intermediate, or grid, electrode.*

The introduction within the exhausted bulb of this third electrode (preferably, but not necessarily, located between the filament and the anode plate) at once placed the Audion in a distinct class, both as to sensitiveness when used as a responsive radio detector, and also as to its mode of operation. No longer, by any form of argument, could the grid Audion be classed as a rectifier, or "vacuum valve." Attempts to confuse it with the so-called Fleming "valve tube" have, I believe, failed to convince any one who has actually compared these two detectors, when each is properly constructed and connected. The difference observed in their operation is more than merely one of degree.

The vacuum valve, as first discovered by Edison, and carefully studied by Elster and Geitel*, has most marked rectifying properties, excelling in efficiency any electrolytic or crystal rectifier.

In this original device the cold electrode (or electrodes), was connected to the positive terminal of the filament. It was itself subjected to what may be, under favorable conditions, a very energetic bombardment of negatively charged corpuscles from the negative portions of the filament, while itself giving off prac-

* Wiedemann's Annalen, XVI, 1882.

tically no negative carriers. This valve rectifies small currents of any frequency, and can therefore be used in connection with a continuous current galvanometer to detect and measure the half-oscillations impressed on a tuned radio frequency transformer, the secondary terminals of which are connected to the anode plate and to one end of the filament respectively.

Starting from my experiments with the rectifying qualities of a gas flame (where the sources of heat energy and of the impressed e. m. f. across the electrodes were distinct and separate), the writer early found that if this incandescent-filament vacuum-rectifier were provided with a second battery, of proper voltage, independent of the lighting source and connected between one leg of the filament and the plate, (which plate must *always* be made the anode); then the device became more than a simple rectifier.

It took on the nature of a true detector, a "wave-responsive" device. In other words, it became to a degree a "trigger" tube, a genuine *relay*, giving responses in the indicating instrument (which was now preferably the telephone receiver) of considerably greater intensity than could be had from the received currents simply rectified. The regular and steady departure of negative ions, or carriers, from the cathode under stress of the "B" battery was found subject to sudden and great variations, following the variations of the applied radio-frequency e. m. f.; the degree of sensitiveness depending, for any given bulb, upon the heat of the filament and the amount of this applied "B" battery voltage.

At once the old vacuum valve took rank in sensitiveness with the electrolytic detector, until that time the most sensitive known.

The next step in advance was, as above stated, the introduction of the electrical impulses into the gaseous medium by means of a third and independent electrode.

This grid and filament, connected across the condenser of a receiving circuit, adds a slight capacity and a very high resistance shunt to that condenser. The additional damping thus introduced in the oscillating circuit is excessively small, its resonant qualities are unimpaired. The Audion thus lends itself, far more than any other form of energy-transforming detector, to sharp tuning. I have found properly designed receiving circuits equally sharply tuned, whether the Audion or the make-and-break contact detector (Poulsen tikker) be used therewith.

There appears to be no lower limit of sensitiveness to the Audion, no minimum of suddenly applied e. m. f., below which the received impulses fail to produce any response. The exciting impulse, it is true, may be be too minute to cause a directly discernible effect in the "B" circuit, but if this effect be amplified through one or more steps, as will be hereafter described, the effect of the original excitation is found to exist.

Even in this day of universally accepted long-distance "wireless records," it is difficult for those who have not actually used the Audion detector to believe some of the receiving feats which are scored to its credit.

The Audion has the further advantage of entire absence of adjustment in the receiver itself. If the two battery potentials are once properly adjusted it requires no further attention. There is no fatigue under any conditions of use. A powerful spark discharge in close proximity may cause the "blue arc," or visible cathode discharge, to pass, but after a second or two this "paralysis" will automatically disappear, while a high resistance path (of the order of several megohms) between plate and grid will generally prevent this blue arcing.

This is therefore a suitable detector to use with a telegrapher's "breaking key." If a key-actuated relay be arranged to open the filament circuit with each depression of the Morse key, the Audion is always ready to catch the distant operator's call, or attempt to "break" in the midst of a message. On the other hand, the sensitiveness of the electrolytic and that of the best of the crystal detectors is frequently destroyed by one violent impulse, unless protected by shunting and disconnecting switches.

If there is any justification in the attempt to classify radio detectors as "current-operated" or "potential-operated" devices, there can be not the slightest hesitancy in classing the Audion as a potential-operated detector. Obviously it is in no sense a contact device, perfect or imperfect. Its response is strictly quantitative, up to the critical point of the voltage-response curve where "saturation" of ions or gas carriers begins. Up to that critical point the change in current passing from plate to filament is essentially proportional to the voltage applied to the grid electrode.

This voltage may be merely that of the hand discharging into the grid. When the lead to this grid is suddenly grasped, a click may be heard in the telephone receiver in the "B" circuit. If the charge thus impressed upon the grid be negative a repulsion or scattering of the negatively charged carriers emanating from the filament occurs. If the impressed charge be positive, then these carriers may be attracted to the grid and discharged there, or delayed in the neighborhood. In either case, therefore, a diminution in the number of ions reaching the plate results, and we observe a diminution in the deflection of a sensitive milliameter or galvanometer in the "B" circuit when a prolonged series of impulses is delivered to the grid. The normal current is usually of the order of a milliampere, and its response diminution may vary from an undiscernible amount to 50 or even 90 per cent. of its full value.

An insight into what forces are at work in the Audion is afforded by experiments with a special circuit, such as that shown in Figure 2. When a negative charge is applied to the grid a click is heard in the telephones. When a positive charge is applied, almost no sound is heard. If G is negatively charged to 20 volts, while a sound from telephonic currents from source S is being received, the sensitiveness of the Audion is practically annulled. When G is positively charged the sound is greatly reduced, but not to the foregoing extent. This experiment seems to show that in the normal operation of the Audion the imposition of a charge, negative or positive, upon the grid acts either to repel from its neighborhood the ionic carriers or to hold them idle there, thus in either case increasing the effective resistance in the filament-to-plate path.

The first establishment of the ionic circuit of an Audion when the filament is suddenly lighted is perfectly silent—no sound is heard in the telephone, altho a galvanometer in series shows that the "make" is practically instantaneous. The establishment of the gaseous conductivity is sufficiently gradual, however, to make no sound. I do not know of any similar rapid circuit-closing device which is thus silent. While this "B" current assumes nearly its maximum amplitude practically instantly, its full value may not be reached for several seconds. A slow and irregular creeping of the milliameter needle reveals how gradually the ions adjust themselves to a completely stable condition. Upon the rupture of the filament circuit, however, the "B" circuit opens practically instantly, invariably accompanied by a click in the receiver telephone.

Referring again to the arrangement in Figure 2: when the grid is positively charged, the voltage of B_2 made nearly equal to that of B_1, and the heating current carefully adjusted, a very intense whistling note may be obtained in the telephone receiver, loud enough to be heard a meter from the instrument.

••• ••• •••

George Squier on Vacuum Tubes in World War I

IN this extract from a longer paper from the *AIEE Transactions* of 1919, the Chief Signal Officer of the Army Signal Corps reviewed the dramatic advance in the production and utilization of vacuum tubes that had been stimulated by military needs. The production rate had been increased to more than one million per year through the use of factory methods and standardization. Squier forecast that vacuum tubes would soon "become widely used in every field of electrical development." He also mentioned that the airplane radiophone had been among the "most spectacular achievements" of the war. See a paper by Paul Clark on the early impacts of communications on military doctrine in the PROCEEDINGS OF THE IEEE for September 1976.

George O. Squier (1865–1934) was born in Michigan and graduated from West Point in 1887. He continued his studies at Johns Hopkins University while stationed at Fort McHenry and received the Ph.D. degree in physics in 1893. He invented the polarizing photochronograph for the measurement of projectile velocity while stationed at Fort Monroe and was transferred to the Signal Corps in 1899. He was commander of a cable ship in 1901–1902 and made several inventions relating to high-speed telegraphy and carrier telephony. He was instrumental in acquiring the first airplane for the Army from the Wright Brothers in 1908. Squier served as military attache in London from 1912 to 1916 until he was recalled to become Chief Signal Officer in 1917. He retired in 1924. See "Electrical Engineers of the Times. Major G. O. Squier," *Electrical World*, vol. 45, p. 928, 1905. Also see *NAS*, vol. 20, pp. 151–159 and *NCAB*, vol. 24, p. 320.

Radio Development Work

IN the question of the engineering achievements of the Signal Corps during the war, the development of radio apparatus forms a large part. Inasmuch as the vacuum tube occupies so prominent a role in almost every kind of radio apparatus, an outline of its development logically precedes discussion of the radio sets.

Vacuum Tubes

The application to radio inter-communication of the vacuum tube—perhaps more properly called the thermionic tube or bulb—is one of the most interesting developments in the whole field of applied science. For not only has it made possible what has been justly heralded as one of the most spectacular achievements of the whole war—the airplane radiophone—but the confidence growing out of the extensive experience with the vacuum tube in warfare, coupled with its extreme adaptability, have resulted in a rapidly increasing amount of radio development involving its use.

Pre-War History

The vacuum tube was known in various forms before the war. Following extensive experiments with the so-called "Edison effect", Fleming, some years ago, produced the well known Fleming valve—a current rectifying device, capable therefore of being used as a detector of radio signals. This device contains two elements: an incandescent filament emitting electrons, and a plate upon which an alternating voltage is impressed, both placed within an evacuated bulb. Later Dr. Lee DeForest introduced an important modification by placing a wire mesh or "grid" between the filament and the plate. A small voltage variation on this grid produces the same current change through the tube as would a much larger voltage variation on the plate, thus adding amplifying properties to the detector characteristics of the Fleming valve. DeForest called his device the "audion". Later, with superior facilities for evacuation available and with a more intimate knowledge of the laws of thermionic emission from hot bodies, improvements and modifications were made in the audion or vacuum tube by both the General Electric Company and the Western Electric

Reprinted from *Trans. AIEE*, vol. XXXVIII, part I, pp. 45–48 (extract of a longer paper), Jan. 10, 1919.

Company, the latter designating their product as "vacuum tube", and the former the "pliotron".

In addition to acting as detectors and amplifiers, as mentioned above, vacuum tubes can function in two other important ways:

1. As Oscillators. In properly designed circuits containing inductance and capacity they will act as radio frequency generators, for use in transmitting or receiving radio signals.
2. As Modulators. By suitable connection to an oscillator circuit or antenna, they can be made to vary the power radiated so that the envelopes of the waves transmitted shall have any desired wave form, as for example, the speech waves from an ordinary telephone transmitter.

The most striking use made of vacuum tubes prior to the time we entered the war was the transmission of speech by radio from Washington to Paris and Honolulu, during the experiments carried out by the American Telephone and Telegraph Company and the Navy Department. Vacuum tubes were used as the radio frequency generator for transmitting, and for detector and amplifier in receiving.

When the United States entered the war, vacuum tubes already were in use by the Allied forces for various signaling purposes. The French particularly had been quick to recognize the military value of vacuum tubes and had, previous to June, 1917, developed very creditable tubes and apparatus. In America, tubes were in limited use as "repeaters" on telephone lines, and as detectors and amplifiers in laboratories and radio stations. The total production, however, in this country did not exceed three or four hundred a week.

Developments During the War

Early in our participation in the war, it became evident that vacuum tubes would be required in very large quantities in order to meet the growing demands for radio communication and signaling. It was equally evident that service conditions, not hitherto anticipated, would require great mechanical strength, freedom from disturbance under extreme vibration, and uniformity of product sufficient to make possible absolute interchangeability of the tubes in sets, without the necessity of readjusting when changing tubes. To these conditions must be added that of minimum size consistent with dependable operation.

To make such a device, with its complicated, yet accurately

constructed metallic system, within a practically perfect vaccum, is no small problem even when made in the laboratory, on the individual unit basis, by a skilled operator who appreciates the delicacy of the job. To turn out tubes by the thousands by factory methods involves almost infinitely greater difficulties. How well certain companies, in collaboration with the Signal Corps, have succeeded in solving these difficulties is indicated by the fact that recently the total rate of production in the United States of high quality standardized tubes was considerably in excess of one million a year. This rate of production could be made many times greater on short notice.

As an example of the difficulties which this quantity production has involved may be mentioned that of evacuation. The degree of vacuum required is such that unusual methods of exhaust are necessary. The heating of the tubes in electric ovens is supplemented by heating of the elements of the tube by excessive filament and plate electrical power input. Molecular pumps are employed, necessitating an extremely large number of pumps to handle quantity production. Special treatment of metal parts prior to assembly is employed to reduce the gas given off by them during the exhaust process.

Another problem is that of making the complicated metallic structure of all tubes exactly alike, in order to insure identical electrical properties. As an indication of progress in this direction, it may be stated that one company is prepared to manufacture, in quantity, a certain tube in which the clearance between filament and grid is only three hundredths of an inch, the allowable variation being of course only a small percentage of this.

Manufacturing in quantity involves careful inspection. The problem of specifying definitely the required performance of tubes, the development of adequate testing specifications, the placing of standardized testing and inspection methods, personnel, and equipment in the various factories so that tubes manufactured at different times and places would, after passing inspection, be uniform and interchangeable—these questions were entirely new and have been solved almost entirely by the Signal Corps Engineers.

Present State of the Art

Tubes developed by the Signal Corps may be divided into two general classes: the tungsten filament types as developed and manufactured by the General Electric Company and the De Forest Radio Telephone and Telegraph Company, and the coated filament or Wehnelt Cathode types as developed and manufactured by the Western Electric Company. The coated filament tubes so far have proven superior to the tungsten filament tubes for Signal Corps use. Both classes have been standardized as regards base, exterior dimensions, filament current and voltage, and in addition, plate voltage and output for transmitting tubes; and amplifying power and detecting power for receiving tubes. Except in certain special cases, the Signal Corps uses two types of tubes, one for transmitting, and another for receiving. The French and the British have been using one type for both transmitting and receiving, but present tendencies of the British are toward different tubes for different duties.

Vacuum tubes are now employed for electric wave detection, radio frequency and audio frequency amplification, radio telephony, particularly in the airplane radiophone, continuous wave radio telegraphy, voltage and current regulators on generators, and for other miscellaneous purposes. However, varied as are the applications at present, the uses, actual and potential, growing out of war development work have proved that the art of Vacuum Tube Engineering, and the application of its products to radio engineering, telephone and telegraph engineering, and particularly to electrical engineering in general, are still in their early infancy. That vacuum tubes in various forms and sizes will, within a few years, become widely used in every field of electrical development and application is not to be denied.

The engineering advancement accomplished in less than two years represents at least a decade under the normal conditions of peace, and our profession will, it is hoped, profit by this particular salvage of war which offers perhaps the most striking example extant of a minimum "time-lag" between the advanced "firing line" of so-called pure physics, and applied engineering.

The Chief Signal Officer considers that the work of standardization and quantity production of vacuum tubes accomplished during the last eighteen months under the pressure of military necessity, represents an advance in the art of electrical engineering which will prove of inestimable industrial and scientific value to this country, and to the engineering world at large.

••• ••• •••

The Varians on the Klystron

THE invention of the klystron was a major event in the history of electronics. Its technical features and the context in which it was invented are discussed in this paper by the Varian brothers. As they point out, the trend toward higher frequencies had approached the practical limits of both conventional circuits and amplifier tubes. Their resolution of the dilemma had been to take advantage of the electron transit time that had seemed previously to be undesirable. See Edward L. Ginzton, "The $100 Idea," *IEEE Spectrum*, pp. 30–39, February 1975. Also see Arthur Norberg's paper on the origins of the electronics industry on the Pacific Coast in the PROCEEDINGS OF THE IEEE for September 1976.

Russell H. Varian (1898–1959) was born in Washington, D.C. and received the B.A. and M.A. degrees from Stanford University in 1925 and 1927, respectively. He worked for Humble Oil Company and for the Farnsworth Radio and Television Company until 1933 when he returned to Stanford to resume his studies in physics. He became a close associate of William Hansen who was developing resonant cavities, and his brother also came to Stanford as a research associate in 1937. The klystron was conceived during the summer of 1937 and verified experimentally later the same year. The Varians, Hansen, and others from the Stanford group worked on klystron development at the Laboratories of the Sperry Gyroscope Company during the war. They returned to California in 1948 to found Varian Associates. See Orrin E. Dunlap, Jr., *Radio's 100 Men of Science* (N.Y., 1944), pp. 272–274 and an obituary in PROCEEDINGS OF THE IRE, vol. 47, p. 15A, Oct. 1959.

Sigurd F. Varian (1901–1961) was born in Syracuse, N.Y. and attended the Polytechnic School at San Luis Obispo, Calif. He learned to fly shortly after World War I and was employed as a pilot by Pan American in 1929. His concern over the need for methods to detect bombers was an important factor in stimulating his brother's interest in developing microwave tubes, and his mechanical skills were also useful in making the first klystrons.

A High Frequency Oscillator and Amplifier

RUSSELL H. VARIAN AND SIGURD F. VARIAN
Stanford University, California

(Received January 6, 1939)

A d.c. stream of cathode rays of constant current and speed is sent through a pair of grids between which is an oscillating electric field, parallel to the stream and of such strength as to change the speeds of the cathode rays by appreciable but not too large fractions of their initial speed. After passing these grids the electrons with increased speeds begin to overtake those with decreased speeds ahead of them. This motion groups the electrons into bunches separated by relatively empty spaces. At any point beyond the grids, therefore, the cathode-ray current can be resolved into the original d.c. plus a nonsinusoidal a.c. A considerable fraction of its power can then be converted into power of high frequency oscillations by running the stream through a second pair of grids between which is an a.c. electric field such as to take energy away from the electrons in the bunches. These two a.c. fields are best obtained by making the grids form parts of the surfaces of resonators of the type described in this Journal by Hansen.

I. INTRODUCTION

IT has long been realized that apparatus for transmitting and receiving electromagnetic oscillations at wave-lengths of the order of 10 cm or shorter should lead to the accomplishment of a large number of useful objectives, because wave-lengths of this order would permit concentration of the radiation into rather sharply defined beams. Such waves have been produced in many ways since the time of Hertz, and can be produced readily by a spark transmitter and in other ways; but the perfection of the three-electrode tube, and in particular the ease with which it obtains great amplifications, has raised our standards so far, at least for low frequencies, as to create a hope for similar apparatus for ultra-high frequencies. It is the purpose of the present paper to describe a new type of electronic device which has made progress in this direction.

The reasons for certain important features of this device may be seen most readily after a brief discussion of some of the difficulties encountered by the triode at very high frequencies. The fundamental difficulties have been of two kinds: with the resonant circuits and with the tube itself.

As a resonant circuit the ordinary coil and condenser become unsatisfactory as the frequency is increased, not only because it becomes too small to be mechanically convenient but even more because the losses become too great and therefore the shunt impedance becomes too low. Another related and yet partially independent difficulty is the decline in frequency stability as the losses increase. These difficulties can be, and are, partially avoided by using such resonators as concentric lines; but because of limitations due to leads and other reasons the resonant circuit often remains a limiting factor. Improvement in resonant circuits is, therefore, of primary importance.

Turning to difficulties with tubes, it is always found, with any given triode, that, as the operating frequency is increased too far, the efficiency declines and eventually the decline becomes so rapid that in a rather short range of frequency the efficiency drops from a usable value to zero. It has been found that this difficulty is associated with the transit time of the electrons. Whenever the time of transit from filament to plate becomes appreciable by comparison with the time of a quarter-cycle, the tube fails to act in the normal way. There are two obvious methods of attack: to increase the plate voltage,

Reprinted with permission from *J. Appl. Phys.*, vol. 10, pp. 321–327, May 1939.

and to decrease the physical size of the tube. The former of these methods cannot be carried far; but the latter method has made tubes now commercially available for work at fairly high frequencies. The difficulties associated with this

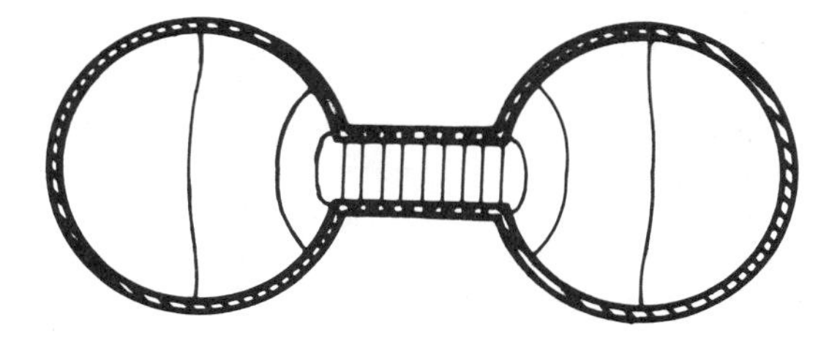

FIG. 1. Cross section of a copper enclosure, useful as a resonator. A qualitative representation of the electric field corresponding to the mode of oscillation used is also given. The horizontal dotted lines represent grids which should be of copper or other good conductor, and which should be designed so as to resist the current flow as little as possible.

method, however, are serious: first because of the mechanical troubles in producing the tube, and, much more important, because each decrease in size decreases the power capacity. It would seem therefore that there is a practical limit to useful results along this line. Accordingly, it becomes necessary to try another plan: not to attempt elimination of the effect of transit time, but to turn it to constructive use. The two objectives in this work, therefore, are first, the application to this problem of a new type of resonator, and second, a constructive use of the transit time.

II. The Resonators, and Excitation by Cathode Rays

Considering first the resonators, we apply here the results obtained by Hansen,[1] who had considered the resonant properties of practically closed metal vessels, like those sometimes called *hohlraums*. In our laboratory, these have been called "rhumbatrons," from the Greek word "rhumba," meaning rhythmic oscillation, and the familiar termination "tron," making the name mean the place where these rhythmic oscillations occur. Hansen obtained mathematical results on frequencies and power consumption for shapes of sufficient variety to make rough estimates possible for many other cases.

For such resonators there are no quantities corresponding exactly to the familiar self-inductance L, resistance R and capacitance C of an ordinary circuit; and there are several essentially different but almost equally good ways of defining quantities to correspond roughly to them. By using any of these ways, and letting $Q=L\omega/R$ as usual,* any simple non-re-entrant rhumbatron, such as a spherical one, has a Q of the order of λ/δ, where λ is the wave-length in free space and δ is the skin depth in the metal of which the rhumbatron is made. The quantity corresponding to shunt impedance, by any reasonable definition, is then expressible in absolute electromagnetic units as something of the order of Qc, where c is the velocity of light, 3×10^{10} cm/sec. Since Q is a pure number, the conversion to practical units affects only c, which becomes 30 ohms. If $\lambda=10$ cm and the metal is copper, $(\lambda/\delta)=5.88\times 10^4$; so such shapes give high values to Q and to the shunt impedance. For reasons to be explained below, the shapes actually used are re-entrant, as, for example, in Fig. 1; and then Q and the shunt impedance are somewhat reduced but their values are still very high.

Hansen proposed to use such resonators, excited by any feasible means, to accelerate electrons by running them through the resonator in the general region and direction of the strongest electric field. So a reversal of phase relations gives the opposite result, the electrons being decelerated and the energy they lose being transferred to the field. Thus to excite our resonator we make holes or grids in two of its opposing walls adjacent to the region of high electric field strength, and pass through these grids and this field a periodically changing stream of electrons, the changes being so timed that more electrons pass when the field is such

[1] W. W. Hansen, J. App. Phys. 9, 654 (1938), and much information imparted in private conversation.

* This may sound as if there are various definitions of Q possible, each leading to a different numerical value. What is intended, rather, is that this is an equation of condition, which, taken along with the equation $LC\omega^2 = 1$ forces us, on choosing some particular definition of, say, L to use with this definitions of C and R which will make the lumped constant "equivalent circuit" have the same frequency and the same damping ratio as the actual rhumbatron. This means that, whatever definition of L one starts with, Q is always defined as 2π times the ratio of the energy stored to the energy lost per cycle.

as to take energy from them than when the field is reversed. The reason why a sphere, for example, is not a suitable shape of rhumbatron is that its wave-length is only about 1.14 times its diameter; so even if an electron had a speed equal to that of light, it would never get through the sphere before the field reversed. For best operation, the path of the electron in the resonator must not be much longer than $\beta\lambda/2$, where $\beta = v/c$ and v = electron velocity, and it seems best to have the path still shorter. We therefore use as a resonator a copper enclosure of shape generally similar to that of Fig. 1.

It will be noted that no use is made of high frequency lead wires, nor is there any solid dielectric. Thus we obtain the full benefit of the desirable qualities of the resonator. Also, it should be specially noted that there is no radio-frequency field outside the resonator, and therefore no radiation, until we add such accessories as are needed for producing the radiation desired for practical uses.

III. Transit Time and Bunched Cathode Rays

The requirement for intermittent cathode rays, just described, cannot be readily met by grid control as in a triode for the reasons stated in our introduction, in connection with transit times. Consequently it leads back to the second

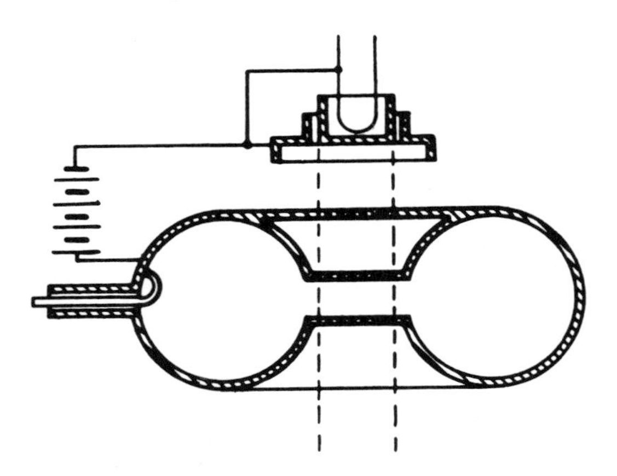

FIG. 2. Diagrammatic representation of a rhumbatron, like that of Fig. 1, a battery, and a cathode which emits a beam of electrons indicated by the parallel dotted lines. Also shown is a concentric line and a coupling loop for supplying radiofrequency power which maintains a radiofrequency field inside the rhumbatron. This field alternately accelerates and retards the electrons passing through.

problem stated at the conclusion of our introduction; namely, that of finding a constructive use for the transit time. The idea of constructive use of transit time, in general, is not new, since for example it appears in the magnetron and Barkhausen-Kurz oscillators. In the present case, however, the method of using this time is different, being more adapted to the present problem.

This method involves the use of an auxiliary rhumbatron, in a way shown diagrammatically in Fig. 2. Here the electrons from the cathode are accelerated by the electrostatic field produced by the battery. In the form shown here, using grids and straight beams, rather than holes and focused beams, the first grid serves to straighten the lines of force of this field. Beyond it the electrons pass through a field-free space to the second grid, and then through the high frequency field of the rhumbatron to the third grid. The latter field is not strong enough to change the speed of an electron by more than a small or very moderate fraction of its speed at the first or second grid. So the current carried by the cathode rays not only has the same value at all points in space from the cathode to the second grid, but also nearly the same to the third grid.

Beyond the third grid, however, the changes of speed have important effects. These are understood best by first considering an electron which passes the center of the high frequency field just as that field is changing from opposing to helping electrons. At the third grid, this electron has practically the same speed as at the second; but another electron which passed the center of the field a few electrical degrees earlier has had its speed reduced; and a third one, passing a few degrees later, is going faster. If there is plenty of field-free space beyond the third grid, these differences in speed cause the electrons ahead and behind the one of unchanged speed to draw nearer to it. Another electron of unchanged speed, a half-cycle earlier or later, has its neighbors draw away from it. Consequently at a suitable distance from the third grid the stream contains bunches of electrons denser than the stream at the third grid and separated by regions less dense.

This stream is of just the sort required for driving the oscillations in the rhumbatron discussed in Section II. Therefore, if this stream is sent through such a rhumbatron as that, and if that rhumbatron is oscillating in the proper

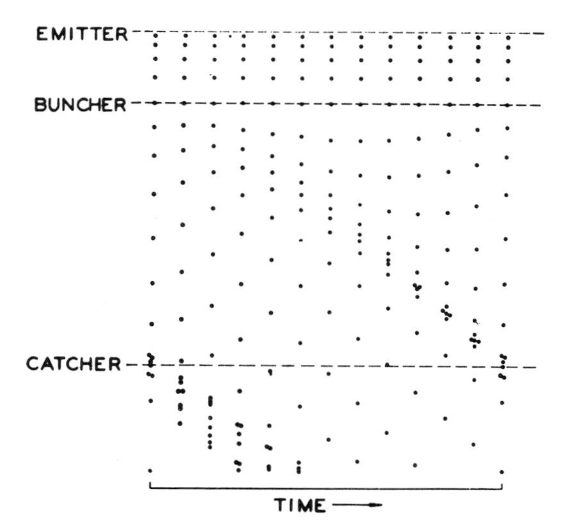

FIG. 3. Sequence of diagrams showing, for 12 times equally spaced throughout one cycle, the positions of chosen typical electrons (dots) in the beam shown in Fig. 2. As to phase, the first vertical row of dots shows the positions of the electrons at a time when the field in the "buncher" is zero and is coming to be in the direction to increase the speeds of the electrons. Horizontal lines show the positions of the cathode and the buncher rhumbatrons. A suitable place for the catcher is designated by the horizontal dotted lines.

phase to retard the electrons in the bunches, and if its oscillating field is strong enough to take more energy from the electrons than was given them by the rhumbatron of Fig. 2, the apparatus as a whole converts part of the d.c. power of the cathode rays into high frequency a.c. power.

Such an apparatus we call a "klystron," from the Greek verb "klyzo," expressing the breaking of waves on a beach. The rhumbatron in which the electrons are given their first modulation of speed is called the "buncher," and the one into which the bunches "break" is called the "catcher" —though it must be noted that the electrons may go through the catcher, if both its surfaces are grids, and that what it "catches" is not primarily current but power.

To show diagrammatically the character of these bunches, in Fig. 3 we have reduced the spacing of the grids in the buncher to zero, and likewise in the catcher, and drawn vertical columns of dots each representing a line of electrons in the places they occupy at one instant, and different columns representing such lines occurring at different instants throughout the cycle. To catch the most power, the catcher must satisfy three requirements: it must be placed where the electrons are in bunches of a certain best form, to be defined later in a paper by D. L. Webster; it must oscillate in the right phase; and the strength of the oscillations must be such as to reduce the electrons in the centers of the bunches just to rest. Under these conditions, as Webster will prove, with ideally perfect grids and other parts, it should convert 58 percent of the power of the cathode rays going through it into high frequency power. Practically, of course, we cannot claim to reach this ideal efficiency, but the efficiency is good.

As the klystron has been described here, with the buncher driven by some independent source of power, such as an antenna receiving a signal, it is evidently an amplifier; but to convert such an amplifier into an oscillator, we need only take a small part of the oscillatory power in the catcher and use it to excite the buncher. The phase relations must, of course, be correct; and if not otherwise adjustable electrically the proper phasing may be simply obtained by

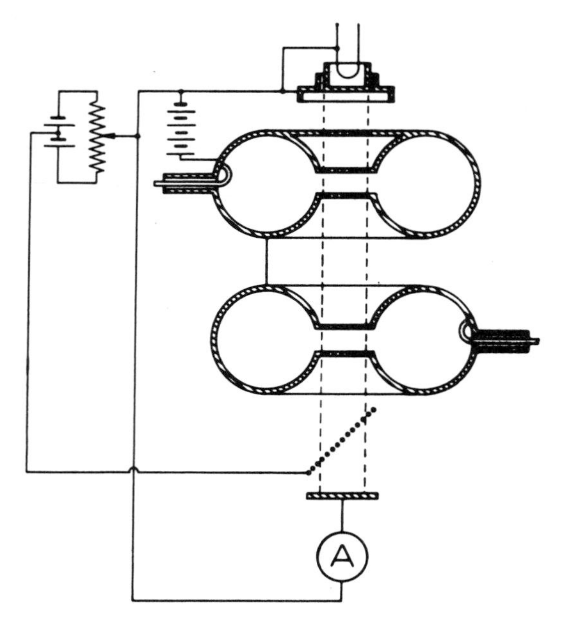

FIG. 4. Diagrammatic representation of radiofrequency amplifier and detector.

varying the mean speed of the electrons, by varying the P.D. between the cathode and the first grid. And, of course, a klystron almost converted from an amplifier to an oscillator makes a good regenerative amplifier.

Another point of interest in connection with the amplifier relates to the energy of the electrons after they have passed through the second rhumbatron. A substantial majority of the electrons give energy to the catcher; therefore they pass the last grid at lower speeds than that at which they passed the first grid. Also, it is unavoidable that a small minority of the electrons go through the catcher during the wrong half of the cycle and acquire more speed. Thus any device which sorts electrons according to their speed may be placed after the last grid of the catcher and be used as a detector. Perhaps the simplest such is a grid nearly at cathode potential. When there is no signal input almost all the electrons go through it. Then if there is a signal to the buncher, the buncher bunches the electrons, which in turn builds up a field in the catcher, and this field slows down some of the electrons so that they can no longer go through the detector grid: this causes a decrease in the current to any collector beyond the grid. Such a device is shown in Fig. 4. Alternatively, the grid may be biased so that with no signal the electrons just fail to pass, and a signal then causes an increase in the collector current.

Thus we have shown how amplification, oscillation, and detection may be achieved. Modulation may be obtained by varying the beam current; also by varying the cathode potential and so making the phase relations come nearer to or farther from the ideal ones noted above; and there are many alternative methods growing out of these.

IV. Practical Details of a Typical Klystron

In this section we describe certain constructional and design details that might not be obvious. For the most part these relate to a particular early oscillator which worked at about 13 cm and which is shown in Fig. 5, but information obtained from other apparatus has been included. Also, in Fig. 6 we have shown a photograph of various parts of interest.

First we may note the means needed to obtain the columnar beam of electrons assumed above. Electrons are drawn from a plane oxide-coated emitting surface A through a plane grid of tungsten wires B. Were it not for space charge, the field between these two planes would be straight and the electrons would all leave in the

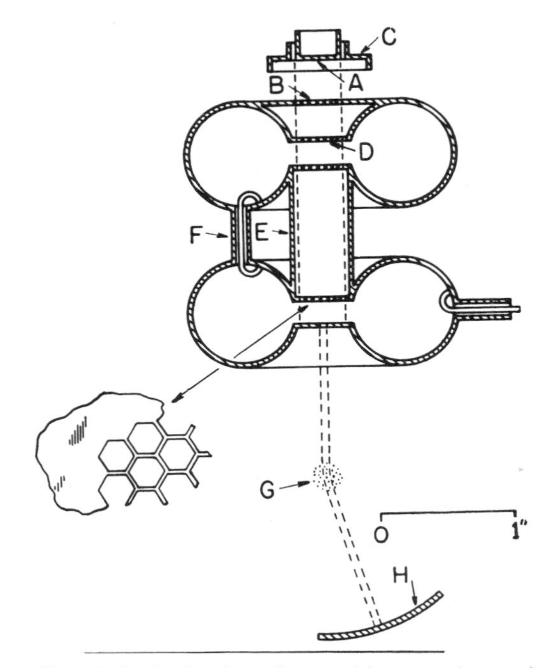

Fig. 5. Scale drawing of essential parts of a typical klystron oscillator. The vacuum envelope, means for tuning, means for varying the feedback and various unimportant details have been omitted.

same direction. But there is space charge, and to compensate for this ring C is added in an endeavor to keep the field due to the electrodes plus that of the space charge approximately straight.

After passing through grid B the electrons pass through three, and in many other cases four, grids, the first of which is D. These grids must have high electrical conductivity to avoid spoiling the resonators; they must have high heat conductivity to dissipate the energy derived from the electrons striking them; and they should have as little projected area as possible to keep the number of electrons intercepted at a minimum. In an effort to satisfy these requirements we have sometimes made the grids of honeycomb form as shown in the detail. They are drilled and filed from copper $\frac{1}{16}''$ thick and are perhaps the most tedious parts to make in the whole apparatus.

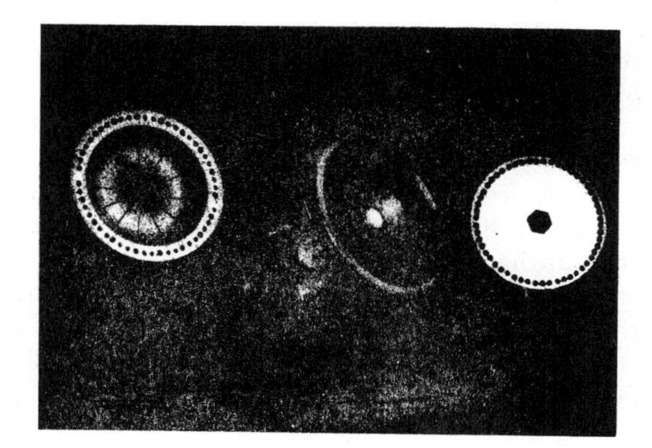

FIG. 6. Photograph of typical cathode, grid and rhumbatron.

The "lattice constant" of the grid is determined by the requirement that the radio-frequency field along the electron path should usually change from zero to full value in a distance short compared to the distance traveled by an electron in one half-cycle. The meaning of "short" will be discussed more precisely in a forthcoming paper by Hansen.

The distance between two grids of the same rhumbatron is usually less than the distance traveled by an electron in one half-cycle. For the buncher this would be $\beta\lambda/2$; but in the catcher some of the electrons are slowed nearly to rest, so that $\beta\lambda/4$ is a closer approximation. In practice the latter spacing is used for both so that the rhumbatrons can be made nearly identical, so as to simplify tuning. Smaller grid spacings can be used, but with some impairment of the shunt impedance.

The two rhumbatrons are connected, mainly for mechanical reasons, by a tube E whose length is determined by a compromise between the following opposing considerations. First, the shorter the tube, the less distance the electrons have in which to bunch, with the result that, for best bunching, they must be given a larger spread in velocity. This, on the other hand, tends to reduce the efficiency because: (a) more power is required to drive the buncher as the buncher voltage is increased; and (b) since no electrons should be thrown backward in the catcher, only the slowest ones can be stopped, and so a larger spread in velocities means more loss of energy due to incomplete stoppage of the faster electrons. On the other hand, a large bunching distance increases the difficulty of getting a large fraction of the beam into the catcher. In practice the compromise chosen is not critical, and the apparatus works well over a large range of distance.

The exact shape of rhumbatron best suited for the present purpose is not known but the shape shown works well and is convenient to manufacture.

Since the Q of the resonators is high (about 1000) it is not easy to build them sufficiently nearly identical to resonate nearly enough to the same frequency without tuning. One of the simplest methods of tuning is that shown, in which the rhumbatron is slotted (along the lines of current flow) and then deformed elastically by a screw. The exact mechanism for doing this is not shown as it is felt that different experimenters will prefer different mechanisms.

The same idea applies to the vacuum envelope and pumping system, which are not shown, and also to the mechanism for rotating the concentric line which carries loops projecting into the rhumbatron. The purpose of this latter is, of course, to feed power back from the catcher to the buncher.

Power may be taken from the catcher for radiation or other uses by a concentric line; or, for radiation, the power may be taken out by simply boring a hole in the catcher.

The energy supply or "plate voltage" has one terminal connected to the cathode parts A and C and the other to the assembly of rhumbatrons, grid B, etc. Choice of a suitable voltage is governed by the following considerations. As the voltage is raised, at constant power, the needed current decreases and the needed shunt impedance increases, perhaps beyond what can be had. On the other hand, as the voltage is decreased, troubles with spreading of the beam by space charge increase. Clearly the proper voltage depends on the power output desired and other factors; but here also conditions are not very critical. For example, it is possible to use many different voltages, from 300 to 4000, on a single klystron oscillator.

Finally we have shown a device that is not essential but very convenient, namely, an oscillation indicator. This consists of a magnetic field G produced by a small permanent magnet (not shown) through which a narrow pencil of electrons from the catcher pass enroute to the fluorescent screen H. When the rhumbatrons are not oscillating the electrons all have the same velocity and make a single spot on the screen. When oscillations occur some of the electrons are retarded and others accelerated by the field in the catcher, and so the spot spreads into a line. This device is exceedingly convenient because of its quick responses. Also, by observing the pattern on the screen one may obtain qualitative information about the bunching.

V. Summary

The most important requisites for a klystron amplifier are as follows:

(1) Efficient resonators, such as rhumbatrons, with holes or grids permitting cathode rays to go through them along the lines of force of their electric fields;

(2) A beam of cathode rays going through both resonators and carrying power enough to make the catcher oscillate more powerfully than the buncher.

For a klystron oscillator, the requirements are those of the amplifier and in addition:

(3) A coupling loop or some practically equivalent line feeding power from the catcher to the buncher;

(4) A P.D. between the cathode and the rhumbatrons such that the bunches arriving at the catcher are not too far out of phase with the field within it.

Note added in proof.—Below are three references closely related to the above article which have come to our attention since the article was written. One of these has appeared since this article was submitted. A. Arsenjewa-Heil and O. Heil, Zeits. f. Physik, **95**, 752–762 (1935); E. Brüche and A. Recknagel, Zeits. f. Physik, **108**, 459–482 (1938); W. C. Hahn and G. F. Metcalf, Proc. I. R. E. **27**, 106–116 (1939).

Gorham on Advances in Electronics During World War II

As had been the case in the first World War (see Paper 52), the impact of almost unlimited funding on the electronics industry was substantial. In this paper, Gorham attempted to summarize the major achievements in the design of electron tubes during the war. He also mentioned silicon and germanium crystal rectifiers that had been widely used in radar receivers, but he did not foresee that they were harbingers of a new solid-state technology. For another view of developments during the war, see Paper 45. Also see Paper 56 for an overview of the history of electron devices. See Charles Susskind's paper on American contributions to electronics in the PROCEEDINGS OF THE IEEE for September 1976.

John E. Gorham (1911–) was born in Moline, Ill. and received the B.S. and M.S. degrees from Iowa State University in 1933 and 1934, respectively. He received a Ph.D. degree from Columbia University in 1938. He worked for the Continental X Ray Company and for Belmont Radio before joining the Signal Corps Engineering Laboratories in 1940. See *Who's Who in Engineering*, p. 928, 1940.

Electron Tubes in World War II*

JOHN E. GORHAM†, MEMBER, I.R.E.

***Summary*—Although the military uses of electronics have been well publicized in technical journals, the improvements in electron tubes that made possible these military innovations have not been fully reported. While this information is known in some detail by the technical people who were engaged in various phases of tube research and development, an over-all summary of the work done by industrial laboratories and Federal agencies has not been available to many engineers and students interested in this field. In this paper, the status of electron-tube development at the close of the war is indicated in broad outline; a more comprehensive picture depends upon more detailed reports from the various laboratories engaged in war activities.**

This summary of wartime advances in electron tubes is based on the knowledge of the vacuum-tube field gained by the engineers of the thermionics branch of the Signal Corps Engineering Laboratories, Bradley Beach, New Jersey, in developing, standardizing, and giving type approval of all tubes procured for the army during the war.

No effort is made to give specific credit either to individuals or industrial organizations. By and large it is a story of common achievement of many people, with industry working hand in hand with the War and Navy Departments to meet the urgent requirements of an ever-expanding demand for new and improved military electronic equipment.

I. General Research and Miscellaneous Related Tube Problems

Cathodes

At THE start of World War II it was the practice to use oxide cathodes in low-power and receiving-type tubes, thoriated-tungsten filaments in medium-power tubes, and pure tungsten in high-power tubes. In general there were few power-pulse requirements. During the war, the use of thoriated filaments had been successfully extended to all types of power tubes, including the highest-power pulsed-oscillator tubes designed. In addition, during the last year of the war, oxide cathodes were used in power tubes capable of delivering 500 or 600 kilowatts peak power, and up to several megawatts peak power in magnetrons. The peak emission of thoriated filaments, for design purposes, has been increased from approximately 100 milliamperes per watt to approximately 200 milliamperes per watt. The peak emission from oxide cathodes has been increased to 30 amperes per square centimeter in production tubes, and as high as 80 amperes per square centimeter for several hundred hours in laboratory tubes. The highest peak emission reported is about 140 amperes per square centimeter. Oxide-cathode direct-current emission has been increased to approximately 0.5 ampere per square centimeter under optimum conditions.

There was little advance in the efficiency and stability of secondary-electron multipliers during the war. Electron multipliers may be considered to have a nominal multiplication factor of approximately 5 per stage for optimum acceleration voltages of the order of a few hundred volts per stage. After about 200 or 300 hours the performance of these multipliers is seriously reduced.

Except for low-power voltage-regulator and mercury-pool type tubes, relatively few tubes having cold cathodes were used in military equipment during World War II because of the lack of satisfactory life from such cathodes. In one type of pulse-modulator tubes, mercury is held in fine iron powder to permit use in aircraft. Mercury-pool ignitrons were used as pulsed modulator tubes to a limited extent.

It has been generally proved that at least 80 per cent of the total emission from the magnetron cathode is largely due to electrons emitted as a result of back-bombardment by electrons that do not reach the anode. As a result of such back-bombardment, considerable power is sent back to the cathode, with resultant heating and evaporation of the cathode coating and even the base metal. This has been overcome to some extent by appropriate reduction in power after the magnetron reaches stable operation. Some higher-frequency magnetron cathodes actually have radiators. More rugged types of coatings have been developed in which the oxides are pressed into a wire mesh which is sintered to a base cylinder, or in which approximately 50 per cent of the coating consists of 3- or 4-micron nickel powder to increase electron and heat conductivity and also to increase to some extent the binding force holding the barium.

Grids

An outstanding advancement during World War II has been the development of alloys and surfaces which overcome the problem of primary grid emission in thoriated-filament tubes, thereby eliminating the phenomenon of grid blocking, which normally leads to destruction of such tubes. These alloys include 4 per cent tungsten-platinum alloy wire, platinum-coated molybdenum-core grid wire, and "mossy"-surfaced tantalum or molybdenum wire.

It has been found that the presence of secondary emission tends to reduce the driving power of conventional grid tubes. The uniformity and stability of such secondary emission are very poor, however, and the current practice is generally to avoid making use of this factor in service tubes.

It has been discovered that there are certain temperatures at which the emission from grid wires is a minimum, and though it is not generally possible to maintain

* Decimal classification: R330. Original manuscript received by the Institute, March 27, 1946; revised manuscript received, July 31, 1946.

† Evans Signal Laboratory, Belmar, New Jersey.

Reprinted from *Proc. IRE*, vol. 35, pp. 295–301, Mar. 1947.

the grids at this temperature, at least one tube type has been put into wide production with a fair degree of success using this principle. A second method used in receiving tubes (and, during the last years of the war, in power tubes) consists of the use of heat conduction to maintain grid temperature sufficiently low to minimize the effect of emission. This has been accomplished in some cases, such as in the 7C22, by the use of nickel cylinders with grid straps punched and rotated 90 degrees to reduce their effective cross section to electron flow but at the same time maintain their cross section for heat flow.

In various other power receiving tubes relatively large copper rods have been fastened at appropriate intervals to help remove heat. Another application makes use of very short grid wires to facilitate conduction to end rings. Still another method employed with moderate success in reducing grid emission involves the use of gold-plated molybdenum wire. It has been found that the gold will dissolve barium for at least 1000 hours in tubes such as the 715C. A solid solution is ultimately formed which apparently draws together and exposes the base metal through the resultant cracks.

Anodes

During World War II the problem of anode heat dissipation had not been a major one, except in planar-type lighthouse tubes. A principal problem in connection with anode design, and for that matter with general design of tubes, has been to reduce lead-reactance effects by providing extremely low-impedance paths at radio-frequency connections. The most important advance in anode design has been the development of various different but essentially similar re-entrant anode designs. These employ large-diameter glass-to-metal seals in both copper and kovar, and result in attendant reduction of lead-reactance effects up to at least 700 megacycles.

Zirconium or zirconium compounds have been sprayed on anodes to make them more nearly perfect black-body radiators, and simultaneously serve as getters. The principal requirement in planar-tube anode design at present is to improve heat dissipation and frequency drift due to warm-up.

Gas Reservoirs

Titanium-hydride reservoirs have been developed which are capable of maintaining the pressure at constant value and appreciably extending the life of hydrogen thyratrons under extreme operating conditions.

II. Magnetron Tubes

The development of the magnetron as an efficient microwave generator took place almost entirely during World War II. During the war, the magnetron advanced from the status of the elementary split-anode variety to the highly perfected and complex multiresonant-cavity type. Operating efficiencies were raised from about 10 to over 50 per cent. Tubes were developed and produced in large numbers for wavelengths as short as approximately 1 centimeter. Representative types of magnetrons are shown in Fig. 1.

Fig. 1—Representative types of magnetrons: (a) 2J31 hole-and-slot pulse type. (b) 2J54 tunable pulse type. (c) 2J64 vane type for pulse communication. (d) 5J31 split-anode continuous-wave type. (e) 3J21 rising-sun pulse type.

During the course of the war, extensive studies were made of mode separation and the manner and efficiency of operation. Methods of eliminating undesirable modes (arising from multiple degeneracy due to the multiresonator anode blocks) were developed, such as strapping and the use of "rising-sun" alternately long and short cavity construction. The usual technique of strapping consists of electrically connecting alternate cavity vanes near the cathode ends by means of metal straps or wires within the tube. This strapping depends on end effects at the top and bottom of the vanes. The rising-sun anode construction consists of making alternate cavities tuned respectively to frequencies above and below the operating frequency of the magnetron, and was originated to avoid straps in super-high-frequency tubes. Lately, the fact that the rising-sun structure does not depend on end effects has been used in designing higher-power, longer-anode magnetrons.

Several mechanical tuning methods were developed. These include internal tuning by means of moving plungers in the resonant cavities ("crown of thorns"), changing the capacitance of the straps to ground and each other, the addition of an external tunable resonator coupled to an internal resonator or strap, and simultaneous application of strap and plunger tuning.

"Packaging" was also introduced, whereby the

magnetron was produced as a complete unit containing or having attached permanent magnets as an integral part of the magnetron instead of depending on the furnishing of proper magnetic fields as part of the operating equipment.

At present, several 25-centimeter pulsed magnetrons of fixed frequency are available with peak powers as high as 1 megawatt. Development has just been completed on a tunable type capable of 600 kilowatts peak power output and 8 per cent tuning range.

At wavelengths of about 10 centimeters, tubes have been produced in quantity with peak powers ranging up to approximately 2 megawatts. Tunable tubes have been made which have approximately 7 per cent tuning range and 1 megawatt peak power output.

The maximum peak power attainable at about 3 centimeters is approximately 1 megawatt from a fixed-frequency magnetron. A variable-frequency magnetron is also available at this frequency capable of 50 kilowatts peak power and 12 per cent tuning range. At about 1 centimeter only two fixed-frequency pulse magnetron types have been produced in quantity. The tubes are capable of peak powers of the order of 50 kilowatts. In general, the life expectancy of pulsed magnetrons is in the neighborhood of 500 hours, except at extremely short wavelengths where life expectancy is about 250 hours.

Continuous-wave magnetrons using split anodes in the high- and ultra-high-frequency bands, and cavities in the higher frequency bands, have been developed primarily as sources of jamming power of from over 1 kilowatt down to about 50 watts. Due to serious back-bombardment of the cathodes, tube life is usually less than 100 hours, although efficiencies are about 40 per cent. Interdigitated magnetrons, having as anodes two cylindrical sets of interlocking teeth, have been made to give about 15 watts output at 7 centimeters.

During the last part of the war, magnetron modulation was investigated to permit communication at all frequencies. One electronic frequency-modulation method consists of varying the current of an electron beam through one of the magnetron cavities. This method has been used at 4000 megacycles to get 4 megacycles total swing at about 25 watts continuous-wave power output. Preliminary tests show that external magnetrons may be used to modulate the magnetron generator tube by virtue of change of electronic reactance, but at present modulation linearity is not as good as that obtained by the former method. Amplitude modulation is not satisfactory at this date, but there are indications that considerable success may be achieved in the near future. Pulse-time modulation is feasible at any frequency and involves transmission at constant power level.

III. Transmit-Receive Tubes

The transmit-receive (TR) tube (Fig. 2) is a switching tube, usually gas-filled, which is generally used in radio-frequency systems (radar, for example) where a transmitter and receiver make use of a common antenna. Its function is to protect the receiver input-circuit elements

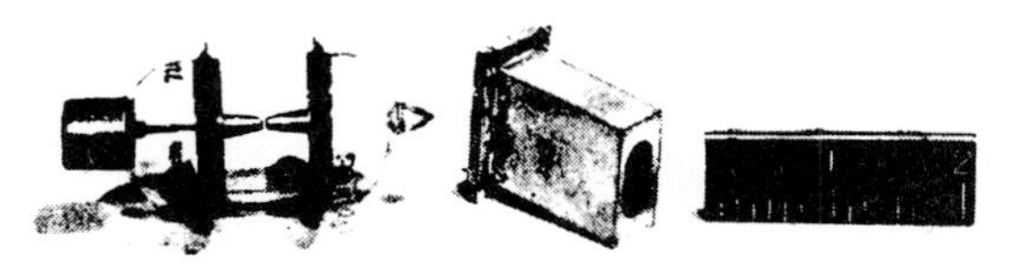

Fig. 2—Transmit-receive tubes: types 721A, 1B37.

during the pulsing of the transmitter and allow the radio-frequency power received by the antenna between pulses to reach the receiver. Antitransmit-receive (ATR) tubes are used in conjunction with TR tubes to reduce the dissipation of receiver signals in the transmitter.

Table I
TR-Tube Performance Characteristics

Type	Application	Wavelength	Power level	Insertion loss	Recovery time	Bandwidth
1B23	TR	20-50 centimeters	50 kilowatts	1 decibel	—	high Q
702A, B	TR	20-50 centimeters	50 kilowatts	—	—	—
721B	TR external cavity	10 centimeters	250 kilowatts	1.0-1.5 decibels	<7 microseconds	high Q
1B27	TR external cavity	10 centimeters	250 kilowatts	1.0-1.5 decibels	<5 microseconds	high Q
1B58	TR fixed-tuned	8-11 centimeters	200 kilowatts	1.0-1.5 decibels	15 microseconds	10 per cent
1B55	TR fixed-tuned	8-11 centimeters	200 kilowatts	1.0-1.5 decibels	15 microseconds	10 per cent
PS3S	TR fixed-tuned	8-11 centimeters	200 kilowatts	1.0-1.5 decibels	15 microseconds	10 per cent
1B44	ATR fixed-tuned	8-11 centimeters	1 milliwatts	1 decibel	-	5 per cent
1B52	ATR fixed-tuned	8-11 centimeters	1 milliwatts	1 decibel	-	5 per cent
1B53	ATR fixed-tuned	8-11 centimeters	1 milliwatt	1 decibel		5 per cent
1B56	ATR fixed-tuned	8-11 centimeters	1 milliwatt	1 decibel		5 per cent
1B57	ATR fixed-tuned	8-11 centimeters	1 milliwatt	1 decibel		5 per cent
1B38	Pre-TR for use with low-power TR	10.7 centimeters	1 milliwatt	0.10 decibel	20 microseconds	
1B54	Pre-TR for use with low-power TR	8.4 centimeters	1 milliwatt	0.10 decibel	20 microseconds	—
1B24	TR tunable self-contained cavity	3 centimeters	60 kilowatts	1.0-1.5 decibels	<3 microseconds	high Q
724B	TR external cavity	3 centimeters	60 kilowatts	1.0-1.5 decibel	<6 microseconds	high Q
1B63	TR broad-band fixed-tuned	3 centimeters	300 kilowatts	<0.8 decibel	<5 microseconds	12 per cent
1B35	ATR fixed-tuned cavity	3 centimeters	60 kilowatts	0.8 decibel	—	6 per cent
1B37	ATR fixed-tuned cavity	3 centimeters	60 kilowatts	0.8 decibel	—	6 per cent
1B26	TR self-contained cavity	1 centimeter	40 kilowatts	0.85-1.5 decibels	<4 microseconds	high Q
1B36	ATR fixed-tuned	1 centimeter	40 kilowatts	0.8 decibel		>2 per cent

Pre-TR tubes are used for added receiver protection during transmitter pulses. These last two tube types have general requirements similar to that of TR tubes (see Table I). TR, ATR, and pre-TR tubes should have low leakage power to the receiver during the transmitter pulse, rapid recovery time immediately following the pulse to enable the maximum received energy to reach the receiver for short range echoes, and satisfactory life. Most tubes were filled either with argon or mixtures of hydrogen and water vapor at pressures in the range of 10 to 25 millimeters.

In general, the recovery time of good tubes at power levels of 30 kilowatts peak is in the order of 4 to 7 microseconds. At higher powers, recovery-time figures are progressively larger. For instance, at line powers of 100 kilowatts the recovery time is approximately 50 per cent greater than at 30 kilowatts. As might be expected, leakage power is also a function of line power. At 30 and 100 kilowatts the leakage powers are of the order of 20 and 75 milliwatts peak, respectively. The insertion loss is approximately one decibel. Recently multicavity fixed-tuned tubes have been made with a frequency coverage of about 12 per cent.

IV. Crystal Rectifiers

Crystal rectifiers (Fig. 3) are used in receiver applications for mixers, video detectors, second detectors, and direct-current restorers. In construction they consist of a semi-conductor, either silicon or germanium, in contact with a cat's whisker of metal, usually tungsten. At present, crystal mixers give the lowest noise figures in receivers above about 1000 megacycles.

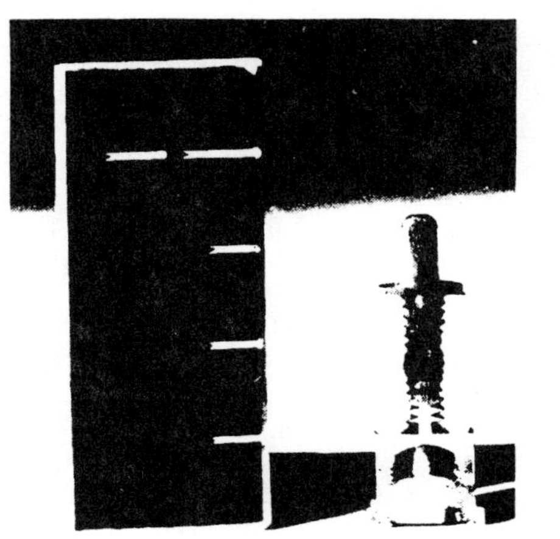

Fig. 3—Crystal detector: 1N21.

In general, microwave crystal converters have conversion losses of the order of about 6.5 to 8.5 decibels, being best at 3000 megacycles and worst at about 30,000 megacycles. They are capable of withstanding pulses ranging from 5 ergs at 3000 megacycles to 0.1 erg at 30,000 megacycles. On the basis of their use with an intermediate-frequency amplifier of 5 decibels noise figure, receiver noise figures are attainable which vary from about 12.7 decibels at 3000 megacycles to 15.2 decibels at 30,000 megacycles.

Germanium crystals, used as second detectors, at present are capable of withstanding 50 or more volts in the back direction, compared with about 5 volts for silicon crystals. In general, they have rectification efficiencies in the same order of magnitude as receiving-type diode tubes. For direct-current restorer applications, germanium crystals have resistances, measured at 1 volt, greater than 0.1 megohm in the back direction and approximately 200 ohms in the forward direction. Germanium crystals are being used at present as second detectors and direct-current restorers for experimental circuit work. Their properties, especially as compared to diodes, are being studied.

V. Klystrons

Development and application of klystrons during World War II has mainly centered about reflex tubes for local-oscillator use, requiring about 20 milliwatts of power output, and signal-generator use, requiring about one watt. Although the theoretical maximum efficiencies are 30 per cent for the reflex klystron and 58 per cent for the two-cavity type, the actual efficiencies thus far attained are only a few per cent for reflex tubes and 5 to 6 per cent for two-cavity types. The best tube in this respect, to date, is the 2K54 for which efficiencies of 10 per cent are obtained under pulsed operating conditions.

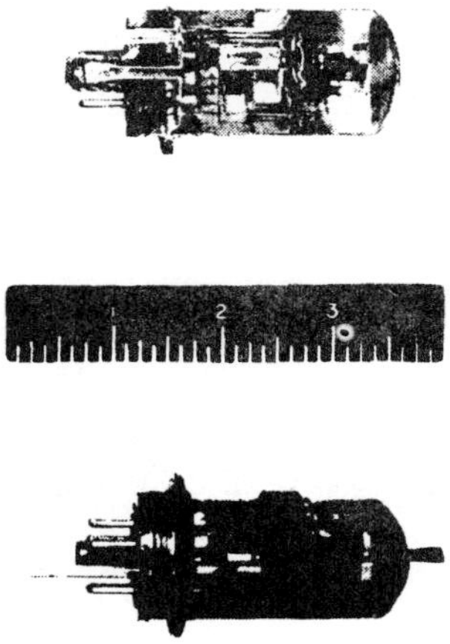

Fig. 4—Thermally tuned reflex klystron, 9000 megacycles: type 2K45.

Tuning of klystrons is generally accomplished by either changing the resonant frequency of the cavity or, in the case of reflex klystrons, by varying the potential of the repeller. Repeller-voltage changes are capable of producing only relatively small frequency changes of the order of 1 per cent. The degree of frequency change attainable by means of cavity variation depends largely on the cavity construction. Klystrons designed to operate with external cavities may have frequency tuning ranges in the order of 2 to 1. Klystrons constructed with cavities which are an integral part of the tube usually are tuned by the motion of a metal diaphragm, which permits variation in the spacing of the resonator grids. This produces changes in grid-to-grid capacitance and consequent shift in the resonant frequency.

Tuning has also been accomplished in some tube types by electronic control of an auxiliary electron source within the same envelope, which heats a thermally sensitive mechanical element attached to the cavity diaphragm. The thermal time constant of such devices varies between 2 and 10 seconds, depending on the type of tube. Tubes with thermal tuning are available in the regions of 10,000 and 25,000 megacycles (Fig. 4).

The power output of reflex klystrons below 3000 megacycles is of the order of 1 watt. Between 3000 and 10,000 megacycles, $\frac{1}{4}$ watt may be attained. Above 10,000 megacycles, available types exist only in the region of about 25,000 megacycles and are capable of approximately 20 milliwatts output. Two-cavity klystrons have been produced in the 2300- to 4000-megacycle region, capable of delivering between 20 and 40 watts of power.

VI. Planar Tubes

Planar-type tubes (Fig. 5) are suitable for high-frequency operation because of (a) reduction in lead inductances by use of disk seals, (b) reduction in interelectrode capacitances by means of small electrode areas and parallel-plane structure, and (c) essentially complete enclosure of the radio-frequency fields permitted by a tube construction suitable for operation in an inclosed cavity. A number of types, all developed during World War II, are now available. These include the 2C40, a low-power triode with 50 milliwatts output at 3370 megacycles; 2C43, a pulse triode with 750 watts peak output at 3370 megacycles; 3C22, a continuous-wave triode with 25 watts output at 1400 megacycles; 2C38, continuous-wave triode with 10 watts output at 2500 megacycles; and the 2C36 and SB846A, British-type disk-seal triodes for low-power use up to about 4000 megacycles.

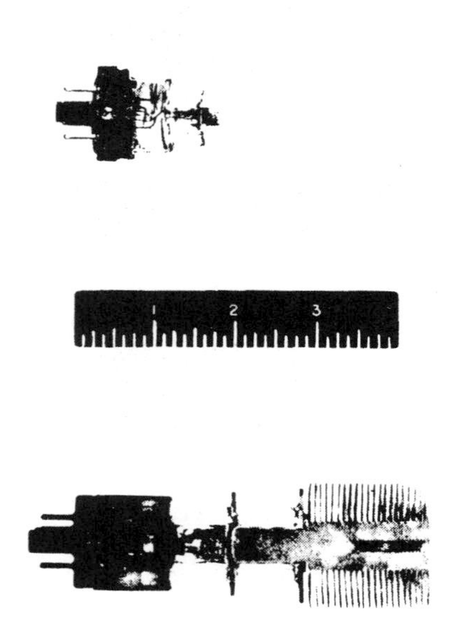

Fig. 5—Cutaway planar tubes: types 2C43, 3C22.

Fig. 6—Cathode-ray-tube screen test.

Present types of planar tubes are constructed with oxide-coated cathodes; tungsten or nickel grids; and steel, molybdenum, or kovar paltes. The glass seals are made to silver-plated steel or kovar. In the case of silver-plated steel, special glass having a thermal-expansion coefficient equal to steel is used. Interelectrode spacings on the 2C40 type are as low as 0.003 inch and 0.010 inch for grid to cathode and grid to plate, respectively. These small spacings, in view of the fact that such tubes are intended for use in accurately machined cavities, require unusually small mechanical tolerances in manufacture.

VII. Indicator and Pickup Tubes

Cathode-ray indicator tubes are used wherever a visible indication of rapidly changing electrical phenomena is required. Because of the almost infinitesimal inertia of the electron beam, these tubes are capable of responses far more rapid than any mechanical indicators.

The transforming of visible or invisible radiation images into electrical signals is accomplished in electronic pickup tubes. Such tubes are designed to have high sensitivity to radiation. By means of very rapid electronic scanning of a photosensitive mosaic, high-resolution electrical transmission of rapidly moving images is accomplished.

During the war, the following improvements were made in electron guns:

(a) In electrostatic-focus types, zero first-anode-current guns were developed in which the first anode did not intercept any beam current, with the result that power-supply requirements were reduced and better focusing control was obtained.

(b) In magnetic-focus types an additional cylinder was added to the high-voltage anode, which aided in alignment of the gun and improved the focus.

(c) Limiting apertures were added in magnetic-focus types to reduce the spot size and improve the focus.

With regard to screens, several new types were developed:

(a) Double-layer screens which have the property of emitting increased intensities of persistent light after successive excitations of the screen. The color of its fluorescent light is different from that of its phosphorescent light.

(b) Dark-trace screens showing a darkening of the normally white screen material, usually potassium chloride, at the point of excitation by the electron beam, were used in projection systems.

(c) Exponential screens having light output which decays at such rate that its instantaneous intensity is proportional to exponential t/t_0, where t_0 is a constant of the screen and t is the time.

Commutator tubes of several varieties were developed during the war for multichannel communication over a single transmission frequency.

Improved tubes suitable for projection purposes were also developed during the war with high light output (6 candle power per watt) and good contrast. Cathode-ray tubes with two or more guns in the same envelope were developed for special applications, eliminating complex switching circuits. Pickup tubes were developed with sensitivities in the infrared. Tubes were also developed capable of converting infrared images directly to visible images by focusing the electron pattern from a photosensitive surface on a fluorescent screen at the opposite end of a cylindrical tube.

At present, cathode-ray tubes with faces from 1 to 12 inches in diameter are available in quantity. These tubes are in some instances focused and deflected by electrostatic methods and in others by magnetic methods.

The various screen types and general information concerning their properties are listed in Table II.

Levels of fluorescent light output vary according to screen types, being about 15 foot-lamberts for tubes of the highest output (nonprojection tubes with P1 screens). Improvements in focusing and line widths were limited and were less than a factor of 2 to 1. Present line widths of from 0.3 millimeter to 1 millimeter are a function of tube size and gun construction (Fig. 6).

Pickup tubes of the orthicon type have been produced with sensitivity in the infrared and in the blue part of the spectrum. Orthicon tubes have been made with a resolution of 1500 lines per frame at the center for high resolution reconnaissance work. For portable systems, tubes have been constructed operating with only a few hundred volts having a sensitivity of 0.03 microamperes per foot-candle.

VIII. Power and Gas Tubes

At the start of World War II radar transmitters were operated at or below about 200 megacycles and used tubes which had thoriated filaments. During the war oxide cathodes came to be used in power-oscillator tubes with a reduction of cathode power by a factor of about five.

Table II

Cathode-Ray-Tube Screen Characteristics

Screen type	Composition	Color	Persistence	Decay time to 1 per cent (seconds)	Applications
P5	$C_aWO_4:(W)$	Blue	short	10^{-5}	Photography of rapid transients (to 60 kilocycles).
P11	α^*—Zns:Ag	Blue	short	0.005	Photography of transients (to 9 kilocycles).
P4	α^*—Zns:Ag + $Zn_8BeSi_5O_{19}:Mn$	White	short	0.005+0.06 (B) (Y)	Television.
·P1	$Zn_2SiO_4:Mn$ (α)	Green	short	0.05	Most cathode-ray oscilloscopes. Rapid-scan radar cathode-ray tube.
P12	$Zn(Mg)F_2:Mn$	Orange	long	0.4	Fire-control radars operating at 4 to 16 scans per second.
P2	ZnS:Cu(Ag) (β^*)	Green	long	0.3	Prewar long-persistence oscilloscopes.
P14	β^*—ZnS:Ag on ZnS(75):CdS:Cu	White ↓ Orange	long	1	Eagle and H2K radars operating at about 1 scan per second.
·P7, (P8)	β^*—ZnS:Ag on ZnS(86):CdS:Cu	White ↓ Yellow	long	3	Most radars operating slower than 1 scan per second.
·P10	KCl	Magenta on White	long	5	Radars operating in high ambient light and slower than 0.2 scan per second.

Several types of $\frac{1}{4}$- to $\frac{1}{2}$-megawatt triode tubes have been developed with the tuned circuits inside the vacuum envelope. By the close of the war, triode oscillator tubes had been developed which gave approximately 0.6 megawatt up to about 700 megacycles. Power-amplifier tubes are available that can handle 100 or 200 watts continuous-wave output up to 700 megacycles with a power gain of about 5 decibels.

Hydrogen thyratrons were originated and put into production during the war to eliminate temperature dependence of mercury tubes. These thyratrons handle powers of from a fraction of a watt to 2 megawatts pulse power. Series and/or parallel operation of thyratrons has been accomplished to allow up to four times the power of a single thyratron. Ignitrons have been used up to 2 megawatts at 20 microseconds pulse width. High-vacuum modulator tubes have been developed to handle a few hundred kilowatts peak power at duty ratios of about 0.0006. Tubes of each of these types are shown in Fig. 7.

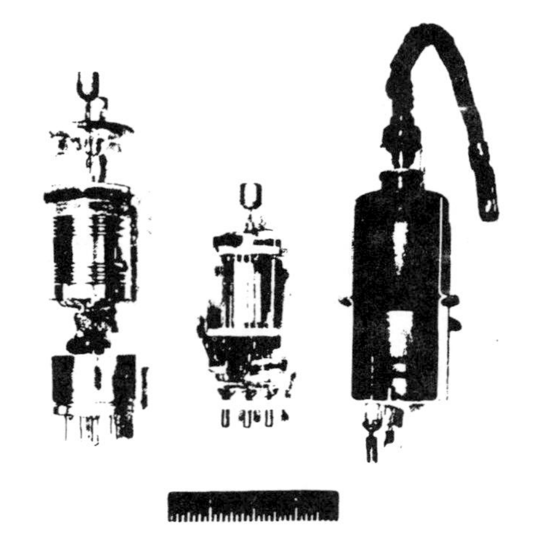

Fig. 7—Pulse modulator tubes: 5C22 hydrogen thyratron; 715C high-vacuum type; 1R21 mercury-pool ignitron.

The resnatron, employed during the war in radar countermeasures to jam German radar, is the most powerful ultra-high-frequency oscillator and amplifier now in existence. It supplies over 50 kilowatts in continuous-wave operation at frequencies ranging from 350 to 650 megacycles, with a plate efficiency of the order of 60 to 70 per cent. Features of this tetrode include beam-forming grids, electron bunching, and self-contained resonant cavities which permit phase-shift compensation for transit-time effects without lowering efficiency.

IX. Receiving Tubes

There are so many types of receiving tubes that it is impossible to begin to describe them here. Consequently only a few practices of a general nature that came into considerably wider employment during the war will be mentioned in this section. The use of standard tubes at low plate and screen voltages was accomplished to allow operation directly from a 24-volt storage battery in place of a high-voltage power supply. Subsequently, tubes with 26.5-volt filaments and a design optimized for 28-volt plate and screen operation were developed. Tubes were "ruggedized" to withstand vibration and shock up to 500 times the acceleration of gravity. Subminiature tubes (T-3 bulbs of $\frac{3}{8}$-inch diameter) were in existence before the war for hearing-aid use. During the war, subminiature types for VT fuzes were developed which could withstand being shot from guns. Size and weight limitations of new radar and allied equipment, along with the need for high peak power output, created the need for receiving-type tubes capable of operating in a pulsed condition at potentials and currents far above their rated values. Fig. 8 shows six different types of receiving tubes.

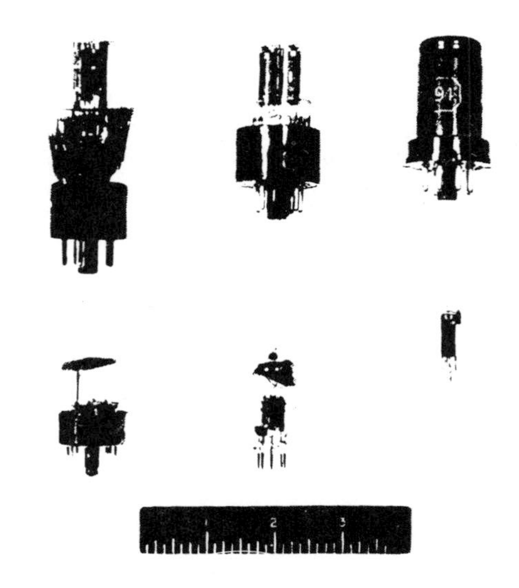

Fig. 8—Receiving tubes having transconductances of 3000 to 5000: G, GT, metal, lockin, miniature, and subminiature tube types 6J5G, 6J5GT, 6J5, 7F8, 6J6, 6K4.

There is now an overabundance of receiving-tube types—one or two thousand, or perhaps more. It is not unusual to find half a dozen or more tubes, substantially equivalent, differing by having several filament voltages, two or three types of bases and bulbs, and different arrangements of pin connections. Almost every metal-tube type is duplicated in a glass version with the same base, and most are also duplicated in lock-in construction under different type designations. Now most of these types are becoming available in miniature bulbs.

X. Acknowledgment

It is a pleasure to acknowledge the aid of my associates in the preparation of this paper: M. E. Crost, K. Garoff, D. R. Gibbons, L. L. Kaplan, B. Kazan, H. L. Ownes, D. E. Ricker, and C. S. Robinson, Jr.

Bardeen and Brattain Discuss the Transistor

LIKE the deForest audion, the transistor marked a major turning point in the history of electronics. In this extract from a longer paper, the co-inventors of the point contact transistor, the first solid-state amplifier, discussed the background of the invention and some of its characteristics. The concerted effort that led to the transistor was initiated by an internal Bell Telephone Laboratories memorandum in July 1945. The first transistor action was observed in December 1947 and the public announcement was made in July 1948. As indicated by the authors of this paper, many others at the Laboratories, including especially William Shockley, were important contributors to the success of the project. Bardeen, Brattain, and Shockley shared the Nobel Prize in Physics for 1956 for their discovery. For an excellent discussion of the prehistory of the transistor, see Charles Weiner, "How the Transistor Emerged," *IEEE Spectrum*, pp. 24–33, January 1973.

John Bardeen (1908–　　) was born in Madison, Wis. where his father was a Professor at the University. He received B.S. and M.S. degrees from Wisconsin in 1928 and 1929, respectively, and then worked as a geophysicist with the Gulf Research and Development Corporation for about three years. He then enrolled at Princeton where he studied solid-state physics and received the Ph.D. degree in 1936. He taught physics at the University of Minnesota until 1941 when he joined the Naval Ordnance Laboratory. He joined the Bell Telephone Laboratories in 1945 and remained until 1951 when he became a Professor at the University of Illinois. His fundamental contributions to the theory of superconductors led to his receiving a second Nobel Prize in Physics in 1973. See a biographical note in *Nobel Lectures* (N.Y., 1963), pp. 342–343. Also see *Engineers of Distinction*, p. 15, 1973.

Walter H. Brattain (1902–　　) was born in China where his father was employed as a teacher. He graduated from Whitman College in Washington in 1924 and received an M.A. degree from the University of Oregon in 1926. He continued his studies at the University of Minnesota where he learned quantum mechanics from John H. Van Vleck. Brattain joined the National Bureau of Standards in 1929 after receiving his doctorate at Minnesota. He moved to the Bell Telephone Laboratories in 1929. See a biographical note in *Nobel Lectures*, pp. 385–386, 1963. Also see *Current Biography*, pp. 68–70, 1957.

Physical Principles Involved in Transistor Action*

By J. BARDEEN and W. H. BRATTAIN

The transitor in the form described herein consists of two-point contact electrodes called emitter and collector, placed in close proximity on the upper face of a small block of germanium. The base electrode, the third element of the triode, is a large area low resistance contact on the lower face. Each point contact has characteristics similar to those of the high-back-voltage rectifier. When suitable d-c. bias potentials are applied, the device may be used to amplify a-c. signals. A signal introduced between the emitter and base appears in amplified form between collector and base. The emitter is biased in the positive direction, which is that of easy flow. A larger negative or reverse voltage is applied to the collector. Transistor action depends on the fact that electrons in semi-conductors can carry current in two different ways: by excess or conduction electrons and by defect "electrons" or holes. The germanium used is n-type, i.e. the carriers are conduction electrons. Current from the emitter is composed in large part of holes, i.e. of carriers of opposite sign to those normally in excess in the body of the block. The holes are attracted by the field of the collector current, so that a large part of the emitter current, introduced at low impedance, flows into the collector circuit and through a high-impedance load. There is a voltage gain and a power gain of an input signal. There may be current amplification as well.

The influence of the emitter current, I_e, on collector current, I_c, is expressed in terms of a current multiplication factor, α, which gives the rate of change of I_c with respect to I_e at constant collector voltage. Values of α in typical units range from about 1 to 3. It is shown in a general way how α depends on bias voltages, frequency, temperature, and electrode spacing. There is an influence of collector current on emitter current in the nature of a positive feedback which, under some operating conditions, may lead to instability.

The way the concentrations and mobilities of electrons and holes in germanium depend on impurities and on temperature is described briefly. The theory of germanium point contact rectifiers is discussed in terms of the Mott-Schottky theory. The barrier layer is such as to raise the levels of the filled band to a position close to the Fermi level at the surface, giving an inversion layer of p-type or defect conductivity. There is considerable evidence that the barrier layer is intrinsic and occurs at the free surface, independent of a metal contact. Potential probe tests on some surfaces indicate considerable surface conductivity which is attributed to the p-type layer. All surfaces tested show an excess conductivity in the vicinity of the point contact which increases with forward current and is attributed to a flow of holes into the body of the germanium, the space charge of the holes being compensated by electrons. It is shown why such a flow is to be expected for the type of barrier layer which exists in germanium, and that this flow accounts for the large currents observed in the forward direction. In the transistor, holes may flow from the emitter to the collector either in the surface layer or through the body of the germanium. Estimates are made of the field produced by the collector current, of the transit time for holes, of the space charge produced by holes flowing into the collector, and of the feedback resistance which gives the influence of collector current on emitter current. These calculations confirm the general picture given of transistor action.

I—Introduction

THE transistor, a semi-conductor triode which in its present form uses a small block of germanium as the basic element, has been described briefly

* This paper appears also in the *Physical Review*, April 15, 1949.

in the Letters to the Editor columns of the Physical Review.[1] Accompanying this letter were two further communications on related subjects.[2, 3] Since these initial publications a number of talks describing the characteristics of the device and the theory of its operation have been given by the authors and by other members of the Bell Telephone Laboratories staff.[4] Several articles have appeared in the technical literature.[5] We plan to give here an outline of the history of the development, to give some further data on the characteristics and to discuss the physical principles involved. Included is a review of the nature of electrical conduction in germanium and of the theory of the germanium point-contact rectifier.

A schematic diagram of one form of transistor is shown in Fig. 1. Two point contacts, similar to those used in point-contact rectifiers, are placed in close proximity (−.005–.025 cm) on the upper surface of a small block of germanium. One of these, biased in the forward direction, is called the emitter. The second, biased in the reverse direction, is called the collector. A large area low resistance contact on the lower surface, called the base electrode, is the third element of the triode. A physical embodiment of the device, as designed in large part by W. G. Pfann, is shown in Fig. 2. The transistor can be used for many functions now performed by vacuum tubes.

During the war, a large amount of research on the properties of germanium and silicon was carried out by a number of university, government, and industrial laboratories in connection with the development of point contact rectifiers for radar. This work is summarized in the book of Torrey and Whitmer.[6] The properties of germanium as a semi-conductor and as a rectifier have been investigated by a group working under the direction of K. Lark-Horovitz at Purdue University. Work at the Bell Telephone Laboratories[7] was initiated by R. S. Ohl before the war in connection with the development of silicon rectifiers for use as detectors at microwave frequencies. Research and development on both germanium and silicon rectifiers during and since the war has been done in large part by a group under J. H. Scaff. The background of information obtained in these various investigations has been invaluable.

The general research program leading to the transistor was initiated and directed by W. Shockley. Work on germanium and silicon was emphasized because they are simpler to understand than most other semi-conductors. One of the investigations undertaken was the study of the modulation of conductance of a thin film of semi-conductor by an electric field applied by an electrode insulated from the film.[3] If, for example, the film is made one plate of a parallel plate condenser, a charge is induced on the surface. If the individual charges which make up the induced charge are mobile, the conductance of the film will depend on the voltage applied to the condenser.

The first experiments performed to measure this effect indicated that most of the induced charge was not mobile. This result, taken along with other unexplained phenomena such as the small contact potential difference between n- and p- type silicon[8] and the independence of the rectifying properties of the point contact rectifier on the work function of the metal point, led one of the authors to an explanation in terms of surface states.[9] This work led to the concept that space charge barrier layers may be present at the free surfaces of semi-conductors such as germanium and silicon, independent of a metal contact. Two experiments immediately suggested were to measure the dependence of contact potential on impurity concentration[10] and to measure the change of contact potential on illuminating the surface with light.[11] Both of these experiments were successful and confirmed the theory. It was while studying the latter effect with a silicon surface immersed in a liquid that it was found that the density of surface charges and the field in the space charge region could be varied by applying a potential across an electrolyte in contact with the silicon surface.[12] While studying the effect of field applied by an electrolyte on the current voltage characteristic of a high-back-voltage germanium rectifier, the authors were led to the concept that a portion of the current was being carried by holes flowing near the surface. Upon replacing the electrolyte with a metal contact transistor action was discovered.

The germanium used in the transistor is an n-type or excess semi-conductor with a resistivity of the order of 10-ohm cm, and is the same as the material used in high-back-voltage germanium rectifiers.[13] All of the material we have used was prepared by J. C. Scaff and H. C. Theuerer of the metallurgical group of the Laboratories.

While different metals may be used for the contact points, most work has been done with phosphor bronze points. The spring contacts are made with wire from .002 to .005″ in diameter. The ends are cut in the form of a wedge so that the two contacts can be placed close together. The actual contact area is probably no more than about 10^{-6} cm^2.

The treatment of the germanium surface is similar to that used in making high-back-voltage rectifiers.[14] The surface is ground flat and then etched. In some cases special additional treatments such as anodizing the surface or oxidation at 500°C have been used. The oxide films formed in these processes wash off easily and contact is made to the germanium surface.

The circuit of Fig. 1 shows how the transistor may be used to amplify a small a-c. signal. The emitter is biased in the forward (positive) direction so that a small d-c. current, of the order of 1 ma, flows into the germanium block. The collector is biased in the reverse (negative) direction with a higher voltage so that a d-c. current of a few milliamperes flows out through the collector point and through the load circuit. It is found that the current in the collector circuit is sensitive to and may be controlled by changes of current from the emitter. In fact, when the emitter current is varied by changing the emitter voltage, keeping the collector voltage constant, the change in collector current may be larger than the change in emitter current. As the emitter is biased in the direction of easy flow, a small a-c. voltage, and thus a small power input, is sufficient to vary the emitter current. The collector is biased in the direction of high resistance and may be matched to a high resistance load. The a-c. voltage and power in the load circuit are much larger than those in the input. An overall power gain of a factor of 100 (or 20 db) can be obtained in favorable cases.

Terminal characteristics of an experimental transistor[15] are illustrated in Fig. 3, which shows how the current-voltage characteristic of the collector is changed by the current flowing from the emitter. Transistor characteristics, and the way they change with separation between the points, with temperature, and with frequency, are discussed in Section II.

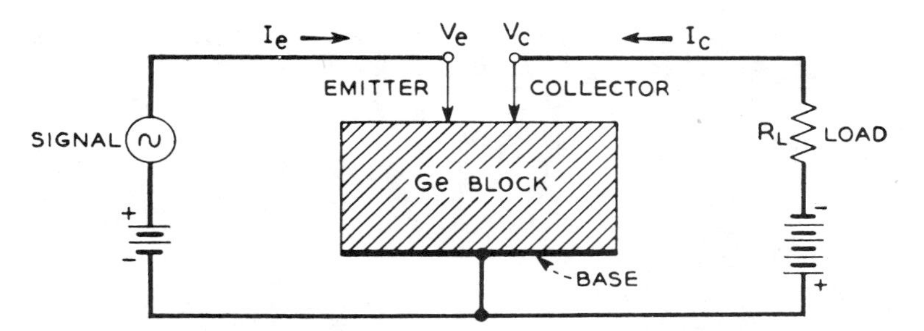

Fig. 1—Schematic of transistor showing circuit for amplification of an a-c. signal and the conventional directions for current flow. Normally I_e and V_e are positive, I_c and V_c negative.

The explanation of the action of the transistor depends on the nature of the current flowing from the emitter. It is well known that in semi-conductors there are two ways by which the electrons can carry electricity which differ in the signs of the effective mobile charges.[16] The negative carriers are excess electrons which are free to move and are denoted by the term conduction electrons or simply electrons. They have energies in the conduction band of the crystal. The positive carriers are missing or defect "electrons" and are denoted by the term "holes". They represent unoccupied energy states in the uppermost normally filled band of the crystal. The conductivity is called n- or p-type depending on whether the mobile charges normally in excess in the material under equilibrium conditions are electrons (negative carriers) or holes (positive carriers). The germanium used in the transistor is n-type with about 5×10^{14} conduction electrons per c.c.; or about one electron per 10^8 atoms. Transistor action depends on the fact that the current from the emitter is composed in large part of *holes*; that is of carriers of opposite sign to those normally in excess in the body of the semi-conductor.

The collector is biased in the reverse, or negative direction. Current flowing in the germanium toward the collector point provides an electric

Fig. 2—Microphotograph of a cutaway model of a transistor

field which is in such a direction as to attract the holes flowing from the emitter. When the emitter and collector are placed in close proximity, a large part of the hole current from the emitter will flow to the collector and into the collector circuit. The nature of the collector contact is such as to provide a high resistance barrier to the flow of electrons from the metal to the semi-conductor, but there is little impediment to the flow of holes into

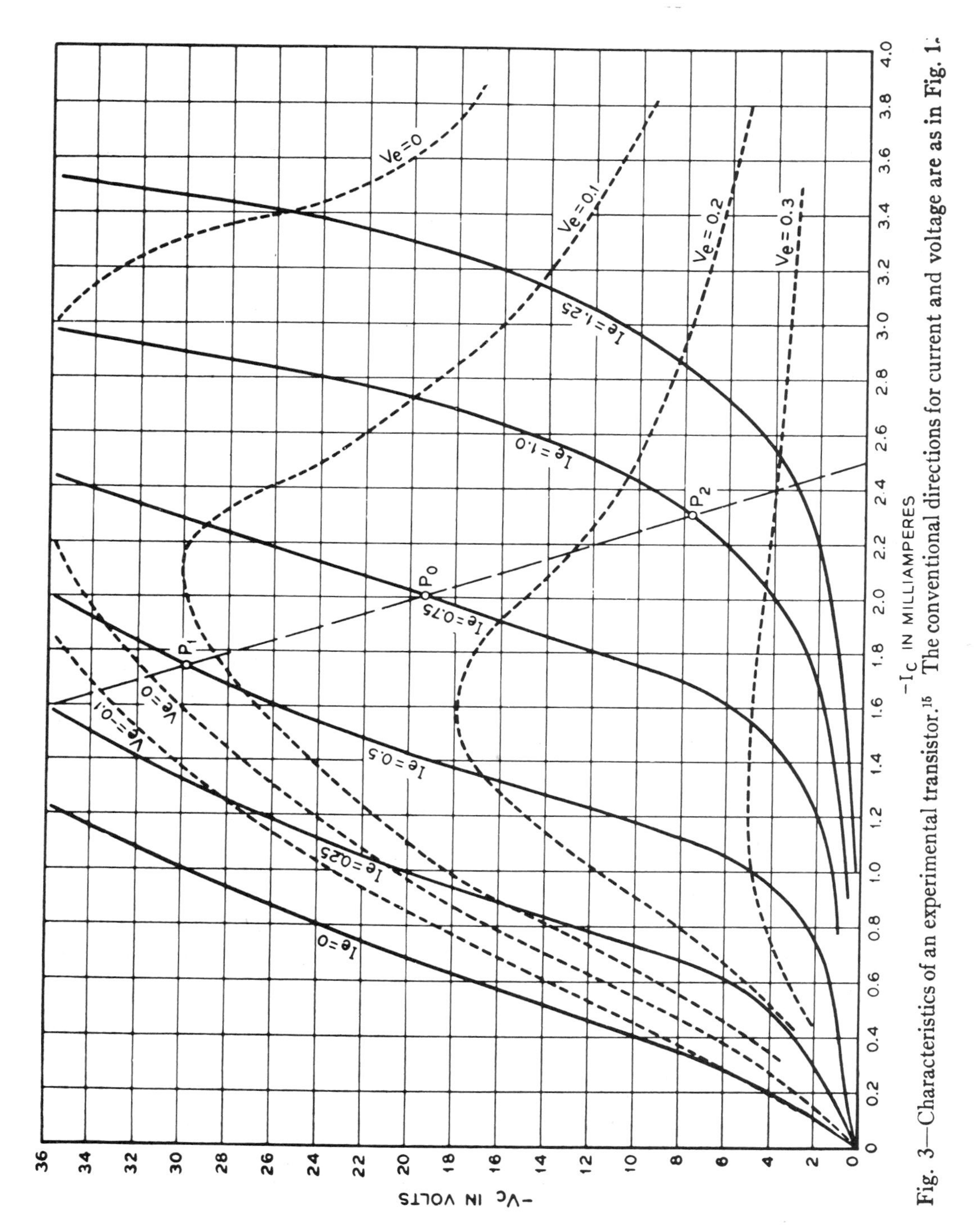

−Ic IN MILLIAMPERES

Fig. 3—Characteristics of an experimental transistor.[15] The conventional directions for current and voltage are as in Fig. 1.

the contact. This theory explains how the change in collector current might be as large as but not how it can be larger than the change in emitter current. The fact that the collector current may actually change more than the emitter current is believed to result from an alteration of the space charge in the barrier layer at the collector by the hole current flowing into the junction. The increase in density of space charge and in field strength makes it easier for electrons to flow out from the collector, so that there is an increase in electron current. It is better to think of the hole current from the emitter as modifying the current-voltage characteristic of the collector, rather than as simply adding to the current flowing to the collector.

In Section III we discuss the nature of the conductivity in germanium, and in Section IV the theory of the current-voltage characteristic of a germanium-point contact. In the latter section we attempt to show why the emitter current is composed of carriers of opposite sign to those normally in excess in the body of germanium. Section V is concerned with some aspects of the theory of transistor action. A complete quantitative theory is not yet available.

There is evidence that the rectifying barrier in germanium is internal and occurs at the free surface, independent of the metal contact.[9, 17] The barrier contains what Schottky and Spenke[18] call an inversion region; that is a change of conductivity type. The outermost part of the barrier next to the surface is p-type. The p-type region is very thin, of the order of 10^{-5} cm in thickness. An important question is whether there is a sufficient density of holes in this region to provide appreciable lateral conductivity along the surface. Some evidence bearing on this point is described below.

Transistor action was first discovered on a germanium surface which was subjected to an anodic oxidation treatment in a glycol borate solution after it had been ground and etched in the usual way for diodes. Much of the early work was done on surfaces which were oxidized by heating in air. In both cases the oxide is washed off and plays no direct role. Some of these surfaces were tested for surface conductivity by potential probe tests. Surface conductivities, on a unit area basis, of the order of .0005 to .002 mhos were found.[2] The value of .0005 represents about the lower limit of detection possible by the method used. It is inferred that the observed surface conductivity is that of the p-type layer, although there has been no direct proof of this. In later work it was found that the oxidation treatment is not essential for transistor action. Good transistors can be made with surfaces prepared in the usual way for high-back-voltage rectifiers provided that the collector point is electrically formed. Such surfaces exhibit no measurable surface conductivity.

One question that may be asked is whether the holes flow from the emitter to the collector mainly in the surface layer or whether they flow through the body of the germanium. The early experiments suggested flow along the surface. W. Shockley proposed a modified arrangement in which in effect the emitter and collector are on opposite sides of a thin slab, so that the holes flow directly across through the semi-conductor. Independently, J. N. Shive made, by grinding and etching, a piece of germanium in the form of a thin flat wedge.[19] Point contacts were placed directly opposite each other on the two opposite faces where the thickness of the wedge was about .01 cm. A third large area contact was made to the base of the wedge. When the two points were connected as emitter and collector, and the collector was electrically formed, transistor action was obtained which was comparable to that found with the original arrangement. There is no doubt that in this case the holes are flowing directly through the n-type germanium from the emitter to the collector. With two points close together on a plane surface holes may flow either through the surface layer or through the body of the semi-conductor.

Still later, at the suggestion of W. Shockley, J. R. Haynes[20] further established that holes flow into the body of the germanium. A block of germanium was made in the form of a thin slab and large area electrodes were placed at the two ends. Emitter and collector electrodes were placed at variable separations on one face of the slab. The field acting between these electrodes could be varied by passing currents along the length of the slab. The collector was biased in the reverse direction so that a small d-c. current was drawn into the collector. A signal introduced at the emitter in the form of a pulse was detected at a slightly later time in the collector circuit. From the way the time interval, of the order of a few microseconds, depends on the field, the mobility and sign of the carriers were determined. It was found that the carriers are positively charged, and that the mobility is the same as that of holes in bulk germanium (1700 cm^2/volt sec).

These experiments clarify the nature of the excess conductivity observed in the forward direction in high-back-voltage germanium rectifiers which has been investigated by R. Bray, K. Lark-Horovitz, and R. N. Smith[21] and by Bray.[22] These authors attributed the excess conductivity to the strong electric field which exists in the vicinity of the point contact. Bray has made direct experimental tests to observe the relation between conductivity and field strength. We believe that the excess conductivity arises from holes injected into the germanium at the contact. Holes are introduced because of the nature of the barrier layer rather than as a direct result of the electric field. This has been demonstrated by an experiment of E. J. Ryder and W. Shockley.[23] A thin slab of germanium was cut in the form of a pie-shaped wedge and electrodes placed at the narrow and wide boundaries of the wedge. When a current is passed between the electrodes, the field strength is large at the narrow end of the wedge and small near the opposite electrode. An excess conductivity was observed when the narrow end was made positive; none when the wide end was positive. The magnitude of the current flow was the same in both cases. Holes injected at the narrow end lower the resistivity in the region which contributes most to the over-all resistance. When the current is in the opposite direction, any holes injected enter in a region of low field and do not have sufficient life-time to be drawn down to the narrow end and so do not alter the resistance very much. With some surface treatments, the excess conductivity resulting from hole injection may be enhanced by a surface conductivity as discussed above.

The experimental procedure used during the present investigation is of interest. Current voltage characteristics of a given point contact were displayed on a d-c. oscilloscope.[24] The change or modulation of this characteristic produced by a signal impressed on a neighboring electrode or point contact could be easily observed. Since the input impedance of the scope was 10 megohms and the gain of the amplifiers such that the lower limit of sensitivity was of the order of a millivolt, the oscilloscope was also used as a very high impedance voltmeter for probe measurements. Means were included for matching the potential to be measured with an adjustable d-c. potential the value of which could be read on a meter. A micromanipulator designed by W. L. Bond was used to adjust the positions of the contact points.

II—Some Transistor Characteristics

The static characteristics of the transistor are completely specified by four variables which may be taken as the emitter and collector currents, I_e and I_c, and the corresponding voltages, V_e and V_c. As shown in the schematic diagram of Fig. 1, the conventional directions for current flow are taken as positive into the germanium and the terminal voltages are relative to the base electrode. Thus I_e and V_e are normally positive, I_c and V_c negative.

There is a functional relation between the four variables such that if two are specified the other two are determined. Any pair may be taken as the independent variables. As the transistor is essentially a current operated device, it is more in accord with the physics involved to choose the currents rather than the voltages. All fields in the semi-conductor outside of the space charge regions immediately surrounding the point contacts are determined by the currents, and it is the current flowing from the emitter which controls the current voltage characteristic of the collector. The voltages are single-valued functions of the currents but, because of inherent feedback, the currents may be double-valued functions of the voltages.

••• ••• •••

Kompfner on the History of Electron Devices

IN this paper the inventor of an important electron device, the traveling wave tube, reviewed the history of electron devices to illustrate how engineering "more than any other profession appears to shape the future." His list of important discoveries and inventions since 1900 is suggestive of opportunities for research by electrical historians. See also Charles Susskind's paper on American contributions to electronics in the PROCEEDINGS OF THE IEEE for September 1976.

For a biographical note on Kompfner and other references, see Paper 48.

Electron devices in science and technology

The many achievements in electron devices, particularly since the turn of the century, can be used as a yardstick to forecast trends in this rapidly expanding field

R. Kompfner *Bell Telephone Laboratories, Inc.*

Whatever one may think about progress in the affairs of mankind, it cannot be denied that great things have happened in the recent past in science and technology, much of it a direct consequence of the invention and development of many different kinds of electron devices. This article speculates on future trends and developments, and the engineer's probable role.

Some of you may remember that the IRE celebrated its 50th year of existence in May 1962. On that occasion the PROCEEDINGS published a jubilee issue, which contained a number of very useful and competent reviews of the state of the art by experts in many fields. It also contained a section entitled "Communications and Electronics—2012 A.D.: A Predictive Symposium by Fellows of the IRE."

I was one of those who were asked to contribute to this section. This was a great and appreciated honor, but I found that any ideas that came to me in trying to foresee what our world would be like in 2012 were so feeble, such obvious extensions of our present-day world, such trivial projections of already existing trends that, in the end, I declined the honor with regrets.

Now, just five years later, I am not sorry that I did. From reading these forecasts now, it is clear that engineers, however eminent they may be in their many fields, are not science fiction writers. They are not in a class with Jules Verne, H. G. Wells, and Jonathan Swift. Obviously much thought had gone into this "Symposium" and there were many ingenious ideas. Still, the overall impression—on me, at any rate—was unconvincing, and I came to the conclusion that engineers are not particularly good at forecasting the future.

Engineering, however, more than any other profession, appears to shape the future. I say this because of what engineers have accomplished in the past. It is tempting to make a simple extrapolation: they will keep on shaping the future. Such forecasting seems to be pretty safe. But is it really?

It is this problem to which I shall address myself. In order to avoid floundering in generalities, I shall restrict the field of inquiry. Naturally, science and technology, one of the "two cultures" of C. P. Snow, is my frame of reference. Within it I shall, however, confine my attention to "electron devices," a small but crucial subfield of electrical and electronics engineering. I shall look at its past, consider the present, and then we shall see what can be said about the future. (It seems clear that the better the understanding of the past the more meaningful will be the appraisal of the future.) If we then should come to a conclusion about the future of electron devices, we may, or may not, apply it to technology in general.

Revised text of a talk presented at the IEEE Electron Devices Meeting, Washington, D.C., Oct. 26–28, 1966.

Reprinted from *IEEE Spectrum*, vol. 4, pp. 47–49, Sept. 1967.

Electron devices in the past

Historically, I suppose, the term "electron devices" goes back to the time when radio engineers were concerned with antennas, with propagation, with circuits, and with components that made it possible for radio to work—namely, the vacuum tubes, the devices in which the electrons were moving around freely. These were clearly so different from other components and clearly so important that a special, although broadly applicable, name was invented: *electron devices*. For my purpose here I propose to define an electron device even more broadly, as a man-made construction in which electrons play a crucial part by virtue of their charge, mass, or spin.

A device may be invented, designed, or developed for a variety of purposes. It may be intended to assist in making discoveries. It may simply serve a useful function, such as amplifying a weak signal. It may be used to generate radiation. It may be built just out of curiosity, so that its behavior can be studied, as a means of studying nature.

With all this in mind I claim that the Geissler discharge tube (actually invented by Plücker in 1858) was the first electron device: a partially evacuated glass tube through which a current can be made to pass. That this current is made up of electrons and ions was, of course, not known at first, but the device was clearly an intriguing one. It demonstrated the power of science in that it combined the science, or art, of pumping air out of a vessel with the new electrical science of the induction coil, to produce an impressive display of glowing color. It was a curiosity—and it provoked curiosity, the desire to discover and learn. Soon someone (I don't know who) discovered that a magnet brought near the discharge tube affected the discharge, and I believe it was Plücker who discovered, after getting a better vacuum than anybody before him, that the envelope could be made to fluoresce and that this fluorescence was due to the impact of rays coming from the cathode—which were called *cathode rays* by Goldstein in 1876.

In 1895 Roentgen was experimenting with such a tube when he discovered X rays. I cannot resist repeating what has been pointed out many times before: that many great discoveries are not planned, or scheduled. Nor was Roentgen's discovery a lucky accident, in the sense that anybody playing with the discharge apparatus would eventually have made the same discovery.* It needed a Roentgen to study this particular combination of apparatus and phenomenon, *and* his perceptiveness *and* his audacity in postulating an invisible radiation penetrating solid bodies other than glass *and* his skill in making clean, comprehensive, and decisive experiments.

Here is an electron device that surely has had an immense impact on our life and civilization. Its most immediate impact, of course, has been on medicine. (Less than three months after their discovery, X rays were put to use in a hospital in Vienna.) With the help of X rays, the structure of the atoms of all the elements has been determined; atomicity itself, the structures of crystals including those of some very important organic compounds such as DNA and RNA—these are only a few of the more recent triumphs.

*The story is told of a laboratory assistant who reported to the Professor of Experimental Philosophy at the Clarendon Laboratory, Oxford, that the photographic plates stored near a Crookes' tube appeared to be darkened by it and was told by the professor to store them elsewhere.

Within a year of Roentgen's discovery, J. J. Thomson had determined that cathode rays are negatively charged particles of a certain charge-to-mass ratio, that they could be accelerated by electric fields and deflected by magnetic fields. This, some people believe, led to the invention of the cathode-ray-tube oscillograph associated with the name of Braun. Several years earlier Heinrich Hertz and Hallwacks had discovered the photoelectric effect; but I think that Elster and Geitel, in the 1920s, found that photoelectric emission took place also in a vacuum. They, presumably, invented the photocell. Edison discovered thermionic emission from hot filaments in his lamp, and thereupon J. A. Fleming in England invented in 1905 what we now know as the vacuum-diode rectifier, which he called a "valve."

I never cease to wonder at the fact that more fuss is not made about the invention, in 1907, of the audion or triode by De Forest. Everybody "knows" that Marconi invented radio, Edison the phonograph, the Wright Brothers the aeroplane, but relatively few people know of De Forest, whose invention really has transformed our civilization. Such a simple thing: to put a grid between the filament–cathode and the plate. And such a momentous result!

Try to picture modern communications, including radio and television; physical sciences, such as nuclear physics, physical chemistry, solid-state physics; life sciences, such as biology and neurophysiology; or defense and space technology, navigation and guidance, and many other fields of human activity—without electronic amplification. (This is now done by transistors, of course.)

You undoubtedly will have noted that I am trying to show just how important electron devices have been; not only to electronics and engineering, but also to science (pure as well as applied) and to human life in all its manifestations. All this, of course, is in support of my thesis that engineers have helped to shape the future.

Notable developments in electron devices

At this point I should like to list a few more of the important discoveries and inventions in electron devices, starting with the period around 1900:

- *J. J. Thomson and Aston:* Positive-ray e/m measurement and mass spectrography
- *Rutherford and Geiger:* Geiger counter

There seems to be a gap of some 20 years during which no electron devices of comparable importance were invented, perhaps as a consequence of World War I. In the 1920s came

- *A. W. Hull, Barkhausen, Kurtz:* Magnetron oscillator and triode transit-time oscillator
- *Zworykin:* Iconoscope
- *Busch and Gabor:* Magnetic electron lens
- *Davisson, Germer, G. P. Thomson:* Electron diffraction tube (which led to the discovery of the wave nature of matter)
- *Langmuir and Hull:* Thyratron
- *Schottky and others:* Multigrid tubes (tetrodes, pentodes)

Later on, in the 1930s, came

- *Cockcroft, Walton, Van de Graaff:* High-energy particle accelerators
- *Lawrence:* Cyclotron
- *Heil and Brüche:* Velocity modulation principle

Before 1890	1890–1900	1900–1910	1910–1920	1920–1930	1930–1940	1940–1950	1950–1960	Since 1960
					Space-charge-limited electron gun			
					Klystron			
				Thyratron	Photomultiplier		Masers and quantum electronics	
				Electron diffraction tube	Electron microscope	Transistors and semiconductor device theory	Tunnel diode	
		Geiger counter	Multigrid tubes	Magnetic electron lens	Science of electron optics	Traveling wave tube and beam wave theory	Diode and electron-beam parametric amplifier	
	Cathode-ray tube	Triode (audion)	Triode oscillator	Iconoscope	Cyclotron	Multicavity magnetron	Strong focusing synchrotron	Gunn-effect oscillator
Geissler tube	X-ray tube	Thermionic diode rectifier	Photocell	Magnetron and Barkhausen-Kurtz oscillator	High-energy particle accelerators	Betatron, synchrotron, and linear accelerator	Ferrite nonreciprocal devices	Optical maser and ramifications

FIGURE 1. Electron-device milestones.

Hahn and Varian brothers: Klystron
Zworykin, Morton, Malter: Photomultiplier
Knoll and Ruska: Electron microscope
Several teams, mostly in Germany: Electron optics

This period marked the beginning of a science of electron devices, which started as the dynamics of particles in conservative fields, and later also in nonstationary fields, including those due to the charged particles themselves. Science by now has come to the aid of electron devices (as a small repayment for the aid electron devices have given to science).

The 1940s saw the following developments:
Pierce: Space-charge-limited electron gun
Kerst: Betatron
W. W. Hansen: Linear accelerator
Oliphant, Veksler, McMillan: Synchrotron
Randall, Boot, Sayers, Millman: Multicavity magnetron
Haeff, Kompfner, Pierce:* The traveling-wave tube and electron beam and wave theory
Bardeen, Brattain, Shockley: The transistor and semiconductor-device theory

Developments in the 1950s included
Hogan: Nonreciprocal ferrite devices
Uhlir and Suhl: Varactor diode and parametric amplification
Esaki: Tunnel diode
Livingston and Christophilos: Strong focusing synchrotron
Townes, Weber, Basov, Prokhorov: Maser (quantum electronics)
Adler: Electron-beam parametric amplifier
Buck: Kryotron

And now in the 1960s:
Townes, Schawlow, Maiman, Javan: Optical maser, and all its ramifications
Franken: Nonlinear optics
Gunn and Read: Gunn oscillator and transit-time effects in semiconductors

The foregoing list is clearly far from complete, nor can I expect to obtain agreement on its general validity. It comprises what I personally consider to be electron-device milestones, identified as far as possible by the invention, and the inventors or major contributors.

* The writer has struggled with the problem of including his own name in the list for some time. It eventually occurred to him that noninclusion would be even more ridiculous than inclusion.

Fubini and Smith Speculate on Limitations in Solid-State Technology

In this paper the formidable problem of how the electronics industry is to cope with seemingly endless innovation is discussed. The authors advance the surprising thesis that research adminstrators need to erect semipermeable barriers to restrict diffusion of information about the latest technical advance to teams engaged in developing new systems or products. They suggest that general laws like that of Shannon in communications theory (see Paper 46) may be found to exist in solid-state systems such as computers.

Eugene G. Fubini (1913–) was born in Italy and received a doctorate degree from the University of Rome in 1935. He worked at the Institute of Electrotechnics in Rome before coming to the U.S. shortly before the war. He was a research engineer at the Harvard Radio Research Laboratory from 1942 to 1945 and joined the Airborne Instruments Laboratory in 1945. He held various positions with the Department of Defense before joining IBM in 1965. He became a private consultant to industry and the government in 1969. See *Engineers of Distinction*, p. 107, 1973.

M. G. Smith received a degree in engineering at the University of Cincinnati in 1950 and an M.S. degree from the University of Colorado in 1956. He worked at the Fort Monmouth Laboratories before joining IBM. See a biographical note in *IEEE Spectrum*, p. 40, May 1967.

Limitations in solid-state technology

With today's exploding technologies, products are technically obsolete by the time they are introduced, a situation that obviously calls for a change. Clearly, technologists must find ways to ascertain the theoretical limits of their capabilities

E. G. Fubini, M. G. Smith

International Business Machines Corporation

In a field in which success is a matter of record some constructive criticism is in order to spur even greater and more meaningful progress. In this vein the present article highlights shortcomings and limitations in the burgeoning area of solid-state technology. One of the main problems is that practice so often outdistances fundamental theory and understanding. Thus it is important that we be able to predict the growth and changes in the technology, but unfortunately our techniques of analysis, simulation, and measurement are inadequate and the number of theoretical limits pitifully small. Undoubtedly, more effort is essential.

Judging by the fantastic achievements that have taken place in the field of solid-state technology, any article on the subject should be a laudatory one. The solid-state engineers have increased computer speeds by two orders of magnitude in ten years. They have made it possible to increase the scope and number of data-processing systems by one order of magnitude and the data processing per dollar by two orders of magnitude. Therefore, congratulations should be in order. However, instead I intend to express my discontent about a number of items that I believe are being neglected. I believe that constructive malcontent serves the role of the mainspring of progress and if I can point the way to some further progress through this approach, this article will have achieved its purpose.

I would like to highlight shortcomings rather than accomplishments, weaknesses rather than strengths, limitations rather than capabilities, and our lack of understanding rather than the great things to come.

The problem of obsolescence

First, let us consider the question of technical obsolescence, and here I will limit myself to digital applications.

If one designs systems in a period of exploding technology, the time span from the completion of a feasibility study to the delivery of machines is an extremely important factor. The length of this span has been surprisingly constant in the past, averaging about four years. If I include research, advanced development, and product engineering and manufacturing, perhaps my four years might seem unduly short. Clearly, this delay will always cause early product obsolescence, especially when technologies explode at the rate that the solid-state technology is today. We are technically obsolete by the time products are introduced. Therefore, we have a critically short system life. Let me call your attention to Fig. 1.

I am not sure that the particular parameter chosen for this diagram is the one that everybody would choose, but I am convinced that whatever parameter we take, the curve would look very much the same: an order-of-magnitude improvement every five or six years.

Reprinted from *IEEE Spectrum*, vol. 4, pp. 55–59, May 1967.

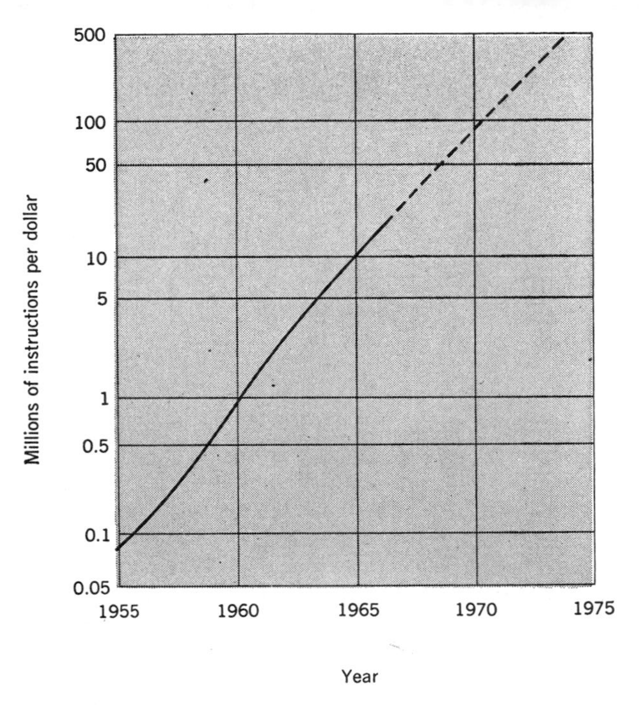

FIGURE 1. Increase in data-processing value.

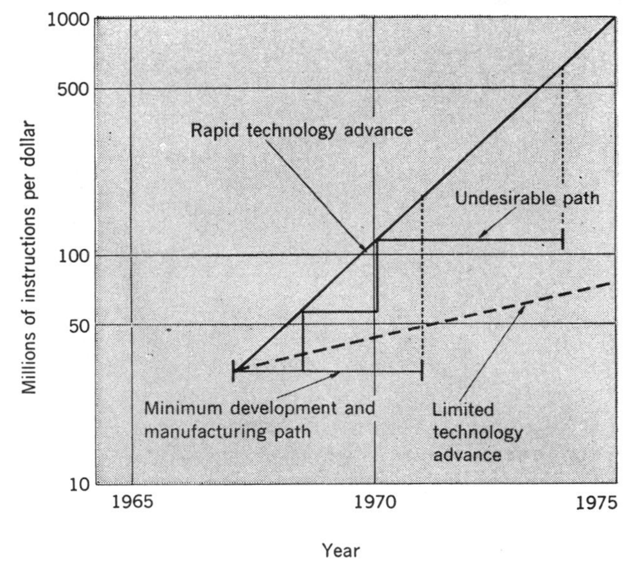

FIGURE 2. Ramifications of introducing a new system in a changing environment.

Figure 2 presents the typical problem of a system designer. Should we anticipate that the rate shown in Fig. 1, that is, an increase of about an order of magnitude every five years, will continue in the foreseeable future? My answer to this question would be "yes" and I would conclude, of course, that a system started today would be nearly an order of magnitude below the performance level for 1971 indicated by our technology advance curve. When a situation such as this arises, a great deal of thinking and care are required in all hardware and software design. We must freeze our design at some early point if we expect to obtain a result.

It is interesting to note that the technology advance is, in its turn, not independent of our efforts and our successes. Technologies are not meaningful if they are not employed, a fact that tends to reduce the true rate of increase of technology by the partial drag of the system considerations. How can we foreshorten our development time in the face of complex technological advances such as large-scale integration (LSI)? We could, of course, try to predict, as shown in Fig. 3. Here we indicate a projected advance not quite as great as a technological advance that actually occurred. We indicate a prediction made regarding this advance, i.e., the penalties that one has to assume in advance because of this difference and the penalties that one may have to pay as a result of mistakes made in assumptions.

Let me emphasize the importance of the "block," shown in Fig. 3. In fact, the equipment will never leave the plant unless we erect a semipermeable wall that will prevent those who do the engineering from observing the exploding technology but will permit those who develop the technology to see what happens to the engineering. As indicated in this illustration, the block extends over the entire development and manufacturing period. If we don't use such a block, the engineers will try to keep up with expanding technology, and the result will be significantly increased costs and relatively little success.

Obviously, the ability to predict technological expansion becomes a most essential part of the system design. Unfortunately, in this area, it is common for practice to outdistance fundamental theory and understanding. And it is to this point that I would like to address most of the balance of this article.

Understanding the limitations

Theory is more difficult to formulate in an environment such as this, in which so many different phenomena and disciplines exist. Our systems are big, interdependent, and complicated. Although this fact may explain why our theoretical achievements are not outstanding, it is also a good reason why more effort is essential. *We cannot continue to permit the distance between practice and fundamental understanding to increase, especially as we press closer to our boundaries.*

Where do we stand and what do we need?

1. *The number of theoretical limits is pitifully small*
2. *The characterization of applications is inadequate.*
3. *The analysis, simulation, and measurement techniques are far too limited.*
4. *There are too few trade-off relationships.*

This situation will have to change. *Technologists must begin to look for the theoretical limits of their capabilities.* The passages in italics present the essence of this article.

In Fig. 3, where we were pointing out the importance of our ability to predict the changes in technology, it would have been extremely useful to know if, at a given time, we are at one tenth of one percent of the theoretical limits or at 80 percent of these limits. If we are close to the limits, the chances of progress are small. If we are far from them, the chances are that we can advance at a constant growth rate.

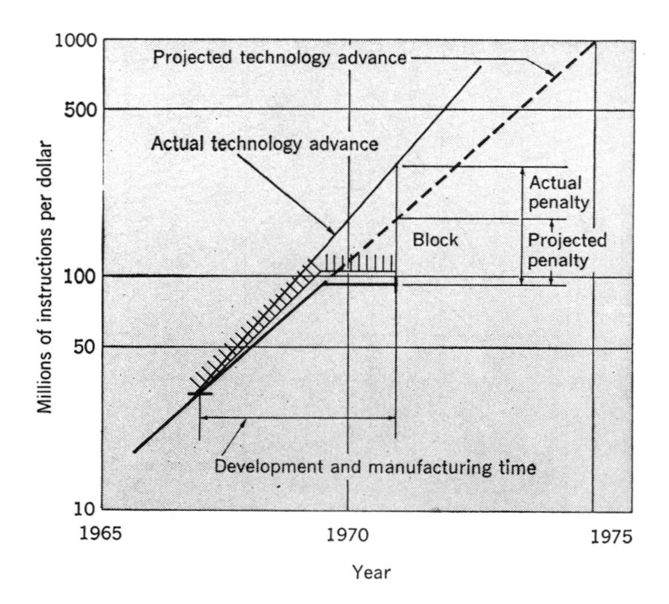

FIGURE 3. Minimizing obsolescence.

We note, for example, that Rolf Landauer, in a paper written in 1961, pointed out that the minimum energy dissipation in switching a specific device must be at least somewhat larger than the energy kT where k is the Boltzmann constant. By introducing a kT/q voltage limit, as was done by Robert Keyes in 1962, we can relate circuit power, circuit delay, and capacitance in an approximate way: $P \approx (C/\tau) \times 10^{-2}$ watt. The following conditions are assumed: voltage must be several times kT/q (valid for transistor circuits as we now understand them); the circuit is switching at a maximum rate as determined by the RC time constant; and delay time $\approx$ time constant.

This is a reasonably generalized expression that is valid for typical semiconductor devices, at least. I will try to use it, not because it is necessarily very good, but because it helps to give an example of the type of information that a system designer needs more and more and that has not been supplied in enough quantity by the theoreticians.

Assuming certain photolithographic limitations, we can postulate that all elements of the circuit, the devices and and interconnections, cannot be smaller than 5–10 μm. We would then conclude from the foregoing expression that the capacitance of the device could not be smaller than 5×10^{-16} farad and that it may be as much as five times greater for a logic gate and 20 times greater for a typical interconnected array of gates. Thus, one would expect, on the basis of the previous formula, a product of power times delay time of about 10^{-16} watt-second.

This fact permits us to make one comparison between the theoretical limitations and what is possible in practice. The typical circuit component used today has 10-ns delay and dissipates 20 mW; thus, the delay–power product is 2×10^{-10}, that is, six orders of magnitude from the postulated theoretical limit. The question arises as to whether a present technology is this far away from the theoretical limits or whether other limits exist that we don't understand as yet. Would they enter into the picture long before the kT limit that we have employed for this comparison? As for the very near future, LSI could give us a reduction of an order of magnitude in capacity, and in the power–delay product, but this is only a small step toward the 10^{-16}-watt-second limitation.

In addition, one should ask what latitude we have in taking advantage of this power–delay time product. The density that we have implied by assuming the one-micrometer limitation would correspond to about 150 000 gates per square centimeter. With this density, what problems will cooling present? If this cooling were restricted to 1 kW/in^2 (150 W/cm^2)—not necessarily a maximum—then the maximum average gate dissipation would be one milliwatt and the minimum gate delay would be about 10^{-13} second. However, propagation delays would dominate and the effective delay τ would be closer to 10^{-12}.

We could use these figures to give a somewhat different picture. Suppose the system could be measured in terms of millions of instructions per second (MIPS) per kilowatt; the 10^{-16}-watt-second limit then would translate to about 100 000 MIPS per kilowatt.

I. MIPS and MIPS per kilowatt for representative machines (for logic only)

Machine Central Processing Unit	MIPS	MIPS/kW
A	0.5	0.3
B	1	0.3
C	1–4	0.2
D	3–12	0.6

If speed is the principal consideration, then, using only the preceding assumption of a 10^{-12}-second gate delay and today's system organization and size (Table I), we could conclude that a maximum performance is of the order of 10 000 to 100 000 MIPS. What would it mean to our present thinking to have machines capable of such performance? A 10 000 MIPS number represents 10^{10} instructions per second. Assume that a user is willing to wait one second for an answer. A machine could, without any of the software complications needed in a shared-terminal teleprocessing system, cope with 10 000 users, each of them asking the machine to finish jobs that require a million instructions before it goes to service the next user. At these speeds, and with this organization, the entire concept of the employment of data-processing machines may have to be modified. I don't know that the theoretical limits that I have set here are indeed the limits, nor do I want to imply that it will ever be possible to achieve them even if they are. I do maintain that it is proper to ask the question, and that it is essential that we work toward finding an answer.

Two further points need to be considered regarding these figures. First, it is not clear at all that MIPS is a good figure of merit for a machine. Second, we have assumed present-day organizations without including such items as parallel processors or pipelining, and greater system-design sophistication will increase the figures by more than an order of magnitude. In addition, we have assumed that the increases in speed can occur not only in the logical devices, but in the memory and in other supporting circuits and equipment as well. Finally, we may have made some unwarranted assumptions regarding the organization of future designs.

Many questions need answers

We have also made mistakes in another direction; we have not allowed for submicrometer dimensions. We must remember that if we can ever go to dimensions corresponding to delays of the order of 10^{-14} second for the logic gate, we may have to consider physical limits, such as the uncertainty principle.

Still I wonder whether there will be limits before the kT limit that will preclude significant advances. Are we really that far from the theoretical limits? Do we have that much growth ahead of us? In designing a system, should we predict a technology growth rate far higher than those rates based on extrapolation of past performance? These are key questions that need to be answered.

I remember being present at the IRE Convention at which Shannon announced his law for the first time. I remember quickly computing with his formula how close a particular communication channel was to the theoretical limit—and finding that it was about a factor of 400 below that theoretical limit. In past years, our channels have come very close to the Shannon limit, and it would not make very much sense for anybody to try to gain the last increment of capacity. If we did not realize this we might waste a lot of time. Why is it that we don't have similar general relationships in computer design?

I am concerned to find that with all our capabilities in complex data-handling problems, we have only very recently ascertained the minimum number of logical levels required for addition. We are functioning in our data-processing environment with limited vision, learning mainly by experience within a particular set of existing conditions. The brilliant intellectual conquests of von Neumann, Turing, and a few others, must be considered as examples to follow.

Quite aside from whether we are or are not approaching fundamental limits, we should be able to characterize all our system designs and performance factors in a general and meaningful way so that we can perform the necessary trade-off exercises. Obviously, trade-off's are required in areas of system work and yet we are pretty far from a satisfactory capability today. I am not sure that we have acquired the level of breadth, perspective, and judgment necessary to tackle problems at this level, particularly since I am not only including machines, but applications as well.

We have methods for examining applications, proceeding from mathematical formulation through numerical analysis and flow charting, and procedures for determining the types and amounts of input and output data and data-processing requirements. Yet, we have not really learned, in general, to characterize the problem formally so that we can predict mathematically the optimum system for a particular problem. We do not have mathematical tools to examine the trade-offs between various approaches and changes in parameters; for example: What should the software be and what do we put into hardware? How large should each memory hierarchy be? How many levels should we have? When do we consider multiprocessing? How should we structure it? Erroneous answers to these questions can have a profound impact on many of our systems.

What type of organization should we use? If we could write algorithms, hardware, and software in the same language, it is conceivable that we could make trade-offs by giving general rules for optimization, which are not dependent upon implementation. Of course, we always include cost and performance in our analyses. In many ways we are quite successful, because our people are competent and have a very good intuitive knowledge of the issue. But I am not really satisfied with our accomplishments.

Even in the areas where we think we are pretty good, we have weaknesses. In circuit design, for instance, we have not seriously tried statistical approaches. In addition, we

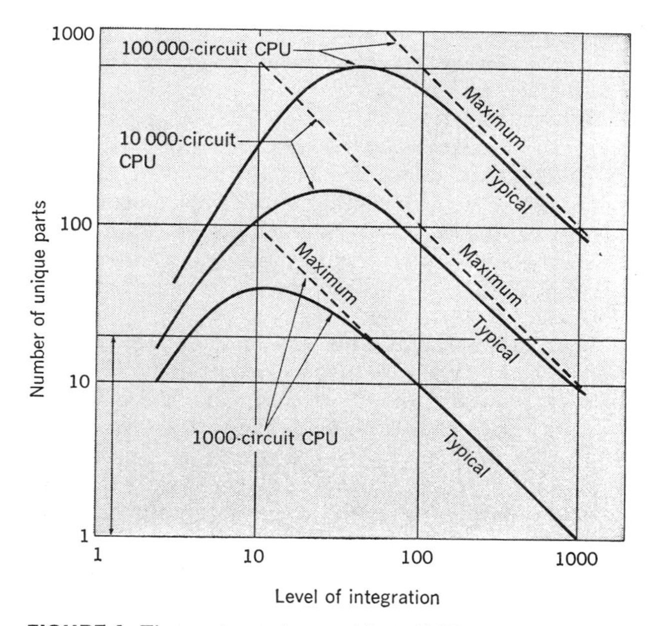

FIGURE 4. The part-number problem (CPU = central processing unit).

FIGURE 5. Integration level for a 20 percent chip yield (photoprocessing defects only) for bipolar and insulated-gate field-effect transistors.

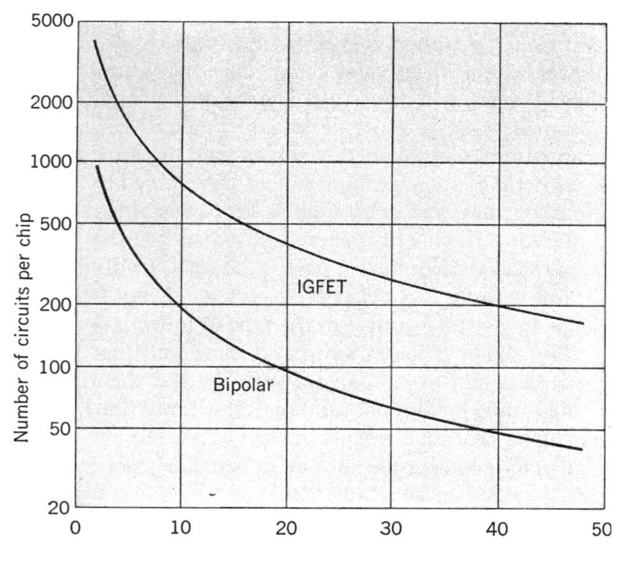

try to forget the fact that high-performance systems constitute highly complex microwave problems. A system may have tens of thousands of nodes, each one different, complicated by a wide variety of loads, stubs, and other discontinuities having a variety of lengths and terminations, including nonlinear loads. None of these can be precisely specified prior to the completion of an extremely exhaustive study of the physical layout. Can we continue to apply a semiempirical design process or will we be compelled by our own success to pause and reset the theoretical basis of our work?

Since we are dealing with very complex problems, we cannot really expect to have a comprehensive and all-inclusive equation of all possible parameters ranging from material properties to applications. This is obviously impossible, so we tend to be happy with the little that we do, but I'm not sure that I can be happy with it. I think we should learn to break up the problem, identify the dominating and fundamental parameters in more detail, and thus reduce our problem to a comprehensible form. I am sure that we would all be embarrassed to know, in our multibillion-dollar solid-state industry, how few people there are who are really trying to pull the whole theory together with all the trade-off problems that are essential to this design—and very few people are identifying and promoting the generation of the necessary fundamental parts of the theory.

The importance of LSI

A substantial part of solid-state technology is concerned with LSI. What is the system impact on LSI? Since the influence of the circuit cost on the ultimate price of a machine is relatively small, a reduction in the circuit cost, even by an immensely large factor, would have little influence on the price to the user. Thus, unavoidably, we will find engineers making use of LSI, not to reduce prices, because they cannot, but to increase capability. And among the capabilities, of course, reliability is one of the most important.

We have assumed low-cost circuitry, and perhaps one can hope that a fantastic reduction in the cost of semiconductors will provide the motivation for the necessary efforts in supporting areas. When will we see people reducing the price of the associated costs (per circuit), such as the cost of generating the layouts and of testing the circuits, packaging them, and, especially, powering them and cooling them? Unless these costs decrease in the same time scale as our semiconductor progress, the effect of LSI will be essentially negated.

Many of us may think that the reason we cannot reduce some of the other costs more significantly is that many hardware components are already much farther out on the learning curves. But I am convinced that this is not so. I think that in many instances we do not devote the resources and intellect to these problems that we devote to the semiconductor aspects. It is a matter of fad, personal taste, and community impact. The interest of the researcher in these areas is not intense because he feels that he is not working at the very forefront of the technology. Of course, there are good reasons why we cannot reduce associated costs. Many components have less uniformity and commonality when we get further from the semiconductor chips, and effort is diluted. On the other hand, I believe it would be technically possible to standardize to fewer component types. For instance, even in the input–output area, we could create a substantial volume if we could do a better job of analyzing all the necessary technical and human factors.

I have heard some people say that if all our components could be converted to solid-state (particularly to semiconductor) components, they would attract the necessary attention. Is it always necessary that research and development people be so heavily influenced by the human desire for conformity?

Another limitation that I would like to consider is a subject that has been discussed a great deal—the part-number problem in LSI. Figure 4 is a rather common representation of what happens to the number of different parts required as a function of the level of integration and of the size of a machine. With certain additional assumptions, the situation in Fig. 4 indicates that the total number of unique part numbers the industry may require for a typical group of different machine types may be as large as 100 000. Can machines be designed to reduce part numbers? Certainly, the hope of reducing part numbers by using extra circuits must be bounded by how much we can waste this way and by the fact that competitive non-LSI approaches, without waste, will prevent this trend from going too far. Would system reorganization be useful to reduce part numbers or might this prove too wasteful?

The problem of generating a new LSI part or a change in a part is much more than just cost per se. The turnaround time becomes an essential element of the picture. Can we set up a manufacturing facility that is optimized for both fast response and low cost? Or should we expect to have two facilities—one for low-cost large production and one for high-cost fast turnaround time?

We must realize that LSI requires a level of process control that may be beyond what we have realized up to now. Figure 5 shows, for instance, the required density level of defects necessary to achieve high-level integration. In this diagram, defects include any defect in the photoprocessing steps. It was assumed that a defect would have a maximum size of two mils and that the number of defects increases more than proportionately to the decrease in defect size. The illustration shows the limits of the level of integration at which one can achieve 20 percent chip yield, assuming no other defects. On the basis of present experience, we would expect, unfortunately, to be on the right side of the curve for both field-effect and bipolar transistors. How far can we expect to go to improve technology? And beyond the yield question, are we going to increase the number of circuits and, at the same time, increase the reliability, or are we not?

Conclusion

The solid-state engineers have pushed so far, so fast, that their empirical ability has outdistanced their theoretical interference. The theoreticians should be encouraged to review the basis of their work. We should try to examine all the items that, from a system point of view, limit our capabilities in cost/performance, in density, in speed, in yield, and so on.

I hope that the limits are far beyond our present level. If this is so, as it may well be, our present technology has a long road ahead for further improvement.

Revised text of the keynote address by Dr. Fubini at the 1967 International Solid State Circuits Conference, Philadelphia, Pa., February 15–17.

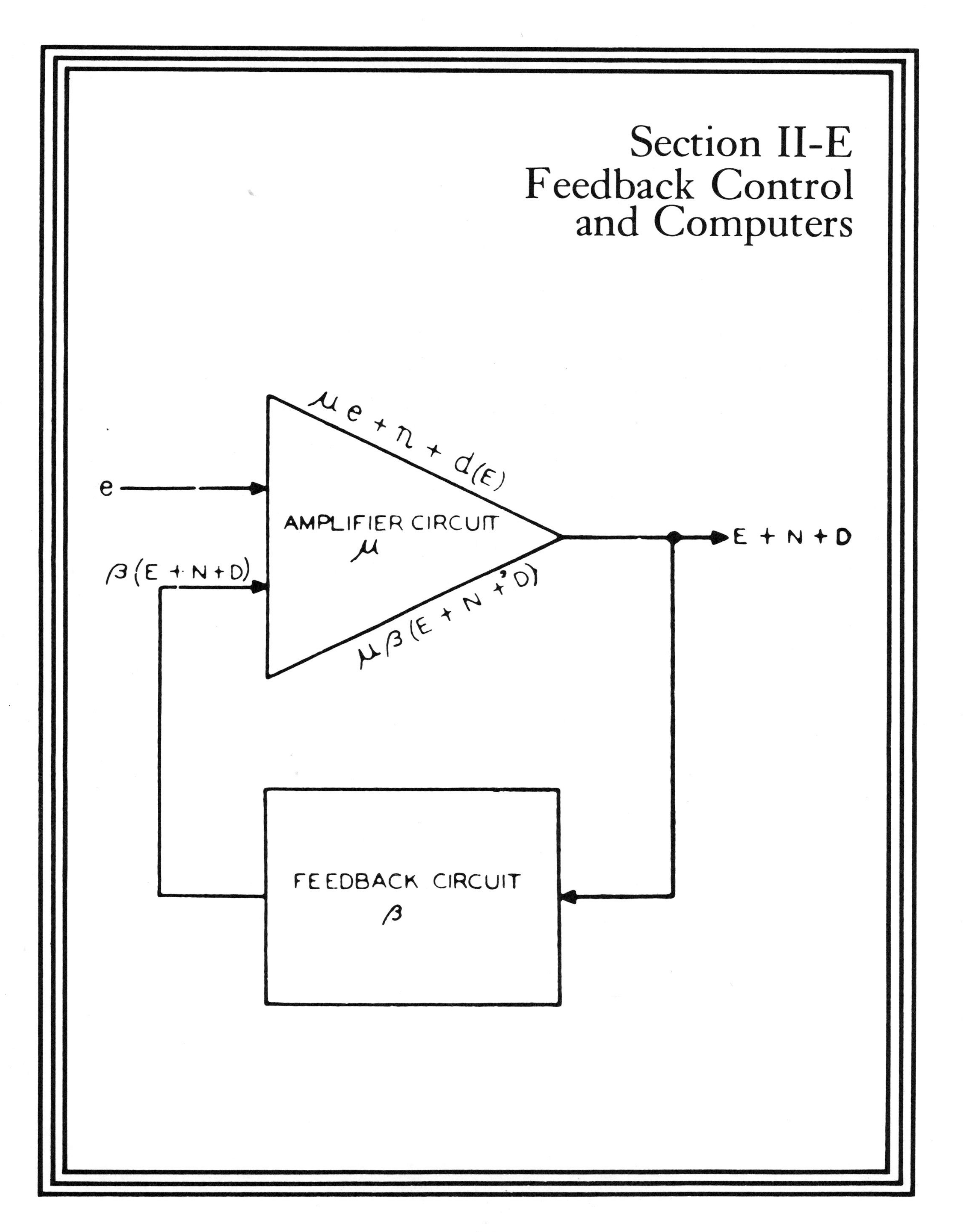

Section II-E
Feedback Control
and Computers
μe + n + d(E)
e
AMPLIFIER CIRCUIT
μ
E + N + D
β(E + N + D)
μβ(E + N + D)
FEEDBACK CIRCUIT
β

Black on Feedback Amplifiers

THIS paper by the inventor of the negative feedback amplifier concerns another of the major innovations in the history of electrical science and technology. M. J. Kelly stated that the negative feedback amplifier ranked with the de Forest audion as the two "inventions of broadest scope and significance in electronics and communications" during the first half of the 20th century. Although it first appeared to many to violate common sense, negative feedback made it possible, in Black's words, "to effect extraordinary improvement in constancy of amplification and freedom from nonlinearity." Black was stimulated to consider an unorthodox approach to amplifier design by a lecture of C. P. Steinmetz early in his career. The essence of the feedback principle came to him while riding the Lackawanna Ferry on his way to work in 1927. He wrote the relevant equations on a page of his newspaper that happened to be blank that day. He received a patent on the invention in 1937.

Harold S. Black (1898–) was born in Massachusetts and received a degree in electrical engineering from Worcester Polytechnic Institute in 1921. He joined the department at Western Electric that later became part of the Bell Telephone Laboratories. Black remained with Bell until 1963. He then became a research scientist with General Precision from 1963 to 1966. Subsequently he has been a communication consultant. See "Harold S. Black, 1957 Lamme Medalist," *Electrical Engineering*, vol. 77, pp. 720–723, 1958. Also see *Engineers of Distinction*, p. 26, 1973.

Stabilized Feed-Back Amplifiers

This paper describes and explains the theory of the feed-back principle and demonstrates how stability of amplification, reduction of modulation products, and certain other advantages follow when stabilized feed-back is applied to an amplifier. The underlying principle of design by means of which "singing" is avoided also is set forth. The paper concludes with some examples of results obtained on amplifiers which have been built employing this new principle.

By
H. S. BLACK
MEMBER A.I.E.E.

Bell Telephone Laboratories, Inc.,
New York, N. Y.

DUE TO ADVANCES in vacuum-tube development and amplifier technique, it now is possible to secure any desired amplification of the electrical waves used in the communication field. When many amplifiers are worked in tandem, however, it becomes difficult to keep the over-all circuit efficiency constant, variations in battery potentials and currents, small when considered individually, adding up to produce serious transmission changes for the over-all circuit. Furthermore, although it has remarkably linear properties, when the modern vacuum tube amplifier is used to handle a number of carrier telephone channels, extraneous frequencies are generated which cause interference between the channels. To keep this interference within proper bounds involves serious sacrifice of effective amplifier capacity or the use of a push-pull arrangement which, while giving some increase in capacity, adds to maintenance difficulty.

However, by building an amplifier whose gain is made deliberately, say 40 decibels higher than necessary (10,000 fold excess on energy basis) and then feeding the output back to the input in such a way as to throw away the excess gain, it has been found possible to effect extraordinary improvement in constancy of amplification and freedom from non-linearity. By employing this feed-back principle, amplifiers have been built and used whose gain varied less than 0.01 db with a change in plate voltage from 240 to 260 volts and whose modulation products were 75 db below the signal output at full load. For an amplifier of conventional design and comparable size this change in plate voltage would have produced about 0.7 db variation while the modulation products would have been only 35 db down; in other words, 40 db reduction in modulation products was effected. (On an energy basis the reduction was 10,000 fold.)

Stabilized feed-back possesses other advantages including reduced delay and delay distortion, reduced noise disturbance from the power supply circuits and various other features best appreciated by practical designers of amplifiers.

It is far from a simple proposition to employ feed-back in this way because of the very special control required of phase shifts in the amplifier and feed-back circuits, not only throughout the useful frequency band but for a wide range of frequencies above and below this band. Unless these relations are maintained, singing will occur, usually at frequencies outside the useful range. Once having achieved a design, however, in which proper phase relations are secured, experience has demonstrated that the performance obtained is perfectly reliable.

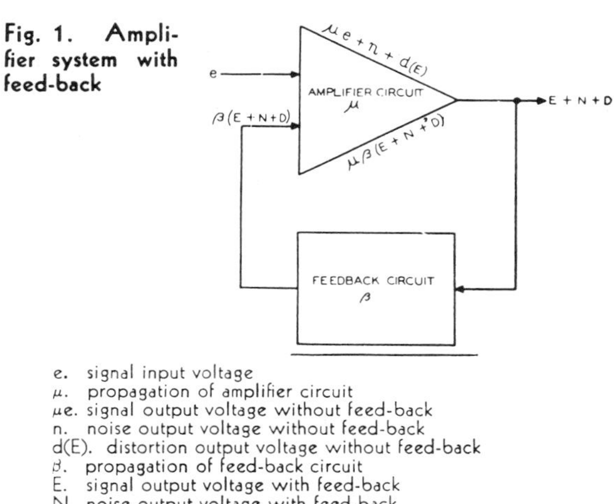

Fig. 1. Amplifier system with feed-back

e. signal input voltage
μ. propagation of amplifier circuit
μe. signal output voltage without feed-back
n. noise output voltage without feed-back
d(E). distortion output voltage without feed-back
β. propagation of feed-back circuit
E. signal output voltage with feed-back
N. noise output voltage with feed-back
D. distortion output voltage with feed-back

The output voltage with feed-back is E + N + D and is the sum of $\mu e + n + d(E)$, the value without feed-back plus $\mu\beta[E + N + D]$ due to feed-back.

$$E + N + D = \mu e + n + d(E) + \mu\beta[E + N + D]$$

$$[E + N + D](1 - \mu\beta) = \mu e + n + d(E)$$

$$E + N + D = \frac{\mu e}{1 - \mu\beta} + \frac{n}{1 - \mu\beta} + \frac{d(E)}{1 - \mu\beta}$$

If $|\mu\beta| \gg 1$, $E \doteq -\frac{e}{\beta}$. Under this condition the amplification is independent of μ but does depend upon β. Consequently the over-all characteristic will be controlled by the feed-back circuit which may include equalizers or other corrective networks.

The carrier-in-cable system dealt with in a recent Institute paper (Carrier in Cables by A. B. Clark and B. W. Kendall. A.I.E.E. TRANS., Dec. 1933, p. 1050) involves many amplifiers in tandem with many telephone channels passing through each

Full text of a paper recommended for publication by the A.I.E.E. committee on communication, and scheduled for discussion at the A.I.E.E. winter convention, Jan. 23–26, 1934. Manuscript submitted March 28, 1933; released for publication December 4, 1933. *Not published in pamphlet form.*

Reprinted from *Elec. Eng.*, vol. 53, pp. 114–120, Jan. 1934.

amplifier and constitutes, therefore, an ideal field for application of this feed-back principle. A field trial of this system was made at Morristown, New Jersey, in which 70 of these amplifiers were operated in tandem. The results of this trial were highly satisfactory and demonstrated conclusively the correctness of the theory and the practicability of its commercial application.

Circuit Arrangement

In the amplifier of Fig. 1, a portion of the output is returned to the input to produce feed-back action. The upper branch, called the μ circuit, is represented as containing active elements such as an amplifier while the lower branch, called the β-circuit, is shown as a passive network. The way a voltage is modified after once traversing each circuit is denoted μ and β, respectively, and the product, $\mu\beta$, represents how a voltage is modified after making a single journey around amplifier and feed-back circuits. Both μ and β are complex quantities, functions of frequency, and in the generalized concept either or both may be greater or less in absolute value than unity: (μ is not used in the sense that it is used sometimes, namely, to denote the amplification constant of a particular tube, but as the complex ratio of the output to the input voltage of the amplifier circuit).

Fig. 2 shows an arrangement convenient for some purposes where, by using balanced bridges in the input and output circuits, interaction between the circuits that connect to the input and output is avoided. Thereby feed-back action and amplifier impedances are made independent of the properties of circuits connected to the amplifier.

General Equation

In Fig. 1, β is zero without feed-back and a signal voltage, e_0, applied to the input of the μ-circuit produces an output voltage. This is made up of

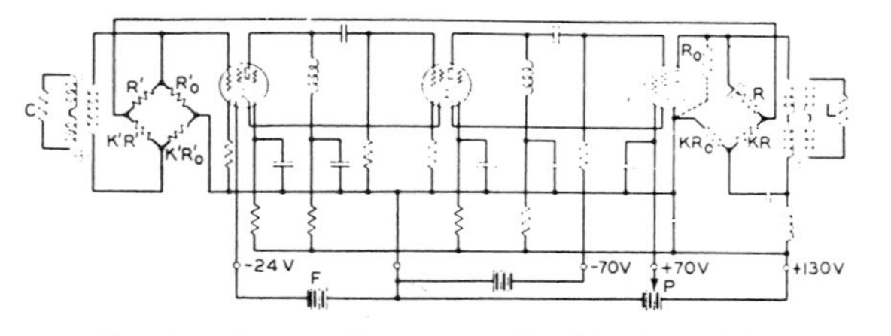

Fig. 2. Circuit of a negative feed-back amplifier

what is wanted, the amplified signal, E_0, and components that are not wanted, namely, noise and distortion designated N_0 and D_0 and assumed to be generated within the amplifier. It is further assumed that the noise is independent of the signal and the distortion of modulation a function *only of the signal output.* Using the notation of Fig. 1, the output without feed-back may be written as:

$$E_0 + N_0 + D_0 = \mu e_0 + n + d(E_0) \tag{1}$$

where zero subscripts refer to conditions without feed-back.

With feed-back, β is not zero and the input to the μ-circuit becomes $e_0 + \beta\ (E + N + D)$. The output is $E + N + D$ and is equal to $\mu[e_0 + \beta\,(E + N + D)] + n + dE$ or

$$E + N + D = \frac{\mu e_0}{1 - \mu\beta} + \frac{n}{1 - \mu\beta} + \frac{d(E)}{1 - \mu\beta} \tag{2}$$

In the output signal, noise and modulation are divided by $(1 - \mu\beta)$, and assuming $|1 - \mu\beta| > 1$, all are reduced.

Change in Gain Due to Feed-Back

From eq 2, the amplification with feed-back equals the amplification without feed-back divided by $(1 - \mu\beta)$. The effect of adding feed-back, therefore, usually is to change the gain of the amplifier and this change will be expressed as

$$G_{CF} = 20 \log_{10} \frac{1}{1 - \mu\beta} \tag{3}$$

where G_{CF} is *db change in gain due to feed-back.* As a quantitative measure of the effect of feed-back $\frac{1}{1 - \mu\beta}$ will be used and the feed-back referred to as *positive feed-back* or *negative feed-back* according as the absolute value of $\frac{1}{1 - \mu\beta}$ is greater or less than unity. Positive feed-back increases the gain of the amplifier; negative feed-back reduces it. The term feed-back is not limited merely to those cases where the absolute value of $\frac{1}{1 - \mu\beta}$ is other than unity.

From $\mu\beta = |\mu\beta|\,\underline{|\Phi}$ and (3), it may be shown that

$$10^{-\frac{G_{CF}}{10}} = 1 - 2\,|\mu\beta|\cos\Phi + |\mu\beta|^2 \tag{4}$$

which is the equation for a family of concentric circles of radius $10^{-\frac{G_{CF}}{10}}$ about the point 1, 0. Fig. 3 is a polar diagram of the vector field of $\mu\beta = |\mu\beta|\,\underline{|\Phi}$. Using rectangular instead of polar coördinates, Fig. 4 corresponds to Fig. 3 and may be regarded as a diagram of the field of $\mu\beta$ where the parameter is db change in gain due to feed-back. From these diagrams all of the essential properties of feed-back action can be obtained such as change in amplification, effect on linearity, change in stability due to variations in various parts of the system, reduction of noise, etc. Certain significant boundaries have been designated similarly on both figures.

For example, boundary A is the locus of zero change in gain due to feed-back. Along this parametric contour line where the absolute magnitude of amplification is not changed by feed-back action, values of $|\mu\beta|$ range from zero to 2 and the phase shift, Φ around the amplifier and feed-back circuits equal $\cos^{-1}\frac{|\mu\beta|}{2}$ and, therefore, lies between -90 deg and $+90$ deg. For all conditions inside or above this boundary, the gain with feed-back is increased; outside or below, the gain is decreased.

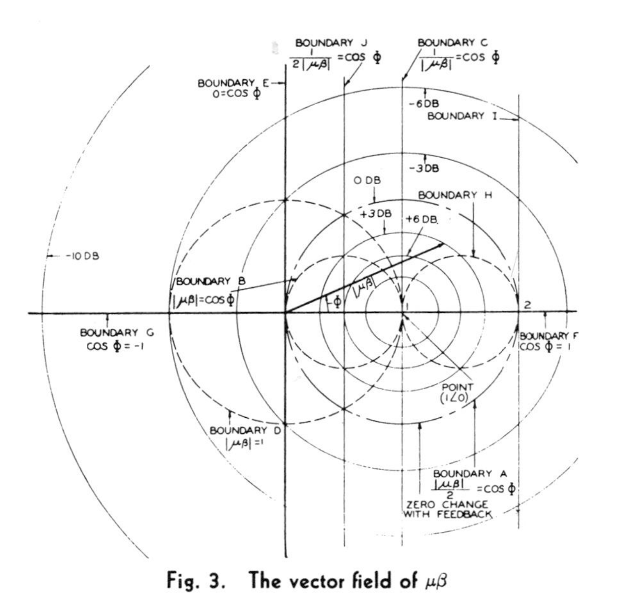

Fig. 3. The vector field of $\mu\beta$

See caption for Fig. 4

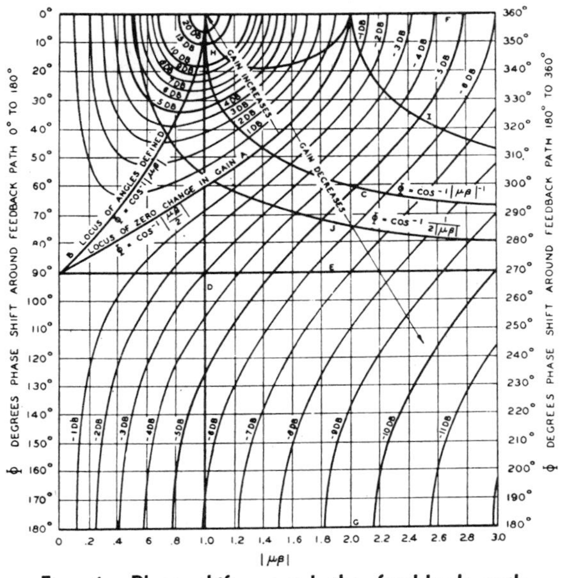

Fig. 4. Phase shift around the feed-back path plotted as a function of $|\mu\beta|$, absolute value of $\mu\beta$

The complex quantity $\mu\beta$ represents the ratio by which the amplifier and feed-back (or more generally μ and β) modify a voltage in a single trip around the closed path

First, there is a set of boundary curves indicated by letters which give either limiting or significant values of $|\mu\beta|$ and ϕ.

Second, there is a family of curves in which db change in gain due to feed-back is the parameter.

Boundaries

A. Conditions in which gain and modulation are unaffected by feed-back.

B. Constant amplification ratio against small variations in $|\beta|$.

Constant change in gain, $\frac{1}{|1-\mu\beta|}$, against variations in $|\mu|$ and $|\beta|$. Stable phase shift through the amplifier against variation in $\Phi\beta$.

The boundary on which the stability of amplification is unaffected by feed-back.

C. Constant amplification ratio against small variations in $|\mu|$.
Constant phase shift through amplifier against variations in $\Phi\mu$.

The absolute magnitude of the voltage fed back $\frac{|\mu\beta|}{|1-\mu\beta|}$ is constant against variations in $|\mu|$ and $|\beta|$.

D. $|\mu\beta| = 1$

E. $\Phi = 90°$. Improvement in gain stability corresponds to twice db reduction in gain.

F. G. Constant amplification ratio against variations in Φ. Constant phase shift through the amplifier against variations in $|\mu|$ and $|\beta|$.

H. Same properties as β

I. Same properties as E

J. Conditions in which $\frac{|\mu|}{|1-\mu\beta|} = \frac{-1}{|\beta|}$ the over-all gain is the exact negative inverse of the transmission through the β-circuit.

Stability

From eq 2, $\frac{\mu e_0}{1-\mu\beta}$ is the amplified signal with feed-back and $\frac{\mu}{1-\mu\beta}$, therefore, is an index of the amplification. It is of course a complex ratio. It will be designated A_F and referred to as the amplification with feed-back.

To consider the effect of feed-back upon stability of amplification, the stability will be viewed as the ratio of a change, δA_F, to A_F where δA_F is due to a change either in μ or β and the effects may be derived by assuming the variations are small.

$$A_F = \frac{\mu}{1-\mu\beta} \tag{5}$$

$$\left[\frac{\delta A_F}{A_F}\right]_\mu = \frac{\left[\frac{\delta\mu}{\mu}\right]}{1-\mu\beta} \tag{6}$$

$$\left[\frac{\delta A_F}{A_F}\right]_\beta = \frac{\mu\beta}{1-\mu\beta}\left[\frac{\delta\beta}{\beta}\right] \tag{7}$$

If $\mu\beta \gg 1$, it is seen that μ or the μ-circuit is stabilized by an amount corresponding to the reduction in amplification and the effect of introducing a gain or loss in the μ-circuit is to produce no material change in the over-all amplification of the system; the stability of amplification as affected by β or the β-circuit is neither appreciably improved nor degraded since increasing the loss in the β-circuit raises the gain of the amplifier by an amount almost corresponding to the loss introduced and *vice versa*. If both μ and β are varied and the variations sufficiently small, the effect is the same as if each were changed separately and the two results then combined.

In certain practical applications of amplifiers it is the change in gain or ammeter or voltmeter reading at the output that is a measure of the stability rather than the complex ratio previously treated. The conditions surrounding gain stability may be examined by considering the absolute value of A_F. This is shown as follows:

Let (db) represent the gain in decibels corresponding to A_F. Then

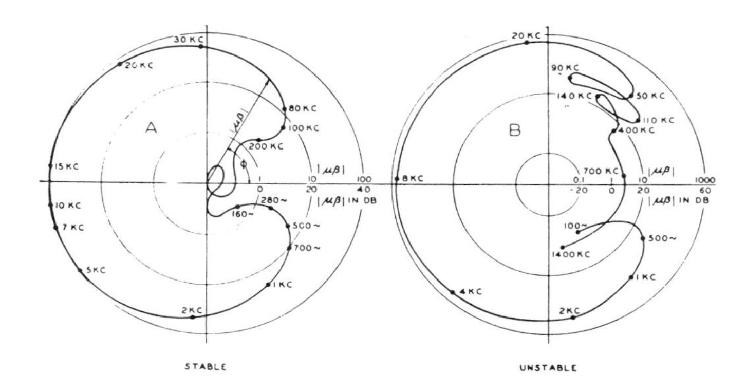

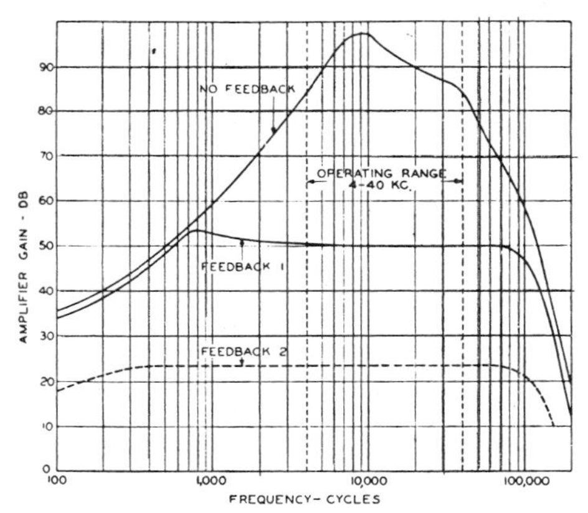

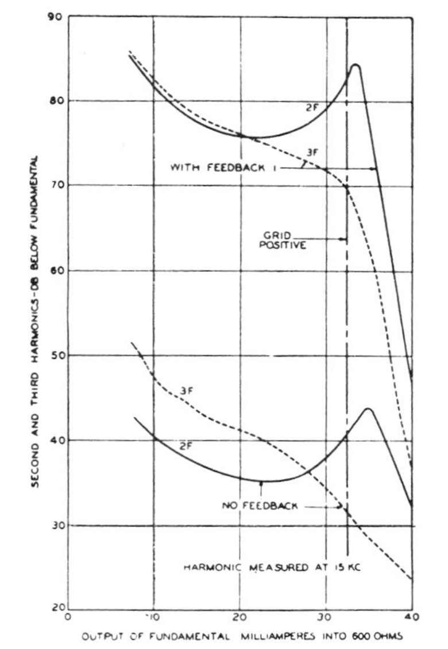

Fig. 5 (above). Measured $\mu\beta$ characteristics of 2 amplifiers

Fig. 6 (above right). Gain-frequency characteristics with and without feed-back of amplifier of Fig. 2

Fig. 7(left). Modulation characteristics with and without feed-back for the amplifier of Fig. 2

$$(db) = 20 \log_{10} | A_F |$$

$$\delta(db) \doteq 8.686 \left[\frac{\delta | A_F |}{| A_F |}\right] \quad (8)$$

To get the absolute value of the amplification:

$$\mu\beta = | \mu\beta | \underline{|\Phi} \quad (9)$$

$$| A_F | = \frac{| \mu |}{\sqrt{1 - 2 | \mu\beta | \cos \Phi + | \mu\beta |^2}} \quad (10)$$

The stability of amplification which is proportional to the gain stability is given by

$$\left[\frac{\delta | A_F |}{| A_F |}\right]_\mu \doteq \frac{1 - | \mu\beta | \cos \Phi}{| 1 - \mu\beta |^2} \left[\frac{\delta | \mu |}{| \mu |}\right] \quad (11)$$

$$\left[\frac{\delta | A_F |}{| A_F |}\right]_{|\beta|} \doteq \left|\frac{\mu\beta}{1 - \mu\beta}\right| \left[\frac{\cos \Phi - | \mu\beta |}{| 1 - \mu\beta |}\right] \left[\frac{\delta | \beta |}{| \beta |}\right] \quad (12)$$

$$\left[\frac{\delta | A_F |}{| A_F |}\right]_\Phi \doteq - \left|\frac{\mu\beta}{1 - \mu\beta}\right| \left[\frac{\sin \Phi}{| 1 - \mu\beta |}\right] [\delta\Phi] \quad (13)$$

A curious fact to be noted from eq 11 is that it is possible to choose a value of $\mu\beta$ (namely, $|\mu\beta| = \sec \Phi$) so that the numerator of the right hand term vanishes. This means that the gain stability is perfect, assuming differential variations in $|\mu|$.

Referring to Figs. 3 and 4, contour C is the locus of $|\mu\beta| = \sec \Phi$ and it includes all amplifiers whose gain is unaffected by small variations in $|\mu|$. In this way it is possible even to stabilize an amplifier whose feed-back is positive, i. e., feed-back may be utilized to raise the gain of an amplifier and, at the same time, the gain stability with feed-back *need* not be degraded but on the contrary may be improved. If a similar procedure is followed with an amplifier whose feed-back is negative, the gain stability theoretically will be perfect and independent of the reductions in gain due to feed-back. Over too wide a frequency band practical difficulties will limit the improvements possible by these methods.

With negative feed-back, gain stability always is improved by an amount at least as great as corresponds to the reduction in gain and generally more; with positive feed-back, gain stability never is degraded by more than would correspond to the increase in gain and under appropriate conditions, assuming the variations are not too great is as good or much better than without feed-back. With positive feed-back, the variations in μ or β must not be permitted to become sufficiently great as to cause the amplifier to sing or give rise to instability as defined in the section devoted to the conditions for avoiding singing.

Modulation

To determine the effect of feed-back action upon modulation produced in the amplifier circuit, it is convenient to assume that the output of undistorted signal is made the same with and without feed-back and that a comparison then is made of the difference in modulation with and without feed-back. Therefore, with feed-back, the input is changed to $e = e_0 (1 - \mu\beta)$ and, referring to eq 2, the output voltage is μe_0 and the generated modulation, $d(E)$, assumes its value without feed-back, $d(E_0)$, and $\frac{d(E)}{1 - \mu\beta}$ becomes $\frac{d(E_0)}{1 - \mu\beta}$ which is $\frac{D_0}{1 - \mu\beta}$. This relationship is approximate because the voltage at the input without feed-back is free from distortion and with feed-back it is not and, hence, the assumption that the modulation is a function *only of the signal output* used in deriving eq 2 is not necessarily justified.

From the relationship $D = \frac{D_0}{1 - \mu\beta}$, it is to be concluded that modulation with feed-back will be reduced decibel for decibel as the effect of feed-back action causes an arbitrary db reduction in the gain of the amplifier; i. e., when the feed-back is negative. With positive feed-back the opposite is true, the modulation being increased by an amount corresponding to the increase in amplification.

If modulation in the β-circuit is a factor, it can be shown that usually in its effect on the output the modulation level at the output due to nonlinearity of the β-circuit is approximately $\frac{\mu\beta}{1 - \mu\beta}$ multiplied by the modulation generated in the β-circuit acting alone and without feed-back.

Additional Effects

Noise. A criterion of the worth of a reduction in noise is the reduction in signal-to-noise ratio at the output of an amplifier. Assuming that the amount of noise introduced is the same in 2 systems, for example, with and without feed-back, respectively, and that the signal outputs are the same, a comparison of the signal-to-noise ratios will be affected by the amplification between the place at which the noise enters and the output. Denoting

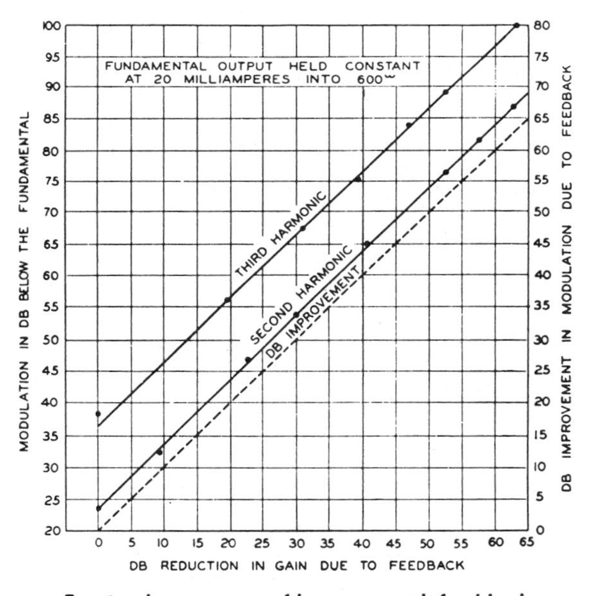

Fig. 8. Improvement of harmonics with feed-back

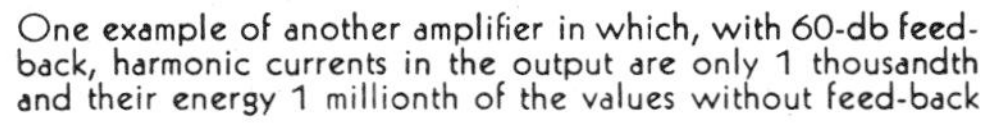

One example of another amplifier in which, with 60-db feed-back, harmonic currents in the output are only 1 thousandth and their energy 1 millionth of the values without feed-back

this amplification by a and a_0, respectively, it can be shown that the relation between the 2 noise ratios is $\frac{a_0}{a}(1 - \mu\beta)$. This is called the *noise index.*

If noise is introduced in the power supply circuits of the last tube, $a_0/a = 1$ and the noise index is $(1 - \mu\beta)$. As a result of this relation less expensive power supply filters are possible in the last stage.

Phase Shift, Envelope Delay, Delay Distortion. In the expression $A_F = \left[\frac{\mu}{1 - \mu\beta}\right] \underline{|\theta}$, θ is the over-all phase shift with feed-back, and it can be shown that the *phase shift through the amplifier with feed-back may be made to approach the phase shift through the β-circuit plus 180 deg.* The effect of phase shift in the β-circuit is not reduced correspondingly. It will be recalled that in reducing the change in phase shift with frequency, envelope delay, which is the slope of the phase shift with respect to the angular velocity, $\omega = 2\pi f$, also is reduced. The *delay* distortion likewise is reduced because a measure of

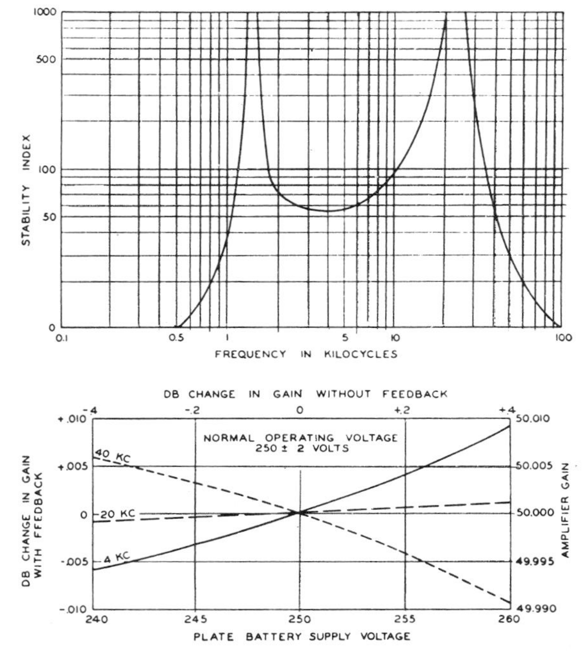

Fig. 9. Representative gain stability of a single amplifier as determined by measuring 69 feed-back amplifiers in tandem at Morristown, New Jersey

The upper figure shows the absolute value of the stability index. It can be seen that between 20 and 25 kc the improvement in stability is more than 1,000 to 1 yet the reduction in gain was less than 35 db

The lower figure shows change in gain of the feed-back amplifier with changes in the plate battery voltage and the corresponding changes in gain without feed-back. At some frequencies the change in gain is of the same sign as without feed-back and at others it is of opposite sign and it can be seen that near 23 kc the stability must be perfect

delay distortion at a particular frequency is the difference between the envelope delay at that frequency and the least envelope delay in the band.

β-Circuit Equalization. Referring to eq 2, the output voltage E approaches $-e_0/\beta$ as $1 - \mu\beta = -\mu\beta$ and equals it in absolute value if $\cos\Phi = \frac{1}{2|\mu\beta|}$

where $\mu\beta = |\mu\beta| \underline{|\Phi}$. Under these circumstances increasing the loss in the β-circuit 1 db raises the gain of the amplifier 1 db, and *vice-versa*, thus giving any gain-frequency characteristic for which a like loss-frequency characteristic can be inserted in the β-circuit. This procedure has been termed β-circuit

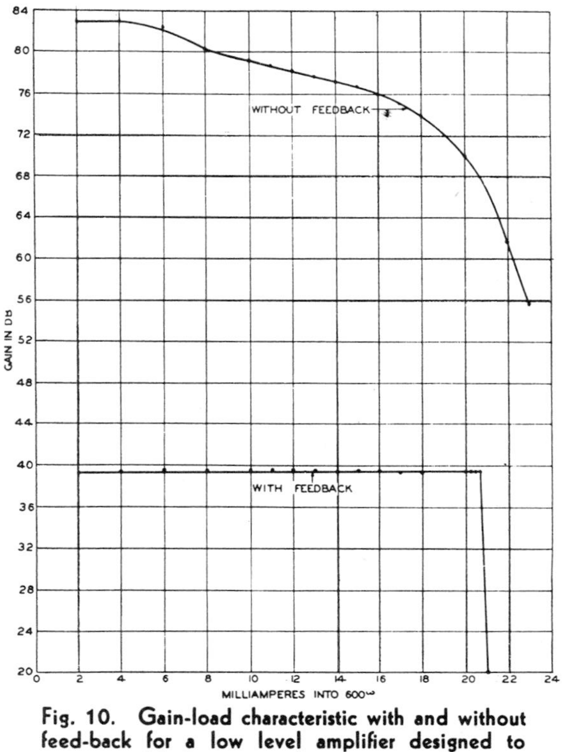

Fig. 10. Gain-load characteristic with and without feed-back for a low level amplifier designed to amplify frequencies from 3.5 to 50 kc

equalization. It possesses other advantages and properties which are beyond the scope of this paper.

Avoid Singing

Having considered the theory up to this point, experimental evidence was readily acquired to demonstrate that $\mu\beta$ might assume large values, 10 to 10,000, provided Φ was not at the same time zero. However, one noticeable feature about the field of $\mu\beta$ (Figs. 3 and 4) is that it implies that even though the phase shift is zero and the absolute value of $\mu\beta$ exceeds unity, self-oscillations or singing will not result. This may or may not be true. When first thinking about this matter it was suspected that owing to practical nonlinearity, singing would result whenever the gain around the closed loop equaled or exceeded the loss and simultaneously the phase shift was zero; i. e., $\mu\beta = |\mu\beta| + \text{jo} \geq 1$. Results of experiments, however, seemed to indicate something more was involved and these matters were described to H. Nyquist who developed a

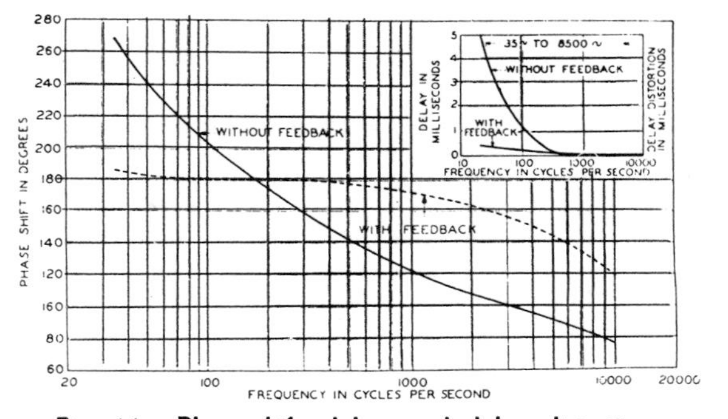

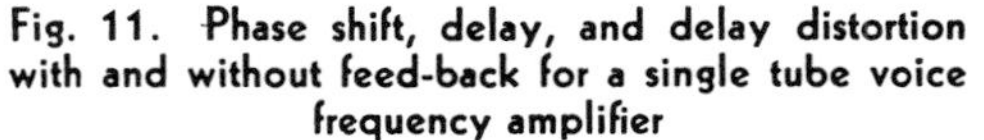

Fig. 11. Phase shift, delay, and delay distortion with and without feed-back for a single tube voice frequency amplifier

more general criterion for freedom from instability applicable to an amplifier having linear positive constants. (For a complete description of the criterion for stability and instability and exactly what is meant by enclosing the point (1, 0), reference should be made to Regeneration Theory, by H. Nyquist. *Bell System Technical Journal*, v. XI, July 1932, p. 126–47.)

To use this criterion, plot $\mu\beta$ (the modulus and argument vary with frequency) and its complex conjugate in polar coördinates for all values of frequency from 0 to $+\infty$. If the resulting loop or loops do not enclose the point (1, 0) the systen will be stable, otherwise not. The envelope of the transient response of a stable amplifier always dies away exponentially with time; that of an unstable amplifier in all physically realizable cases increases with time. Characteristics A and B in Fig. 5 are results of measurements on 2 different amplifiers; the amplifier having $\mu\beta$ characteristic denoted A was stable, the other unstable.

The number of stages of amplification that can be used in a single amplifier is not significant except in so far as it affects the question of avoiding singing Amplifiers with considerable negative feed-back have been tested where the number of stages ranged from 1 to 5, inclusive. In every case the feed-back path was from the output of the last tube to the input of the first tube.

Experimental Results

Figs. 6 and 7 show how the gain-frequency and modulation characteristics of the 3-stage impedance coupled amplifier of Fig. 2 are improved by negative feed-back. In Fig. 7 the improvement in harmonics is not equal exactly to the decibel reduction in gain. Fig. 8 shows measurements on a different amplifier in which harmonics are reduced as negative feed-back is increased, decibel for decibel over a 65-db range.

That the gain with frequency practically is independent of small variations in $|\mu|$ is shown by Fig. 9. This is a characteristic of the Morristown amplifier, described in the paper by Clark and Kendall referred to previously, which meets the severe requirements imposed upon a repeater amplifier

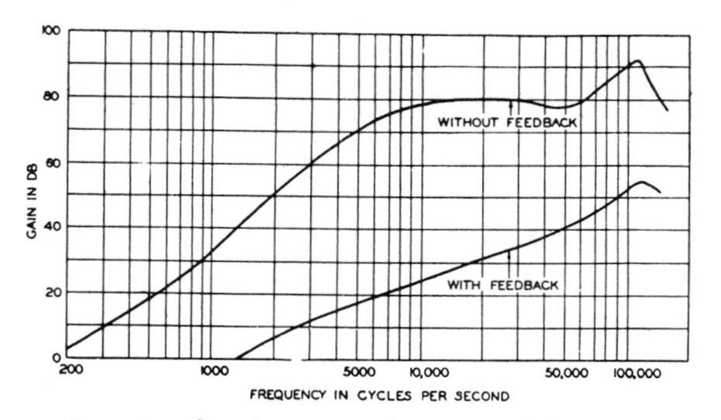

Fig. 12. Gain-frequency characteristic of an amplifier with an equalizer in the β-circuit

This was designed to have a gain frequency characteristic with feed-back of the same shape as the loss frequency characteristic of a nonloaded telephone cable

for use in cable carrier systems. Designed to amplify frequencies from 4 kc to 40 kc the maximum change in gain due to variations in plate voltage does not exceed $\frac{7}{10{,}000}$ db per volt and at 20 kc the change is only $\frac{1}{20{,}000}$ db per volt. This illustrates that for small changes in $|\mu|$, the ratio of the stability without feed-back to the stability with feed-back, called the *stability index*, approaches $\frac{|1 - \mu\beta|^2}{1 - |\mu\beta| \cos \Phi}$ and gain stability is improved at least as much as the gain is reduced and usually more, and is theoretically perfect if $\cos \Phi = \frac{1}{|\mu\beta|}$.

In Fig. 10 is indicated the effectiveness with which the gain of a feed-back amplifier can be made independent of variations in input amplitude practically up to the overload point of the amplifier. These measurements were made on a 3-stage amplifier designed to work from 3.3 kc to 50 kc.

As shown in Fig. 11, the negative feed-back may be used to improve phase shift and reduce delay and delay distortion. These measurements were made on an experimental 1-tube amplifier, 35–8,500 cycles, feeding back around the low side windings of the input and output transformers.

In Fig. 12 is given the gain-frequency characteristic of an amplifier with and without feed-back when in the β-circuit there is an equalizer designed to make the gain-frequency characteristic of the amplifier with feed-back of the same shape as the loss-frequency characteristic of a nonloaded telephone cable.

Conclusion

The feed-back amplifier dealt with in this paper was developed primarily with requirements in mind for a cable carrier telephone system, involving many amplifiers in tandem with many telephone channels passing through each amplifier. Most of the examples of feed-back amplifier performance naturally have been drawn from amplifiers designed for this field of operation. In this field, vacuum tube amplifiers normally possessing good characteristics with respect to stability and freedom from distortion are made to possess superlatively good characteristics by application of the feed-back principle.

However, certain types of amplifiers, in which economy has been secured by sacrificing performance characteristics, particularly as regards distortion, can be made to possess improved characteristics by the application of feed-back. Discussion of these amplifiers is beyond the scope of this paper.

Terman, Hewlett, Buss, and Cahill on Applications of Negative Feedback

THIS paper well illustrates the enormous importance of the negative feedback principle (see Paper 58) outside the field of telephone amplifiers. The authors give numerous examples of improved laboratory instruments that were made possible through the use of feedback. They concluded, however, that these examples merely scratched the surface of possibilities. Dr. Terman, who provided me with a helpful commentary on the selection of papers for this volume, made the following comment on this paper: . . ."the resistance–capacitance tuned variable frequency audio oscillator described in this paper was William Hewlett's contribution to the paper, and was furthermore the foundation on which the Hewlett–Packard Company was built. At the time the company was formed in 1939, this oscillator was their entire product line, and its success in the market place made it possible for H–P to develop other products and hence to grow. Today H–P's annual sales are approaching a billion dollars per year, and still include a line of resistance–capacitance tuned variable frequency audio oscillators! Thus the contents of this 1939 IRE paper have more historic value than might appear on the surface." See also Norberg's paper on the origins of the electronics industry on the West Coast in the PROCEEDINGS OF THE IEEE for September 1976.

Frederick E. Terman (1900–   ) was born in English, Ind. and received an A.B. degree in chemical engineering and a degree in electrical engineering from Stanford in 1920 and 1922, respectively. He received a doctorate degree from M.I.T. in 1924 where he worked under Vannevar Bush. He returned to Stanford to teach in 1925 and remained in various capacities including Department Head, Dean of Engineering, Provost, and Vice President until his retirement in 1965. During World War II he organized and directed the Radio Research Laboratory at Harvard where electronic countermeasure systems were developed. He was President of the IRE in 1941 and the author of the famous *Radio Engineers' Handbook*. See the biographical note on Terman in the special issue of the PROCEEDINGS OF THE IEEE for September 1976.

Robert R. Buss (1913–   ) was born in Provo, Utah and received an A.B. degree from San Jose State in 1935. He received a degree in electrical engineering from Stanford in 1938 and the Ph.D. degree from Stanford in 1940. He worked for Heintz and Kaufman, Litton, and at the U.S. Naval Air Station before joining Terman's group at Harvard in 1942. After the war he taught at Northwestern University from 1946 to 1951 before returning to Stanford in 1951. See *Who's Who in Engineering*, p. 252, 1964.

William R. Hewlett (1913–   ) received a B.A. degree from Stanford and an M.S. degree from M.I.T. He was a co-founder of the Hewlett–Packard Company in 1939 and has remained with the Company since in various administrative capacities. He was IRE President in 1954. See *Engineers of Distinction*, p. 138, 1973 and "W. R. Hewlett receives WEMA Medal," *IEEE Spectrum*, p. 119, September 1971.

Francis C. Cahill (1914–   ) was born in Dixon, Ill. and received an A.B. degree from Stanford in 1936 and a degree in electrical engineering in 1938. He was a radio engineer with Heintz and Kaufman and with RCA before joining the Radio Laboratory at Harvard in 1942. He became an engineer with the Airborne Instruments Laboratory in 1945, becoming chief engineer in 1952. See *Who's Who in Engineering*, p. 257, 1964.

Some Applications of Negative Feedback with Particular Reference to Laboratory Equipment*

F. E. TERMAN†, FELLOW, I.R.E., R. R. BUSS†, STUDENT, I.R.E.,
W. R. HEWLETT†, ASSOCIATE, I.R.E., AND F. C. CAHILL†, STUDENT, I.R.E.

***Summary**—The application of feedback to an entire amplifier rather than just to the final stage makes it possible to realize the characteristics of a perfect amplifier over wide frequency ranges. The use of such amplifiers to give direct-reading audio-frequency voltmeters with permanent calibration and any desired sensitivity is described.*

Negative feedback can be used to reduce the distortion in the output of laboratory oscillators for all loads from open circuit to short circuit by the expedient of throwing away a part of the output power in a resistive network.

Means are described for applying feedback to tuned radio-frequency amplifiers so that the amplification depends only upon the constants of the tuned circuit and is independent of the tubes and supply voltages.

The use of negative feedback to develop a stabilized negative resistance substantially independent of tubes and supply voltages is considered, and various applications described.

High selectivity can be obtained by deriving the feedback voltage from the neutral arm of a bridge, one leg of which involves a parallel resonant circuit. It is possible by this means to obtain an effective circuit Q of several thousand, using ordinary tuned circuits, and the selectivity can be varied without affecting the amplification at resonance. The use of these highly selective circuits in wave analyzers is considered.

Feedback can be used to give improved laboratory oscillators. These include resistance-stabilized oscillators, in which the amplitude-limiting action is also separated from the amplifier action, and oscillators in which the frequency is controlled by a resistance-capacitance network. Such resistance-capacitance oscillators represent a simple and inexpensive substitute for beat-frequency oscillators, and have comparable performance.

An amplifier with negative feedback is an ordinary amplifier in which a voltage is derived from the output and superimposed upon the amplifier input in such a way as under normal conditions to oppose the applied signal voltage. The presence of feedback then reduces the amplification and output distortion by the factor $1/(1-A\beta)$, where A is the amplification in the absence of feedback, and β is the ratio of voltage superimposed on the amplifier input to the output voltage of the amplifier.[1,2] The quantity $A\beta$ determines the magnitude of the feedback effect, and can be conveniently termed the *feedback factor*. It will be noted that when $A\beta$ is large compared with unity, that the amplification approaches $-1/\beta$.

Laboratory Audio-Frequency Amplifiers with Negative Feedback

Although feedback is usually employed in audio-frequency amplifiers for the purpose of reducing the distortion in the power stage, there is much more to be gained in the case of laboratory amplifiers by applying feedback to the entire amplifier. By making the feedback factor $A\beta$ much larger than unity, and arranging matters so the fraction β of the output voltage that is superimposed upon the amplifier input is independent of the tube characteristics, the amplification depends primarily on β and is substantially independent of tube replacements, electrode voltages, aging of tubes, etc. It is then possible to engrave an accurate calibration on the gain control, since the gain calibration becomes as permanent as the characteristics of a small milliammeter. Furthermore, if $A\beta$ is large and the feedback circuit is such as to make β independent of frequency, then the amplification is practically independent of frequency, the phase shift is practically zero over the normal frequency range of the amplifier, and the range for reasonably flat response is greatly increased.

The extent of the improvements obtainable in the performance of an amplifier can be realized by considering Table I, which compares performances with and without feedback in a hypothetical case.

TABLE I

COMPARISON OF RESISTANCE-COUPLED AMPLIFIERS WITH AND WITHOUT FEEDBACK

	No Feedback	With Feedback
Voltage gain (middle-frequency range)	2500	2500
Voltage gain with tube or supply potential change that increases A 25 per cent	3125 (+1.94 db)	2520 (+0.07 db)
Distortion with full output	2%	0.04%
Variation of gain over range 15–30,000 cycles	−50% (−6 db)	+4% (+0.33 db)
Frequency range for gain variation of 50 per cent	15 to 30,000 cycles	5 to 95,000 cycles
Phase shift over range 15–30,000 cycles	90°	4°40′

NOTE: Amplifier without feedback is two-stage resistance-coupled. Amplifier with feedback is two such sections in tandem with each section having $A\beta=-49$ in mid-frequency range.

* Decimal classification: R363. Original manuscript received by the Institute, November 22, 1938; abridged manuscript received by the Institute, August 1, 1939. Presented, Thirteenth Annual Convention, New York, N. Y., June 16, 1938, and Pacific Coast Convention, Portland, Ore., August 11, 1938.

† Stanford University, California.

[1] H. S. Black, "Stabilized feedback amplifier," *Elec. Eng.*, vol. 53, pp. 114–120; January, (1934).

[2] H. Nyquist, "Regeneration theory," *Bell Sys. Tech. Jour.*, vol. 11, pp. 126–147; January, (1932).

Reprinted from *Proc. IRE*, vol. 27, pp. 649–655, Oct. 1939.

It is apparent that whenever flatness of response, reproducibility of gain, low distortion, or low phase shift are of importance, an amplifier cannot be considered as being properly designed unless full use is made of feedback. This is especially true in amplifiers used in measuring equipment and for oscillograph purposes.

Feedback can also be used to improve the balance between the two sides of a push-pull class A amplifier. Typical circuits for doing this are shown in Fig. 1.

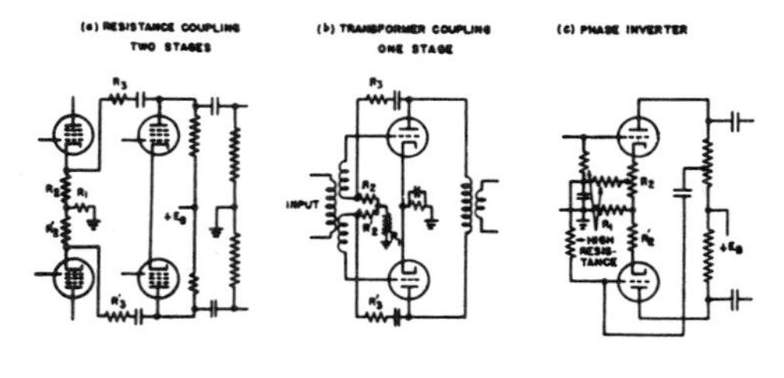

Fig. 1—Use of feedback to maintain balance between outputs of the two sides of a push-pull class A amplifier or phase inverter.

In these arrangements unbalance produces a current through the resistance R_1, resulting in the development of a feedback voltage that is applied to the tubes in such a manner as to reduce the difference in the outputs of the two sides. The use of feedback in this way makes it possible to maintain extremely accurate balance without the necessity of using carefully matched tubes in each push-pull stage. It also makes possible almost perfect balance when phase inverters are used. The great practical value of the result in laboratory push-pull amplifiers requiring accurate balance is obvious.

Use of Feedback Amplifiers in Voltage Measurements

The stability of amplification that results when a large amount of feedback is employed in an audio-frequency amplifier is comparable with the stability of the small d'Arsonval meters commonly used in laboratory work. This, coupled with the very uniform response that can be obtained over a wide frequency range opens up many possibilities in measuring equipment.

The instruments shown in Figs. 2 and 3 are indications of what can be done. The first of these consists of a two-stage amplifier with a large amount of feedback, delivering its output to a vacuum thermocouple. This gives the equivalent of a square-law vacuum-tube voltmeter but has a permanent calibration and requires no zero adjustment. Furthermore by proper design the final tube can be made to overload at slightly above full-scale deflection, so that no matter how large a voltage may accidentally be applied, the thermcouple cannot be burned out.[1,2] The instrument with the circuit proportions given in Fig. 2 has been in use for several years and found to be highly satisfactory. It has an input resistance of 1 megohm, gives full-scale output with an input of 3 volts, and can be used as a direct-reading instrument in the same manner as an ordinary direct-current voltmeter. The stability and flatness of the frequency response are indicated by the performance tests reported in Table II.

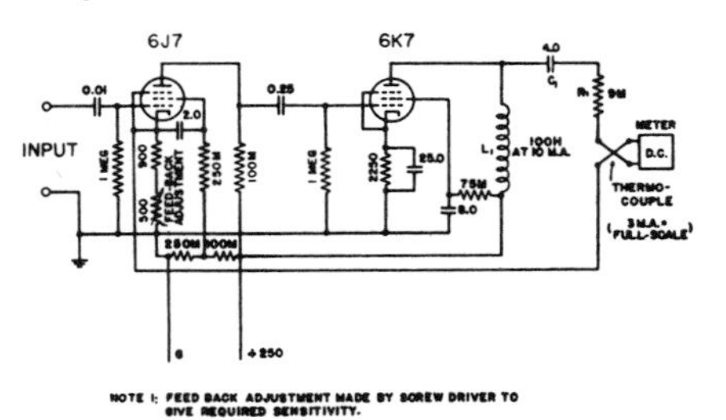

Fig. 2—Circuit diagram of a feedback voltmeter designed for audio-frequency service.

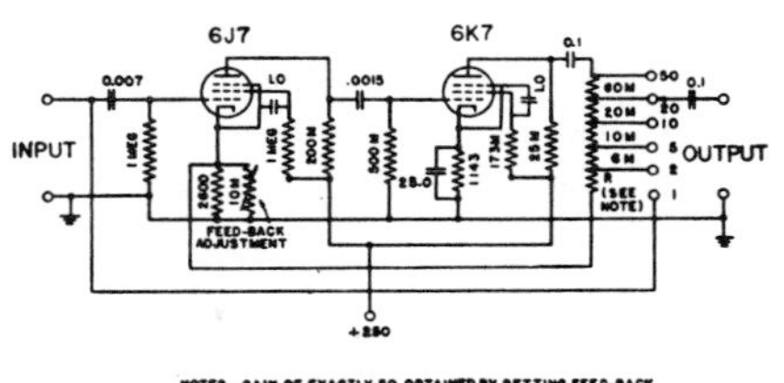

Fig. 3—Circuit diagram ot standard-gain amplifier for increasing the sensitivity of vacuum-tube and feedback voltmeters at audio frequencies. The maximum gain is 50, and the output is tapped as shown so that gains of 20, 10, 5, and 2 are likewise available.

TABLE II

Characteristics of Feedback Measuring Instruments

	Feedback Voltmeter of Fig. 2	Standard-Gain Amplifier of Fig. 3
Drop in response at 40 cycles	under −1%	under −1%
Drop in response at 20,000 cycles	under −1%	under −1%
Change in gain when plate-supply voltage is varied from 250 volts to		
140 volts	under −1%	under −0.5%
400 volts	under +1%	under +0.5%
Change in gain when heater voltage varied from 6 volts to		
7.5 volts	under 0.5%	under 1%
5.3 volts	under 0.5%	under 1%

The instrument of Fig. 3 is an amplifier for extending the range of the feedback voltmeter of Fig. 2 and of ordinary vacuum-tube voltmeters. This consists of an amplifier made stable and given a good frequency response by introducing a large amount of feedback. This feedback is set by means of a screw-driver adjustment to give a gain of exactly 50, and the output resistance is then tapped as indicated so that output voltages that are 2, 5, 10, 20, or 50 times the input voltage can be obtained across a 1-megohm load according to the switch position. The proportions are such that an output of approximately 3 volts effective can be obtained on any range without overload. The design indicated in Fig. 3 was intended for audio-frequency service, and as seen from Table II has excellent stability and practically an ideal frequency response.

Application of Feedback to the Output Amplifier of Laboratory Oscillators

In a well-designed laboratory oscillator the major part of the distortion occurring in the output results from harmonics generated in the output amplifier. The use of negative feedback to reduce this distortion is complicated by the fact that the load impedance to which the oscillator output is delivered may vary from short circuit to open circuit under different conditions of use. The situation can, however, be handled by throwing away a fraction of the output in a resistive network as shown in Fig. 4. Here R_2 prevents the output of the power tube from ever being short-circuited, while the combination R_1+R_2 introduces feedback that makes the voltage E_0 a substantially distortionless reproduction of the input signal E_s.

A quantitative analysis of Fig. 4(a) shows that for maximum output power delivered to the load, the load should be a resistance equal to R_3, while the resistance formed by R_1+R_2 in parallel with R_3 should equal the plate-load resistance which gives maximum power output from the tube operated as

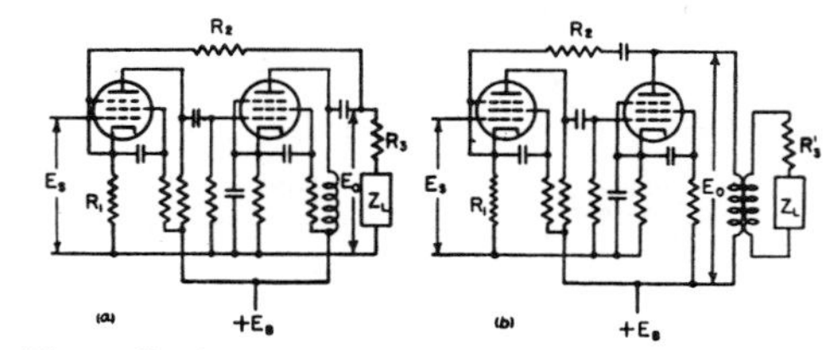

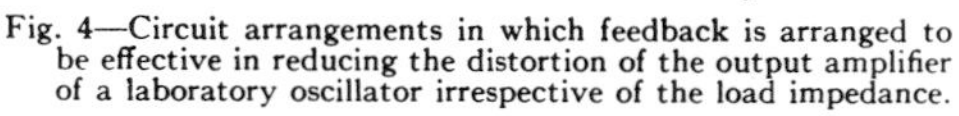

Fig. 4—Circuit arrangements in which feedback is arranged to be effective in reducing the distortion of the output amplifier of a laboratory oscillator irrespective of the load impedance.

an ordinary amplifier. Under these conditions the maximum power that can be delivered to the load is $(P_0/4)\times(R_1+R_2)/(R_1+R_2+R_3)$, where P_0 is the maximum undistorted output which the tube is capable of developing. In the usual case where $(R_1+R_2)\gg R_3$, the output obtainable hence approaches $P_0/4$.

Tuned Amplifiers Employing Negative Feedback

The amplification of a tuned amplifier can be made substantially independent of the tube and the supply voltages by means of the circuit shown in Fig. 5. Here the current that the tube delivers to the tuned output circuit also flows through a resistance R_1 across which is developed a feedback voltage that is proportional to the current passed through the tuned circuit and is *independent of frequency.* When the feedback factor obtained in this way is large, the voltage developed across the resistance R_1, and hence the current through the tuned circuit, is stabilized. The amplification is then determined solely by the

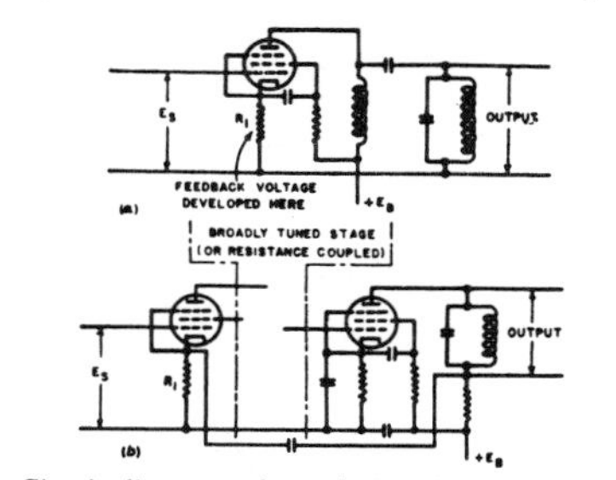

Fig. 5—Circuit diagrams of tuned amplifiers in which feedback is used to make the gain independent of tube conditions. These circuits can be easily modified for band-pass action.

tuned circuit, ***and becomes independent of the tube or electrode voltages.*** The two circuits shown in Fig. 5 accomplish the same result, but the arrangement at (b) is by far the better because it gives appreciable gain even when the feedback factor $A\beta$ is large.

Arrangements of the type shown in Fig. 5 can be used to advantage in the intermediate-frequency and radio-frequency stages of field-strength-measuring equipment. *It is possible in this way to avoid the necessity of frequent calibration, and in fact it is entirely feasible to make a calibration of the field strength in terms of gain-control setting, with the assurance that the only factors that will affect the calibration appreciably are temperature effects and misalignment.*

High Selectivity by Means of Negative Feedback

Negative feedback provides some remarkable possibilities for obtaining the equivalent of a high-Q tuned circuit. One method of doing this is to use feedback to provide a stabilized negative resistance that can be used for regeneration. Another method of approach is to provide a feedback amplifier in which the feedback network is a circuit having a transmission characteristic that depends upon frequency.

Stabilized Negative Resistance

The circuit of Fig. 6 gives a negative resistance across the terminals *aa* that is substantially independent of the tubes and supply voltages, and which can be made constant over a wide range of frequencies. This arrangement can be analyzed by assuming a signal voltage E_s is applied to the input, and then evaluating the ratio E_s/I_s, where I_s is the current that flows into the input terminals *aa*. Assuming that the grid of the first tube is not allowed to go positive, and referring to Fig. 6, one can write

$$I_s = \frac{E_s - E_0}{R} = \frac{E_s - AE_s}{R} = \frac{E_s}{R/(1-A)} \qquad (1)$$

where E_0 is the amplified voltage, and A is the ratio E_0/E_s. If the amplified voltage E_0 has the same phase as E_s, then the resistance which the terminals *aa*

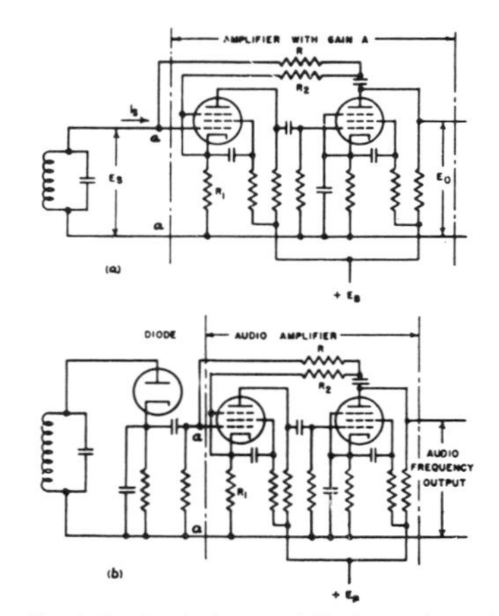

Fig. 6—Circuit for developing a stabilized negative resistance by using a negative-feedback amplifier to give stabilized regeneration, together with several applications.

offer to the voltage E_s is obviously a negative resistance having an absolute magnitude $|R/(A-1)|$. By using a large amount of negative feedback the amplification A can be made substantially independent of tube conditions and supply voltages, and can be made constant over a wide frequency range. The negative resistance under such conditions is correspondingly stabilized.

Such stable negative resistances have a number of uses. If placed in parallel with a tuned circuit as shown in Fig. 6(a), the result is equivalent to reducing the equivalent resistance of the tuned circuit, and hence raising the effective Q. This is a form of regeneration, but unlike ordinary regeneration there is little or no possibility of instability being introduced by variations in the tube or supply voltages. Thus tests in an actual case using a tuned circuit having $Q=100$ at 10 kilocycles showed that with $A\beta=100$ and sufficient negative resistance to raise the effective Q to 2000 when the plate-supply potential was 150 volts, an increase to 400 volts raised the effective Q only 10 per cent.

Another use of a stabilized negative resistance is in the improvement of the (alternating-current) / (direct-current) impedance ratio of diode detectors, by shunting the negative resistance across the diode output as shown in Fig. 6(b). This eliminates the principal cause of distortion in diode detectors.

High Selectivity by Means of Frequency-Selective Feedback Circuits

This method of obtaining a high effective Q makes use of a feedback network such that there is no feedback at some particular frequency, but increasing feedback as the frequency is increased or reduced. A typical circuit arrangement is shown in Fig. 7(a). Here the combination of R_3, R_4, R_5, and LC in the amplifier output constitutes a bridge which is balanced at the resonant frequency of the tuned circuit. The feedback voltage, which is derived from the neutral arm, is then zero at the resonant frequency but increases rapidly as the frequency departs from resonance. Since the amplification becomes less the greater the feedback, it is apparent that the amplification is maximum at the frequency for which the bridge is balanced and less at other frequencies, in spite of the fact that the amplifier itself is resistance-coupled. If the circuits are proportioned so that the feedback factor is large the amplification drops to a small value even when the bridge is only slightly unbalanced. The result is a very selective action. An exact analysis shows that when the output voltage is derived from the tuned circuit, the effective Q of the response curve is $(1+kA)$ times the actual Q of the tuned circuit, where $k=R_5/(R_4+R_5)$. Since it is readily possible to make $(1+kA)$ have values of the order 10 to 30, while the actual Q may readily exceed 100, values of Q from 2000 to 5000 are easily realizable at audio and low radio frequencies.

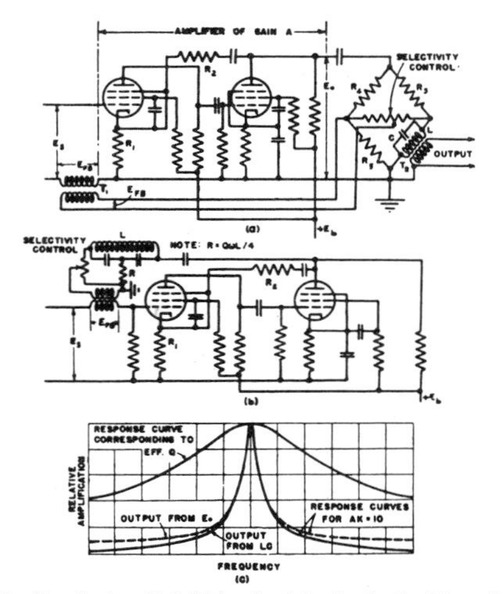

Fig. 7—Circuits in which high selectivity is obtained by using a frequency-selective feedback network.

When the output voltage is taken from the plate electrode of the amplifier tube instead of from the tuned circuit, the response curve no longer has the shape of a resonance curve. In the immediate vicinity of resonance it approximates a resonance curve rather closely, with the effective Q being the same as with the output derived from the tuned circuit, but at frequencies differing appreciably from resonance the output is substantially constant at a value very nearly $1/(1+Ak)$ of the value at resonance. This is indicated by the dotted curve in Fig. 7(c).

The circuit of Fig. 8(a) can be modified by replacing the bridge in the output by a bridged-T arrangement as illustrated in Fig. 7(b). By giving the resistance R the value indicated in the figure, the circuit will have zero transmission at the resonant frequency, and so is equivalent to a bridge, but has the advantage of being a 3-terminal network.

By using the potentiometer to control the feedback in the circuits of Fig. 7, *the selectivity obtainable can be varied without changing the amplification at resonance.* This possibility of obtaining variable selectivity without affecting the gain is possessed by no other tuned amplifier, and is of particular value in wave analyzers, as discussed below.

A New Wave Analyzer Based upon Negative Feedback Circuits

The arrangement shown in Fig. 7 for obtaining high selectivity can be made the basis of a simple, inexpensive, and yet very effective wave analyzer. A schematic arrangement of such an instrument is shown in Fig. 8. The wave to be analyzed is applied to a balanced modulator, using a phase inverter to transform from unbalanced input to balanced output. At the modulator the wave is heterodyned with a locally generated oscillation that is adjusted to give a predetermined difference frequency with the desired component. This difference frequency is then

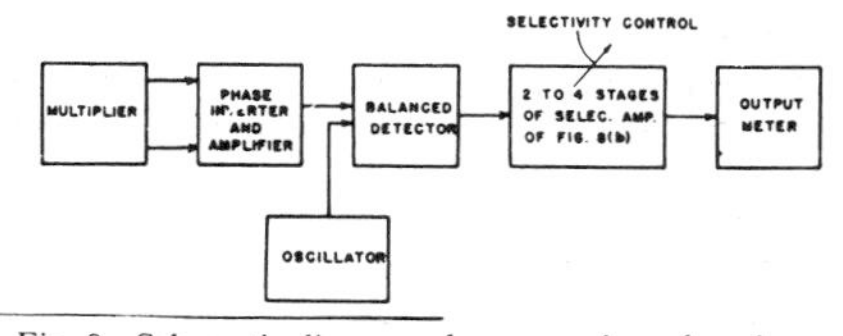

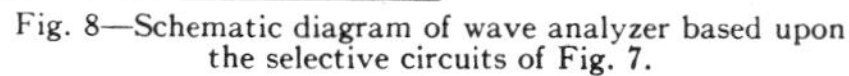

Fig. 8—Schematic diagram of wave analyzer based upon the selective circuits of Fig. 7.

selected from other components that may be present in the modulator output, using a selective system consisting of two to four sections of the type shown in Fig 7(b). By employing coils having cores that are permalloy dust, or better yet, molybdenum-permalloy dust, it is possible to give the fixed frequency a value of the order of 10 to 15 kilocycles, and still obtain adequate selectivity for analyzing waves having the lowest fundamental frequencies commonly encountered.

The selectivity of a wave analyzer of the type shown in Fig. 8 can be varied without changing sensitivity by providing each section of the selective system with a potentiometer for controlling the feedback as in Fig. 7(b), and then ganging the individual potentiometers to give a single-dial control of the overall selectivity. *This gives a wave analyzer having variable selectivity but constant gain, a feature possessed by no other analyzer now available.*

By making generous use of negative feedback throughout the analyzer of Fig. 8 to stabilize the gain of individual stages, the sensitivity and hence the

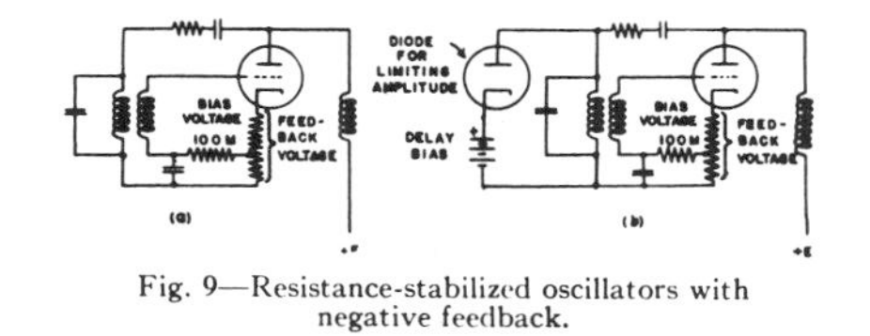

Fig. 9—Resistance-stabilized oscillators with negative feedback.

calibration can be made less dependent on tube changes or supply-voltage variation than is customary. Feedback is also preferably used in the inverter stage to maintain balance.

Laboratory Oscillators Making Use of Negative Feedback

Negative feedback can be used to advantage in a variety of ways in laboratory oscillators. A few illustrations of the possibilities are given below.

Resistance-stabilized Oscillators Employing Negative Feedback[3]

Negative feedback can be introduced in a resistance-stabilized oscillator as shown in Fig. 9(a), resulting in advantages of improved wave form, and higher frequency stability.

Where the ultimate is desired in performance, particularly with regard to wave form, it is desirable to separate the amplifying action required to produce oscillations from the nonlinear action that is necessary to stabilize these oscillations at a definite amplitude. An arrangement for doing this is shown in Fig. 9(b), and involves essentially a regenerative amplifier tube with a large amount of negative feedback, operation on a straight-line part of the tube characteristic, no grid current, and with only a small net voltage applied between the grid and cathode of the tube. The amplitude is then limited by the nonlinear action of the shunting diode and is controlled by the delay bias. With this arrangement most of the distortion in wave form that occurs results from the nonlinear action of the diode, and this will be small if the circuit is adjusted so that the amplifying action is only slightly more than required to start the oscil-

[3] For a discussion of ordinary resistance-stabilized oscillators see pages 283–289 of F. E. Terman, "Measurements in Radio Engineering," McGraw-Hill Book Company, New York, N.Y., (1935).

lations. What small distortion is introduced by the diode is readily calculated by taking advantage of the fact that since the current through the diode flows in the form of pulses of very short duration, then the second-harmonic component of these pulses has substantially the same amplitude as the fundamental component. If the effective resistance to the fundamental frequency which the diode must shunt across the tuned circuit in order to stabilize oscillations is $\alpha(Q\omega_0 L)$, where α is a constant and $Q\omega_0 L$ is the parallel impedance of the tuned circuit at the resonant frequency $\omega_0/2\pi$, then it can be readily shown that

$$\frac{\text{second harmonic voltage}}{\text{fundamental voltage}} = \frac{2}{3\alpha Q}. \quad (2)$$

In a practical case Q will usually be in the range 50 to 200, while with large feedback to stabilize the amplifying action, it is entirely practicable to operate with values of α as high as 100. The resulting distorrion is then of the order of 0.01 per cent, giving a remarkably pure wave.

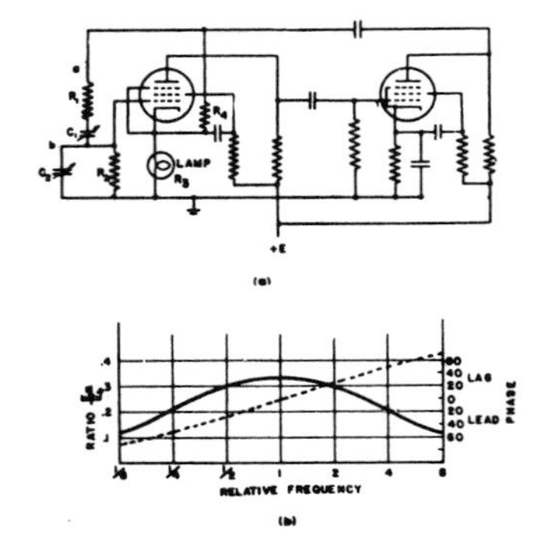

Fig. 10—Oscillator using resistance-capacitance tuning.

Two-Terminal Oscillators

The circuit of Fig. 6(a) can be made to operate as an oscillator by making the negative resistance less than the parallel resonant impedance of the tuned circuit. The amplitude of the oscillations in such an arrangement can be limited by allowing the amplifier to overload, or by using an auxiliary diode in the manner of Fig. 9(b).

Oscillator with Resistance-Capacitance Tuning

The use of negative feedback makes possible a practical sine-wave oscillator in which the frequency is determined by a resistance-capacitance network. An example of such a resistance-capacitance tuned oscillator is shown[4] in Fig. 10. Here $R_1C_1R_2C_2$ provide the regenerative coupling between the input and output circuits of the amplifier that is necessary to maintain oscillations. By proportioning this resistance-capacitance network so that $R_1C_1 = R_2C_2$ the ratio of voltage at b to voltage at a varies with frequency in a manner similar to a resonance curve, as indicated in Fig. 10. At the maximum of this curve the frequency f_0 is $1/2\pi\sqrt{R_1R_2C_1C_2}$ and the voltages at a and b have the same phase. Oscillations hence tend to occur at the frequency f_0.

For such an oscillator to be satisfactory it is necessary that the amplifier associated with the resistance-capacitance network have a phase shift that is independent of changes in supply voltage, etc., and furthermore there must be some means of controlling the amplitude of oscillations so that they do not exceed the range over which the tubes will operate as class A amplifiers. The constant amplifier phase shift is necessary in order to insure a stable frequency. This comes about because the phase angle of the transfer impedance of the resistance-capacitance network from a to b varies only slowly with frequency. Hence a small change in amplifier phase shift such as could be produced by a variation in supply voltage requires a very large change in oscillation frequency to produce a compensating phase shift in the resistance-capacitance coupling system. A stable phase characteristic can be readily obtained in the amplifier by employing a large amount of negative feedback, such as is obtained in Fig. 10 by suitably proportioning the resistance combination R_3R_4.

Amplitude control is obtained by a nonlinear action in the amplifier circuits that prevents the oscillations from building up to such a large amplitude that distortion occurs. It is possible to employ any one of a variety of systems, but the one shown in Fig. 10 is recommended as being both simple and effective. Here the resistance R_3 is supplied by a small incandescent lamp, and the operating conditions so adjusted that the current through this lamp is such that the filament operates at a temperature where the resistance varies rapidly with current. As a result, an increase in oscillation amplitude increases the lamp resistance. This makes the negative feedback larger, so decreases the gain of the amplifier and reduces the tendency to oscillate. Similarly, as the oscillations decrease in amplitude the current through the lamp is reduced, lowering the lamp resistance, reducing the negative feedback, and thereby increasing the tendency to oscillate. The result is that a constant amplitude is maintained, with no tendency to distort the wave shape.

In practical oscillators of this type it is most con-

[4] This oscillator somewhat resembles that described by H. H. Scott, in the paper "A new type of selective circuit and some applications," Proc. I.R.E., vol. 26, pp. 226–236; February, (1938), although differing in a number of respects, such as being provided with amplitude control and having the frequency adjusted by variable condensers rather than variable resistors. The latter feature makes the impedance from a to ground constant as the capacitance is varied to change the frequency, and so greatly simplifies the design of the amplifier circuits.

venient to make $R_1 = R_2$, and $C_1 = C_2$. Under these conditions the frequency of oscillation is

$$\text{frequency} = \frac{1}{2\pi R_1 C_1}. \quad (3)$$

It will be noted that this frequency is inversely proportional to capacitance, instead of inversely proportional to the square root of capacitance as is the case in ordinary tuned circuits. Accordingly if the frequency is varied by means of an ordinary gang-tuning condenser such as used in broadcast receivers, a frequency range of 10 to 1 can be covered on a single dial. Decimal multiplying factors for frequencies can be obtained by changing resistances R_1 and R_2 in decimal values.

The arrangement of Fig. 10 provides an inexpensive and yet highly satisfactory laboratory oscillator capable of performing most of the functions of a beat-frequency oscillator. One version that has been constructed employs a four-gang broadcast condenser with sections paralleled in pairs for tuning and covers the frequency range 20 to 20,000 cycles in three subdivisions (20 to 200, 200 to 2000, and 2000 to 20,000 cycles) by employing three sets of resistances. The output voltage is constant within approximately 10 per cent over the entire frequency range, and has only about 0.25 per cent distortion. A few checks on frequency stability indicate negligible frequency shift (less than 0.1 per cent) with large variations in supply voltage.

Experimental oscillators of this type have been built that operate at frequencies exceeding 2 megacycles.

Conclusion

The various applications of negative feedback that have been described in this paper by no means exhaust the possibilities that this new technique opens up, but rather merely suggest the important part that feedback is bound to play in the measuring and laboratory equipment of the future. Merely scratching the surface as has been done in this paper does, however, bring to view such interesting devices as improved forms of vacuum-tube voltmeters with unlimited sensitivity and a permanent calibration, detectors with the main cause of distortion removed, circuits with amazingly high Q, field-strength-measuring sets with a sensitivity that does not vary with tube conditions or supply voltages, new types of wave analyzers, new types of oscillators, etc.

Acknowledgment

The authors wish to express their appreciation for suggestions and helpful discussions contributed by Edward L. Ginzton of the Stanford Communication Laboratory group.

Bode on the Theory of Feedback Amplifier Design

IN this introduction to a longer paper on the fundamental principles of feedback, Bode called attention to the existence of inexorable mathematical laws that imposed "limits to what can and cannot be done in a feedback design." These limits meant, as Bode wrote on another occasion, that it was necessary "for the individual components to become qualitatively better as the system as a whole becomes quantitatively more ambitious." The history of feedback amplifier design is a striking example of the dynamic interaction between theory and practice that resulted in the formulation of general principles that proved applicable to a much broader range of phenomena than electronic amplifiers.

Hendrik W. Bode (1905–) was born in Madison, Wis. and received the B.A. and M.A. degrees from Ohio State University in 1924 and 1926, respectively. He joined the Bell Telephone Laboratories in 1926 and remained until his retirement in 1967. He received a doctorate from Columbia University in 1935. During World War II Bode worked on the design of gun control systems and later became Vice President with responsibility for military systems engineering. After retiring from BTL, he became a Professor of Systems Engineering at Harvard University. See *Engineers of Distinction*, p. 29, 1973 and *Who's Who in Engineering*, p. 169, 1964. See also Preston C. Mabon, *Mission Communications—The Story of Bell Laboratories* (Murray Hill, N.J., 1975), pp. 53–54.

Relations Between Attenuation and Phase in Feedback Amplifier Design

By H. W. BODE

Introduction

THE engineer who embarks upon the design of a feedback amplifier must be a creature of mixed emotions. On the one hand, he can rejoice in the improvements in the characteristics of the structure which feedback promises to secure him.[1] On the other hand, he knows that unless he can finally adjust the phase and attenuation characteristics around the feedback loop so the amplifier will not spontaneously burst into uncontrollable singing, none of these advantages can actually be realized. The emotional situation is much like that of an impecunious young man who has impetuously invited the lady of his heart to see a play, unmindful, for the moment, of the limitations of the $2.65 in his pockets. The rapturous comments of the girl on the way to the theater would be very pleasant if they were not shadowed by his private speculation about the cost of the tickets.

In many designs, particularly those requiring only moderate amounts of feedback, the bogy of instability turns out not to be serious after all. In others, however, the situation is like that of the young man who has just arrived at the box office and finds that his worst fears are realized. But the young man at least knows where he stands. The engineer's experience is more tantalizing. In typical designs the loop characteristic is always satisfactory—except for one little point. When the engineer changes the circuit to correct that point, however, difficulties appear somewhere else, and so on ad infinitum. The solution is always just around the corner.

Although the engineer absorbed in chasing this rainbow may not realize it, such an experience is almost as strong an indication of the existence of some fundamental physical limitation as the census which the young man takes of his pockets. It reminds one of the experience of the inventor of a perpetual motion machine. The perpetual motion machine, likewise, always works—except for one little factor. Evidently, this sort of frustration and lost motion is inevitable in

[1] A general acquaintance with feedback circuits and the uses of feedback is assumed in this paper. As a broad reference, see H. S. Black, "Stabilized Feedback Amplifiers," *B. S. T. J.*, January, 1934.

Reprinted with permission from *Bell Syst. Tech. J.*, vol. 19, pp. 421–423 (extract of a longer paper), July 1940.

feedback amplifier design as long as the problem is attacked blindly. To avoid it, we must have some way of determining in advance when we are either attempting something which is beyond our resources, like the young man on the way to the theater, or something which is literally impossible, like the perpetual motion enthusiast.

This paper is written to call attention to several simple relations between the gain around an amplifier loop, and the phase change around the loop, which impose limits to what can and cannot be done in a feedback design. The relations are mathematical laws, which in their sphere have the same inviolable character as the physical law which forbids the building of a perpetual motion machine. They show that the attempt to build amplifiers with certain types of loop characteristics *must* fail. They permit other types of characteristic, but only at the cost of certain consequences which can be calculated. In particular, they show that the loop gain cannot be reduced too abruptly outside the frequency range which is to be transmitted if we wish to secure an unconditionally stable amplifier. It is necessary to allow at least a certain minimum interval before the loop gain can be reduced to zero.

The question of the rate at which the loop gain is reduced is an important one, because it measures the actual magnitude of the problem confronting both the designer and the manufacturer of the feedback structure. Until the loop gain is zero, the amplifier will sing unless the loop phase shift is of a prescribed type. The cutoff interval as well as the useful transmission band is therefore a region in which the characteristics of the apparatus must be controlled. The interval represents, in engineering terms, the price of the ticket.

The price turns out to be surprisingly high. It can be minimized by accepting an amplifier which is only conditionally stable.[2] For the customary absolutely stable amplifier, with ordinary margins against singing, however, the price in terms of cutoff interval is roughly one octave for each ten db of feedback in the useful band. In practice, an additional allowance of an octave or so, which can perhaps be regarded as the tip to the hat check girl, must be made to insure that the amplifier, having once cut off, will stay put. Thus in an amplifier with 30 db feedback, the frequency interval over which effective control of the loop transmission characteristics is necessary is at least four octaves, or sixteen times, broader than the useful band. If we raise the feedback to 60 db, the effective range must be more than a hundred times the useful range. If the useful band is itself large these factors

[2] Definitions of conditionally and unconditionally stable amplifiers are given on page 432.

may lead to enormous effective ranges. For example, in a 4 megacycle amplifier they indicate an effective range of about 60 megacycles for 30 db feedback, or of more than 400 megacycles if the feedback is 60 db.

The general engineering implications of this result are obvious. It evidently places a burden upon the designer far in excess of that which one might anticipate from a consideration of the useful band alone. In fact, if the required total range exceeds the band over which effective control of the amplifier loop characteristics is physically possible, because of parasitic effects, he is helpless. Like the young man, he simply can't pay for his ticket. The manufacturer, who must construct and test the apparatus to realize a prescribed characteristic over such wide bands, has perhaps a still more difficult problem. Unfortunately, the situation appears to be an inevitable one. The mathematical laws are inexorable.

Aside from sounding this warning, the relations between loop gain and loop phase can also be used to establish a definite method of design. The method depends upon the development of overall loop characteristics which give the optimum result, in a certain sense, consistent with the general laws. This reduces actual design procedure to the simulation of these characteristics by processes which are essentially equivalent to routine equalizer design. The laws may also be used to show how the characteristics should be modified when the cutoff interval approaches the limiting band width established by the parasitic elements of the circuit, and to determine how the maximum realizable feedback in any given situation can be calculated. These methods are developed at some length in the writer's U. S. Patent No. 2,123,178 and are explained in somewhat briefer terms here.

◆◆◆ ◆◆◆ ◆◆◆

Alexanderson, Edwards, and Bowman on the Amplidyne

THIS paper illustrates and supports Alexanderson's conviction that invention occurs most frequently at the boundary between fields and that the engineer–inventor should avoid overspecialization. His own experience in radio electronics many years earlier (see Paper 36) had been directly applicable to the invention of the amplidyne. As the authors point out, the amplidyne was in principle similar to a two-stage electronic amplifier. Although they mention only the potential industrial application, the device had been conceived during efforts to achieve more effective control of naval guns. That became its major application during World War II.

For a biographical note on Alexanderson, see Paper 36.

Martin A. Edwards (1905–   ) was born in Kansas and received degrees in both electrical and mechanical engineering from Kansas State College in 1928 and 1929, respectively. After working as a plant engineer with the National Refining Company, he joined the General Electric Company. By 1959 he had received 91 patents, primarily in the area of power rectifiers and industrial control systems. See *Who's Who in Engineering*, p. 1699, 1959.

Kenneth K. Bowman (1904–   ) was born in Kansas and received B.S. and M.S. degrees in electrical engineering from Kansas State College in 1926 and 1927, respectively. He joined General Electric in 1927. He worked for the Gurney Elevator Company from 1930 to 1932 and for the Westinghouse Electric Elevator Company from 1932 to 1934 before returning to G.E. He received the Coffin Award at G.E. in 1940. See *Who's Who in Engineering*, p. 189, 1964.

Dynamoelectric Amplifier for Power Control

E. F. W. ALEXANDERSON
FELLOW AIEE

M. A. EDWARDS
NONMEMBER AIEE

K. K. BOWMAN
NONMEMBER AIEE

THE USE of amplifiers has become common knowledge in radio, but on the other hand, the term amplification has seldom been applied to processes in power engineering. Strictly speaking, we may say that a radio amplifier is only a form of control because we always tap a new source of power and the function of the amplifier is to control this power so as to reproduce the changes of energy flow at a higher power level. On this ground we might say that an ordinary d-c generator is an amplifier because we control the power output by the current in the field winding. Such a terminology would, however, be rather misleading because when we say amplification, we imply something more specific than when we say control. An amplifier should give accurate reproduction both of intensity and time intervals, whereas, controlled power circuits used so far have not met these requirements. In the radio-frequency amplifier we are dealing with a time element of less than a millionth of a second. In an audio amplifier the time element is about $^1/_{5,000}$ of a second. The only device we know of that will respond in such a short time is the high-vacuum electron tube. In power circuits, on the other hand, the time element is seldom less than one-tenth of a second.

When it became apparent that amplifiers were needed in power engineering we naturally turned to the vacuum tube. It was found, however, that the high-vacuum tube used in radio did not lend itself so well to high power circuits as the mercury-vapor and gas-filled tube with grid control known as the thyratron. The thyratron amplifier has found many successful uses in power applications and it was in fact this demand that led us to consider what other means may be available to solve similar problems. In the thyratron amplifier we have a device with an extremely high ratio of amplification so that for most practical purposes only a single stage is needed.

A thyratron amplifier may be controlled from a small regulator and may be used to furnish excitation of large synchronous machines. Installations of this type have proved very successful in service and are found superior to the conventional exciter plant in giving quick response to sudden changes of load. From an engineering standpoint, this type of installation is highly successful. Unfortunately, in the present stage of its development, the costs are such that it is not justified economically.

The type of dynamoelectric amplifier which will be described was developed to meet new control functions in industry where a high rate of amplification must be combined with a quick and accurate response. As an approach to this problem we were again guided by the history of radio amplification. It was found in the design of radio amplifiers that the ratio of amplification or the sensitivity can be increased within certain limits by sacrifice of the quickness of response. Thus an improvement of ten to one in sensitivity might be gained by permitting the process to take ten times as long. The same general principle applies to the dynamoelectric amplifier. In radio we have found, however, that by using amplification in several stages we can gain insensitivity in geometric progression without sacrifice of speed of response. The dynamoelectric amplifier which for brevity has been called "amplidyne" generator, has been developed on this principle. It is a two-stage amplifier incorporated in one dynamoelectric machine. In its physical structure it resembles the Winter-Eichberg motor, the Rosenberg generator, and the Pestarini metadyne, characterized by a pair of short-circuited brushes at right angles to the power brushes. In its functions, it is quite different. The first stage of amplification is from the control field to the short-circuited brushes and the second stage from the short-circuited brushes to the power brushes.

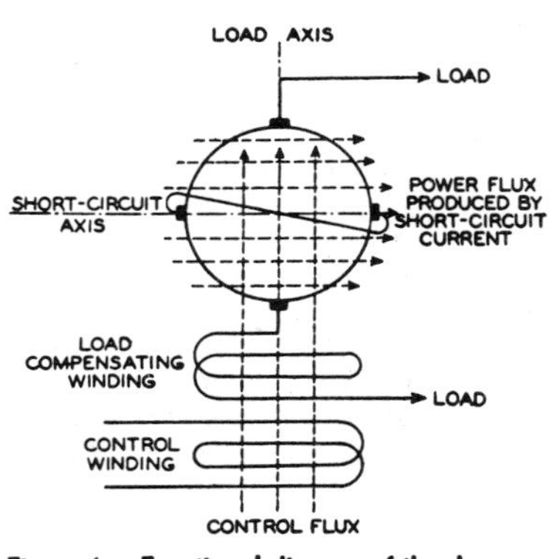

Figure 1. Functional diagram of the dynamoelectric amplifier (amplidyne generator)

Figure 1 shows a functional diagram of the dynamoelectric amplifier (amplidyne generator). The principal elements are:

- A control field
- A short-circuit axis on the armature
- An output circuit from the armature including a compensating winding

A high ratio of amplification of the order of 10,000 to 1 can be realized by this system. The over-all amplification is the product of an amplification of about 100 to 1 between the input winding and the short-circuit axis and another of 100 to 1 between the short-circuit axis and the output axis. The process of amplification in two stages can be further explained by assuming that the magnetic flux which generates the voltage in the short-circuit axis may be made only one-tenth as strong as flux in the power axis. Therefore the volt-amperes required to produce the control flux need be only $^1/_{100}$ as great as if the power field were excited directly from the control circuit. The control circuit may therefore contain enough resistance in proportion to its reactance to realize the desired short time constant.

The diagram also shows that the output circuit is on the same magnetic axis as the control circuit. There may be thus a direct magnetic back coupling from the output circuit to the control circuit with the tendency to regeneration and oscillations with which we are familiar in radio amplifiers. The object of the compensating winding is to neutralize this back coupling. That this is a rather delicate adjustment may be realized when it is remembered that the volt-amperes in the output circuit may be more than 10,000 times as great as the volt-amperes in the control circuit.

We also find that a compensation which is correct for direct current in the steady state may not be right for transients or alternating current. In the design of an amplifier system we must therefore take into account not only the characteristics of the machine itself but also the circuits

Paper 40-7, recommended by the AIEE committee on industrial power application, and presented at the AIEE winter convention, New York, N. Y., January 22–26, 1940. Manuscript submitted November 8, 1939; made available for preprinting November 29, 1939; released for final publication March 15, 1940.

E. F. W. ALEXANDERSON is consulting engineer, General Electric Company, Schenectady, N. Y.; M. A. EDWARDS and K. K. BOWMAN are with the same company.

Reprinted from *Trans. AIEE*, vol. 59, pp. 937–939, 1940.

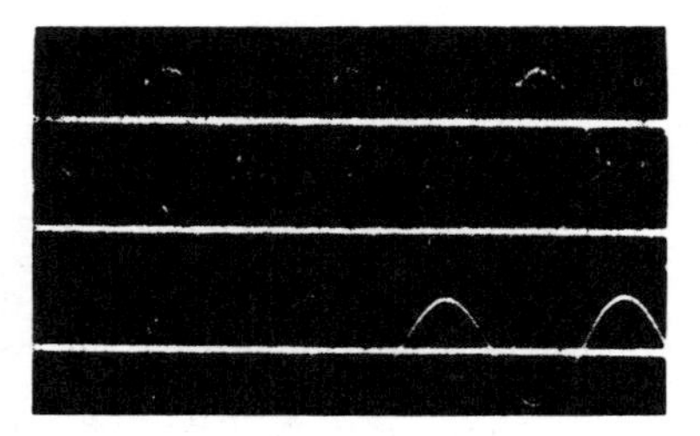

Figure 2. Correctly compensated amplidyne generator shows no back coupling when the load circuit is closed

A—Control-field volts, 60-cycle timing wave
B—Load volts
C—Load current

with which it is associated and we must analyze the performance of the system from the point of view of a-c and transient stability as well as d-c stability.

The reaction of the output circuit upon the input circuit with alternating current is under certain circumstances opposite to the reaction from direct current. This statement may seem surprising and need some further explanation. Let us assume that the machine is so designed that the back-coupling action for direct current is negative which produces stable operation for the steady state. Let us now assume that we use this same machine for amplification of alternating current and that we deliver the power to an inductive circuit like the field of a synchronous machine. Under those circumstances the back-coupling effect for alternating current may be positive, and cause regenerative oscillations at a critical frequency. The reason for this is that the alternating current in the output circuit is displaced more than 90 degrees from the current in the control circuit, thus reversing the back coupling from negative to positive. This positive reaction is regenerative and may result in continuous oscillations, as shown on oscillogram figure 4.

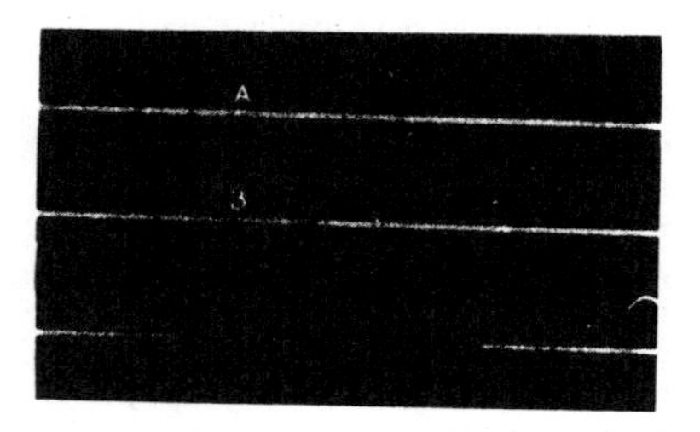

Figure 3. Effect of back coupling in undercompensated amplidyne generator when closing load circuit

A—Control-field volts, 60-cycle timing wave
B—Load volts
C—Load current

Figure 2 shows a correctly compensated machine feeding an inductive load. No back coupling between the load circuit and the control circuit is noticed when the load circuit is closed. The operation is perfectly stable and there is no tendency to oscillate.

In contrast to this, figure 3 shows the behavior of the same machine when undercompensated. The action of the load current upon the control flux is shown by a substantial increase of the output voltage due to regeneration. This is in contrast to a reduction in output voltage when a steady d-c load is applied. When the control field is open-circuited the machine passes from forced oscillations at four cycles per second to free oscillations at seven cycles, as shown in figure 4.

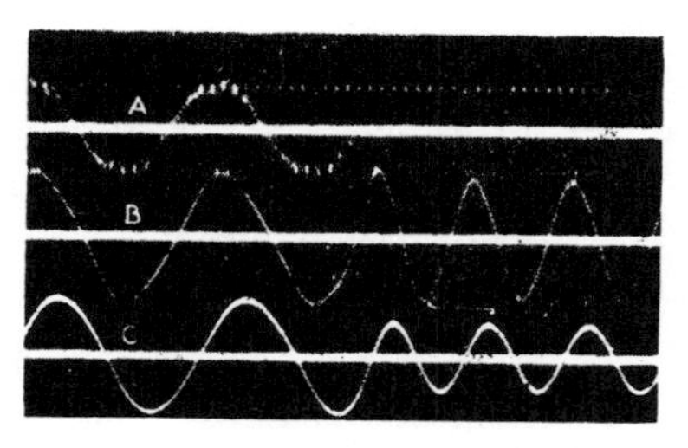

Figure 4. Free oscillations in undercompensated amplidyne generator when opening control circuit

A—Control field volts, 60-cycle timing wave
B—Load volts
C—Load current

Such tests with alternating current are convenient for analyzing and defining the amplifier characteristics. In most cases of practical application we are however dealing with rapid control transients in a generally unidirectional flow of current. Figure 5 shows an oscillogram taken of a typical machine. In this test the control field voltage was applied instantaneously. The field current did not build up instantaneously due to the inductance of the control field. However, the short-circuit-axis current and load voltage are nearly in phase with the field current. Hence, the time constant of the machine is only slightly larger than that of the control field and is not the numerical sum of the time constants of the control field, short-circuit axis, and load. This fact has been proved both experimentally and mathematically. The overall time constant of the amplifier shown on this record is $^1/_{20}$ of a second.

The tendency toward regenerative oscillations is present when the dynamoelectric amplifier feeds a highly inductive load such as the field winding of a large machine. If the desired combination of d-c stability and a-c stability has not been realized in the design of the machine it is possible to overcome the difficulty by adding an external back coupling which functions only at transient and alternating current. In most cases, however, this expedient is not needed. No case has however been found where it was not possible to adjust the system for stable operation in one way or another.

A theory of oscillations and the methods for avoiding them to be complete should take into account the effect of capacity as well as inductance in the output circuit. Electrostatic capacity in low-voltage circuits is so small that it may be neglected, but effects equivalent to capacity are apt to be present to an extent that becomes important in the functioning of the system. The mechanical inertia of a d-c armature reacts upon the circuit like an electrostatic capacity with the only difference that it is usually so much greater that it may be measured in farads instead of microfarads. Electrostatic capacity in conjunction with electromagnetic inductance produces tuning effects at a natural period in accordance with the well-known formula

$$\text{frequency} = \frac{1}{2\pi\sqrt{LC}}$$

When we use a dynamolectric amplifier to deliver power to a d-c motor we have a practical case of that sort. The inductance of the armature winding of the generator plus the inductance of the motor winding create a natural period of resonance in conjunction with the electromechanical capacity which is determined by the inertia attached to the motor shaft. In practical systems this natural period is of a magnitude of a few

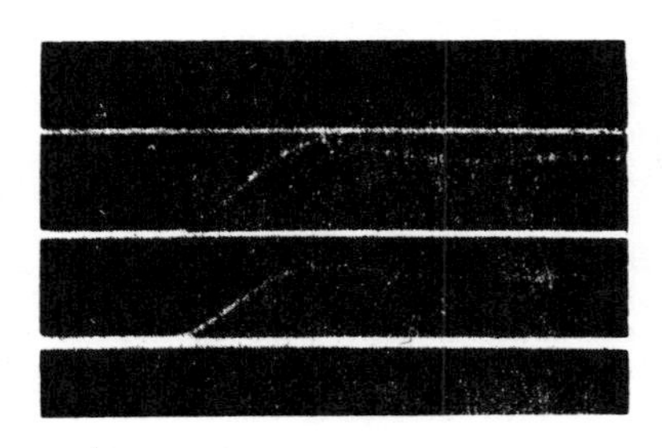

Figure 5. Response of an amplidyne generator to control transients

A—Control-field current, 60-cycle timing wave
B—Short-circuit axis current
C—Load volts

cycles per second. This is likely to be a frequency which is representative of the control function for which the system is to be used, and it has been found that in such cases the electromechanical resonance is apt to be a very real problem, which must be foreseen in the design. In such cases it is found to be desirable to make the inductance of both generator and motor as low as possible by distributed compensating windings at the same time as the mechanical inertia is kept as low as possible. Both these factors tend to place the electromechanical resonance frequency above the frequency of the control functions, thus avoiding any tendency to mechanical oscillations.

When a control system is desired which demands a still higher ratio of amplification than is attainable in a single dynamoelectric amplifier it is possible to introduce a first stage of electron tubes of the type used in radio sets. Such tubes are adapted to an output sufficient to control an amplidyne whereas the primary control energy of such tubes is measured in microwatts. We have thus established simple means for industrial purposes to amplify a minute energy such as a beam of light into a quick and accurate control of power flow of many horsepowers. It is hoped that this will open up many new possibilities in the control of industrial processes.

Other aspects of the dynamoelectric amplifier are discussed in a paper by A. Fisher, dealing with machine design (pages 939–44) and a paper by D. R. Shoults, M. A. Edwards, and F. E. Crever, dealing with practical applications (pages 944–9).

Bibliography

D-C Machines With Armature Excitation:

Principles of D-C Machines (a book), **A. S. Langsdorf.** Rosenberg generator, page 555.

Theory of the Metadyne, **J. M. Pestarini.** *Revue Generale De L'Electricity*, 1930.

The Metadyne, **C. Trettin.** *E. T. Z.*, April 14, 1938.

Brainerd and Sharpless on the ENIAC

THIS paper contained the first general description and assessment of the first all-electronic general-purpose computer to appear in an engineering journal. The ENIAC was not only the most complex electronic machine built prior to 1945, but marked the effective beginning of the computer age. As pointed out by the authors, electrical engineers had pioneered in the development of complex computing systems such as the ac calculating board and the differential analyzer. ENIAC was a far more ambitious undertaking, and there were many who doubted that any machine that employed 18 000 vacuum tubes could be a practical success. However, the "atmosphere of pressure" generated by the war made it seem a worthwhile risk to the Army Ordnance Department that encouraged the Moore School engineers to undertake the project and funded it. See John G. Brainerd, "Project PX-The ENIAC," *The Pennsylvania Gazette*, pp. 16–17 and 32, March 1946. Also see Thomas Parke Hughes, "ENIAC: Invention of a Computer," *Technikgeschicte*, vol. 42, no. 2, pp. 148–165, 1975; Henry Tropp, "The Effervescent Years: A Retrospective," *IEEE Spectrum*, pp. 70–79, February 1974; John W. Mauchly, "Mauchly on the Trials of Building ENIAC," *IEEE Spectrum*, pp. 70–76, April 1975.

John G. Brainerd (1904–) was born in Philadelphia, Pa. and received a B.S. degree in electrical engineering and a doctorate degree from the University of Pennsylvania in 1925 and 1934, respectively. He remained at the Moore School of Electrical Engineering and was its Director from 1954 to 1970. He was Assistant State Director of the Power Division of the Public Works Administration in 1935–1936 and Supervisor of the ENIAC project from 1943 to 1946. He is currently serving as President of the Society for the History of Technology. See *Engineers of Distinction*, p. 34, 1973 and *Who's Who in Engineering*, 1954.

Thomas K. Sharpless (1913–) was born in Philadelphia, Pa. and received an A.B. degree from Haverford College in 1936. He received a B.S. degree in electrical engineering from the University of Pennsylvania in 1943 and was a research engineer on the ENIAC project. He later joined the Technical Engineering Company in Philadelphia. See *Who's Who in Engineering*, p. 1686, 1964.

The ENIAC

J. G. BRAINERD — FELLOW AIEE
T. K. SHARPLESS — ASSOCIATE AIEE

The ENIAC is the only electronic large-scale general-purpose digital computing device now in operation. Its speed of operation compares favorably with other electric and mechanical computers. Developed under wartime pressure, it has been of value not only in producing results but in pointing the way toward improvements for future designs.

THE ENIAC (electronic numerical integrator and computer) is a large-scale device, adapted to problems requiring a large amount of work for their solutions, and particularly to problems which involve the repetition of a large number of similar types of computations to achieve a result. The ENIAC does not reach a point of practical usefulness until applied to a problem which is such that a large amount of repetitious computation is necessary to obtain numerical answers. Such problems often involve the solution of differential equations, the evaluation of series, or the preparation of mathematical tables. For example, the equation:

$$\frac{d^2y}{dt^2}+\epsilon(1+k\cos t)y=0$$

(where ϵ and k are parameters) is one which arises in several electrical problems, and has extensive use elsewhere. To solve it for values of $\epsilon=1, 2, 3, \ldots 10$ and $k=0.1, 0.2, \ldots 1.0$ requires 100 results, each of which is a table of y versus t. Each separate complete solution applies to one value of ϵ and one of k. To get this the equation is solved by a corresponding difference equation, and if $\Delta t=0.0004$, then about 7,850 "lines" are called for in the range $0<t<\pi$. In carrying out each line of work, approximately 10 multiplications and many more additions and subtractions were needed. Thus, in one solution of the equation for the given values of ϵ and k and for the range $0<t<\pi$, there were 78,500 multiplications and many more additions or subtractions. Multiplying these by 100 (as this number of complete solutions was desired), there resulted 7,850,000 multiplications and many more additions and subtractions. The work of obtaining two separate solutions for y in each instance, corresponding to two different sets of initial conditions, required 15 continuous hours of ENIAC operating time. This included making a test run after every 5 regular runs, thus increasing the work outlined by 20 per cent. Ten digit numbers were used throughout, (the reason for using such "large" numbers is discussed briefly later) and each result was recorded for $t=0.0, 0.1, 0.2, \ldots$ up to $t=3.14$ after which the values for $t=3.1408$, 3.1412, 3.1416, 3.1420, and 3.1424 were added to enable values of the results at π to be obtained accurately by interpolation. This brief summary of machine operations and time used in solving an actual problem indicates the orders of magnitudes of these two items as they are related to the ENIAC.

Electrical engineers in the United States have had a major interest in the development of large-scale computing devices. Probably the most extensively used such instrument in the world is the a-c calculating board, but although this is a large scale device it is hardly a general purpose computer even though it has been used by Kron of the General Electric Company in such problems as solving certain partial differential equations.

Another large-scale device is the differential analyzer, which originated in the electrical engineering department of Massachusetts Institute of Technology. Of the five now in the United States, two are at Massachusetts Institute of Technology, one at the Moore school of electrical engineering of the University of Pennsylvania, and one at the General Electric Company.* Differential analyzers have been used extensively in connection with machinery, stability, and similar problems, as well as in many fields outside electrical engineering, and all the analyzers now in the United States are in almost continuous use. The primary purpose of differential analyzers is to solve sets of ordinary differential equations, and, although they have been put to numerous other uses, this remains their prime objective.

Among other large scale computing devices are the Bell Telephone Laboratories relay computer and the International Business Machines automatic sequence controlled calculator at Harvard University.[1] Particular note should be made of these machines, because in some respects it may be said that operations carried out by

J. G. Brainerd is professor and **T. K. Sharpless** is research engineer at the Moore school of electrical engineering of the University of Pennsylvania, Philadelphia, Pa.

Considerable credit should go to Colonel Paul N. Gillon, Colonel Leslie E. Simon, and Major H. H. Goldstine of the Army Ordnance Department for their backing of the project, which was carried out under an Ordnance Department research and development contract. The ENIAC was developed and built at the Moore school of electrical engineering of the University of Pennsylvania. J. P. Eckert, Jr., was chief engineer and was primarily responsible for design; Doctor J. W. Mauchly was research engineer as was the junior author of this paper (Doctor Mauchly had much to do with the original proposals): and numerous others contributed to the development including particularly the following engineers: Arthur Burks (A '42), Joseph Chedaker (A '43), Chuan Chu, James Cummings, Leland Cunningham (astronomer whose war work was in large-scale computations), John Davis, Harry Gail, Robert Michael, Frank Mural, and Robert Shaw. The senior author of this paper was the project supervisor.

* The fifth is substantially a duplicate of the one at the Moore school, and is in the Ballistic Research Laboratory of the Army Ordnance Department, Aberdeen Proving Ground, Md.

Reprinted from *Elec. Eng.*, vol. 67, pp. 163–172, Feb. 1948.

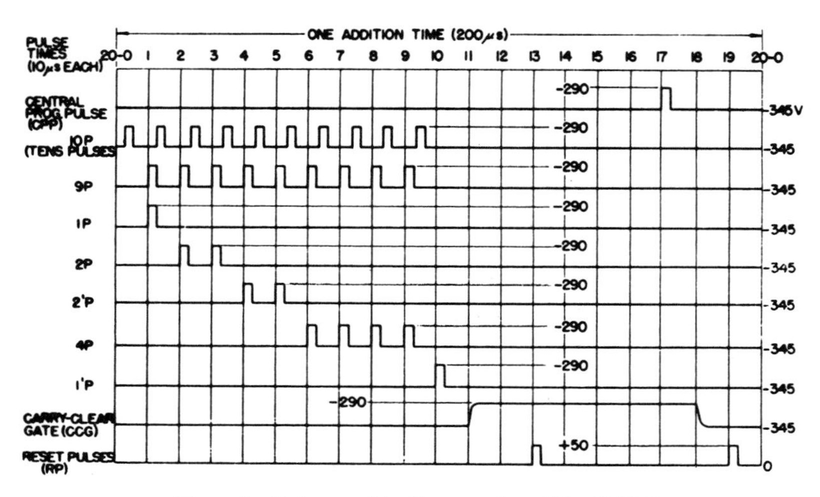

Figure 1. Pulses used in the operation of the ENIAC

the ENIAC by electronic methods are performed in them by mechanical means. (The converse is not necessarily true, as will be explained subsequently.)

Another, and important, large-scale general-purpose computing "unit" is a group of business machines such as those of the International Business Machines Corporation. These devices can be used for addition, subtraction, multiplication, recording, and so forth, and by transferring punched cards from one to another, the sundry numbers in a problem can be operated on as desired. The Watson computing laboratory at Columbia University contains such a unit, as do numerous other places. Engineers might be interested to know there are units at the General Electric Company and in engineering departments of the Massachusetts Institute of Technology and University of Pennsylvania.

TERMS

A discussion of some terms used in connection with computing devices will simplify a description of the ENIAC:

Large-Scale. Although the afore-mentioned large-scale devices all require a moderately large room to contain them, size is not necessarily an accompaniment of large scale. Indeed, one of the purposes behind new electronic machines now under development is to reduce physical size and complexity. Large-scale refers rather to the magnitude of the problems which may be placed on a device so labeled. A desk calculator of any of the common types is a small-scale device; the ENIAC (and others) which without human intervention may perform thousands of additions, multiplications, and so forth in the proper sequence and with numbers evolved in the operations as the work proceeds, is a large-scale device.

General Purpose. A general purpose machine is one which will handle many types of problems, in contrast with specialized devices. A desk calculator is a general-purpose small-scale machine; an a-c calculating board is a special-purpose large-scale calculator.

Continuous (Analogue) Versus Digital. A continuous variable or analogue type of computer is one such as a differential analyzer, where the angular displacements of rotating shafts or other devices give direct measures or analogues of results at each instant. The continuous motions of the shafts are to be contrasted with a desk calculator where continuous variation is impossible because adjacent keys in a column differ by a unit, and it is not possible to go between the values given by unit change in the right-hand column. The distinction between analogue and digital devices does not imply that they are to be applied to separate fields. A large-scale general-purpose digital device such as the ENIAC and others for most practical purposes can differentiate and integrate by using extremely small intervals of the independent variable ($\Delta t = 0.0004$ was cited in a foregoing example). It is true that the independent variable will not change uniformly, but rather in steps; nevertheless, the steps may be taken so small as to make over-all errors small. A-c calculating boards, differential analyzers, slide rules, and so forth, are continuous variable devices; most development now under way on large-scale computers is confined to digital types. The reason for this is that higher accuracy is obtainable with the digital device. For example, the ENIAC can handle numerical quantities of 10 significant figures and with a minor change can handle 20 significant figures. On the other hand, differential analyzers or other analogue machines yield at best 4 or 5 significant figures, and to increase this would require complete and revolutionary design changes.

Electronic Versus Mechanical. By an electronic computing device is meant one in which the arithmetic and control procedures are performed in the machine by electronic circuits. The ENIAC is the only electronic large-scale general-purpose computer now in operation. In contrast, mechanical means such as relays are used in other existing large-scale digital instruments, and the basic arithmetic device in all differential analyzers is the mechanical integrator. A-c calculating boards are electric devices; the word electronic is used because the new electric computers (ENIAC and those under development) use electron tube circuits extensively.

Amplitude Versus Step Mechanisms. The distinction to be attempted here between amplitude and step mechanisms has to do with the internal operation of calculators and not with the question of continuous (analogue) versus digital machines. An amplitude mechanism is one in which a result is indicated by the amplitude of some quantity, as, for example, the voltage on a capacitor. The voltage may be read as closely as possible, or the device might be used in a digital system in which any voltage from 7.5 to 8.5 indicated the number 8. A step mechanism is, like a relay, one characterized by an on-off or an open-closed state. If ten relays are arranged in a column and labeled 0, 1, 2, . . . 9, then if all are open except that labeled 8, the number 8 is indicated. The definiteness of on-off or open-closed devices like relays and certain tube circuits has led to their use almost exclusively in large-scale digital computers. A basic elementary circuit in the ENIAC is a pair of tubes (trigger circuit) so arranged that when one (number 1) is conducting the other (number 2) is not. If, for any cause, number 2 becomes conducting and number 1 not, a reversal has occurred. This reversal can be made to correspond to the change from open to closed or conversely of a relay, or, in more general terms, from a normal to an abnormal state.

Synchronous Versus Sequential. If as soon as one operation such as a multiplication is completed a signal is given which initiates immediately the following operation, a calculator is said to have sequential operation. An individual operating a desk calculator would work on a sequential basis. If on the other hand no operation can begin except at an integer multiple of some fixed time interval after a previous operation has begun, the calculator is said to work synchronously. The ENIAC is of the latter type, its operation being controlled by a group of pulses which are repeated every 200 microseconds. A new operation can start only at the beginning of one of these 200-microsecond intervals.

Series Versus Parallel Operation. A large-scale digital calculator is said to have parallel operation if two or more arithmetic operations (two additions, or an addition and a multiplication) can be carried out simultaneously. It has series operation if no two arithmetic processes can be carried out at the same time. The ENIAC has a limited amount of parallel operation, but because of the high speed of the electronic machines the tendency in new development is toward series operation wherever that mode reduces complexity.

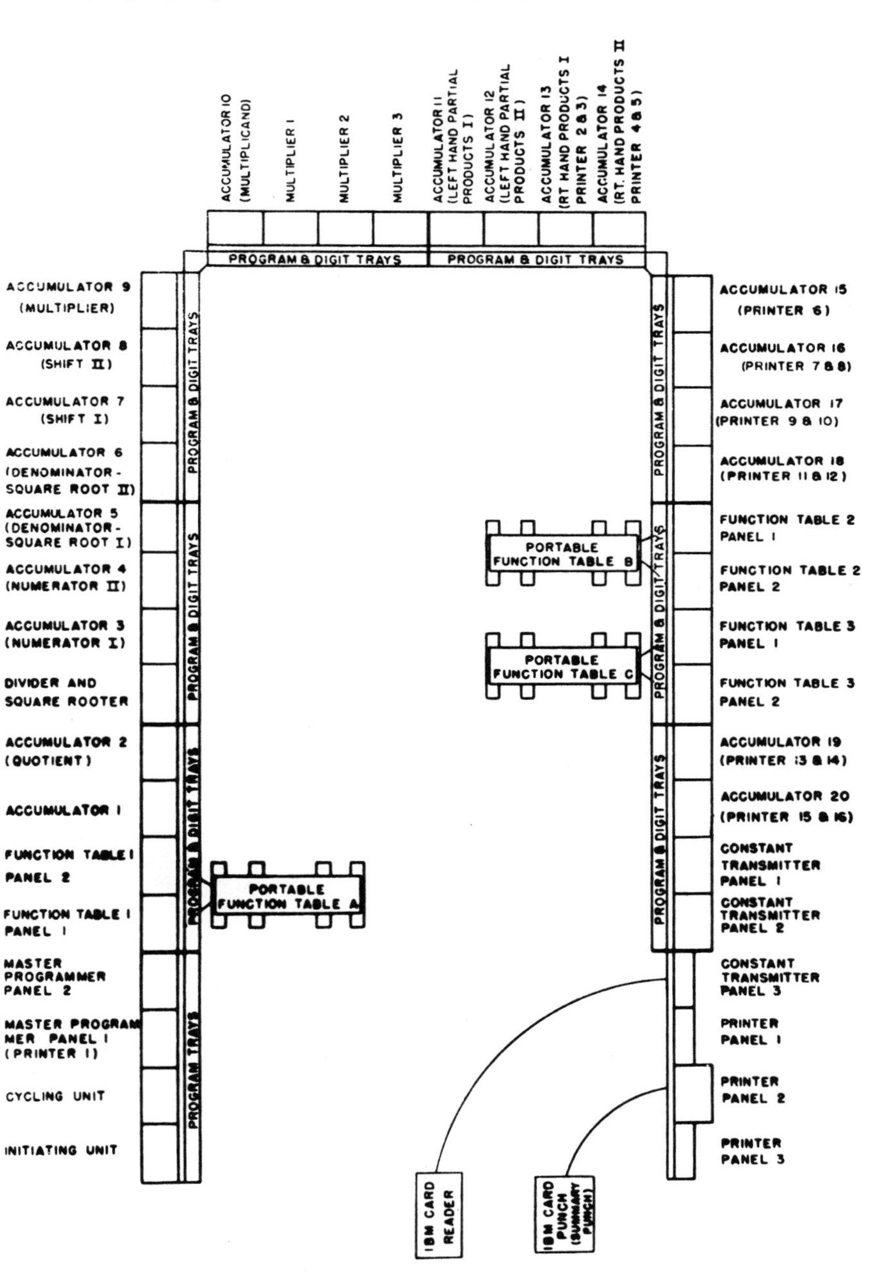

Figure 2. ENIAC floor layout

Initiating unit—Controls for power, operation controls

Cycling unit—Source of pulses used in operation of ENIAC; name is derived from cycle of pulses shown in Figure 1 and does not refer to cycles of arithmetic operations

Master programmer—Controls the cycles of arithmetic operations and performs other programming functions

Function table panels—Used in conjunction with function table to call up at high speed numbers "set" in function table

Accumulator—Performs additions and subtractions, stores results, and so forth

Divider—Performs divisions and also can be used to take square roots

Multiplier—Performs multiplications

Constant transmitter—Receives at low speed information from the input device (an International Business Machines card reader, as indicated) or other device, and supplies this information at high speed when called for (note that the input device is at what appears to be the next to the last position rather than at the "beginning" of the ENIAC; this is of no significance). The constant transmitter also has a limited capacity for storing numbers set on switches on its front panel

Printer—Receives at high speed results to be recorded and transmits them to the relatively slow-speed output device (an International Business Machines card punch as indicated)

Digit trunks—Special transmission lines into which connection can be made to transmit numbers from one part of the ENIAC to another

Program trunks—Special transmission lines into which connection can be made to transmit program orders (electric pulses) from one part of the ENIAC to another

Trunk from cycling unit—Permanently connected to other units to which it supplies the group of pulses shown in Figure 1 every 200 microseconds when ENIAC is in continuous operation

Decimal Versus Binary. This heading lists two of the large number of possible number codes which can be used in connection with a digital machine. In the ENIAC numbers appear in the usual way they are used, and this representation is called a decimal one. On the other hand, the open–closed or on–off characteristic of relays, certain tube circuits, and the like, has led to extensive consideration of the use of the binary, or base two, system. In this case a number would be "translated" from its common expression in the decimal system to its expression in the binary system of numbers. It would be retained in this latter system, and most or all operations would be performed on it in this system, until a desired result is achieved, in which instance it would be translated from the binary to the decimal system either before or after recording. Other number systems besides the binary may be considered—the choice of a system to use internally is determined by saving in equipment, simplification in circuits, magnitude and complexity of translation equipment, weight given to simplicity of understanding for maintenance men and other nontechnical personnel, and so forth. As the ENIAC uses the decimal system, no further discussion of this topic will be included here, except to note that other systems such as the binary and the biquintic are in use.

MACHINE COMPONENTS

Most large-scale computing devices can be broken down into several components. Despite the fact that in the ENIAC these components are mixed almost inextricably with one another, it is convenient to use them for a brief functional outline.

1. The Arithmetic Component consists of 20 accumulators for addition or subtraction, one multiplier, one divider square rooter, and three function tables on each of which can be set values of a known function to be called up in the course of the solution. (The calling up is arithmetic; the settings are memory as next described.)

2. The Memory Component consists of the same 20 accumulators, any one of which can be used to "hold" or "store" a number so long as that accumulator is not used for other purposes, the same three function tables which are memory devices yet at the same time arithmetic, and finally an unlimited memory obtained by sending numbers to be remembered to the output device, and having them available for recall through the input device.

3. The Control Component may be considered to consist of two parts: control of basic operations without regard to the problem on the machine, and control of the sequence of operations for a particular problem. The latter usually is known as programming and is achieved on the ENIAC by external connections inserted by hand between the panels of the various arithmetic and other devices. In order to carry out such processes as to have a cycle of operations repeated, another cycle begun and repeated, and then the first one again carried out a certain number of times, and to do many other programming jobs, there is a master programmer available for control. Other large-scale general-purpose digital machines now in operation have programming done automatically, and the electronic computers now under development likewise will have this feature. The control of basic operations, independent of the particular problem on the machine, is obtained in the ENIAC (disregarding power supply and auxiliary equipment controls and control of the numerous direct voltages for tubes) by a series of pulses generated at the cycling unit and repeated every 200 microseconds. Figure 1 shows the group of pulses so required in each 200-microsecond interval. Some idea of the uses of the pulses will be contained in the description of the operation of an accumulator.

4. Input and Output Devices. The questions of how data, such as initial values of variables, values of parameters, are supplied to a device such as the ENIAC, and how results are to be taken out of the machine, are to a large extent independent of the computing device. Thus, data may be recorded originally on a punch tape, on a magnetic tape, on punch cards such as are used in business machines, or otherwise, and an appropriate mechanism devised to insert into the computing machine the electric or other type of signals needed to inform the machine of the numbers being supplied to it. Likewise, a result which it is desired to record may appear somewhere in the machine, and this result may be brought out to a mechanism which will translate the machine result (given by indications in certain circuits in the case of the ENIAC) to a punch tape, a magnetic tape, an electric typewriter, a punch card, an indication on a film, or other medium. It is interesting to note that the speed at which input and output devices operate may be so low, relatively, that, on occasion, they may be the limiting factor in determining the over-all time in which a problem can be done. This limitation is particularly evident in the case of a high-speed electronic computer such as the ENIAC, and might be illustrated by reference to the problem cited previously. The equation there given was solved on the machine 100 times, and in each instance it was solved for some 7,850 values of t, differing in steps of 0.0004 from zero to π. If each one of these results for a given set of values of ϵ and k were tabulated, the machine would obtain the solution and then wait until the slow-speed recording of the previous solution was completed. Actually, every hundredth result was recorded, and thus the machine ran through 100 lines while the output device was recording the final result of the previous 100 lines. In this way the machine had plenty of work to do during the time required for recording.

In the ENIAC, input and output are by International Business Machines punch cards, and there are other input methods as well. Consider, for example, the case in which it is desired to record a result in the machine. This result, appearing in electrical form, is translated to a set of mechanical relays which, in turn, cause an International Business Machine card punch (usually called the "printer") to punch on a card the result. This is a relatively long-time procedure, but meanwhile the ENIAC is proceeding without interruption, unless it should happen that it produces a new result for recording before recording of the first one is completed, in which instance the ENIAC pauses in its work until it receives an appropriate signal from the card punch.

Current developments in large-scale general-purpose digital computing devices are devoted to a considerable extent to obtaining speedier input and output mechanisms. It may be noticed that the function tables mentioned in "Memory" may be used to insert arbitrary data into the ENIAC at high speed, and in addition there are built-in facilities of limited extent in the ENIAC (in the unit called "constant transmitter") for inserting at high speed numbers (such as π) which frequently may be used in a particular problem.

BASIC CHARACTERISTICS

Large-scale computing devices often are compared on the bases of flexibility, accuracy, speed, reliability, and capacity. These characteristics are so intertwined with one another that no absolute clean-cut distinction between factors affecting each can be maintained, but in a broad way they serve as an introduction to the comparison and evaluation of the large-scale machines.

Fundamentally the desideratum is to solve a given problem to the accuracy desired in the least possible time and at minimum cost. "Time of solution" is a term which has been used in numerous ways; it may include all the time from first tackling the problem with pencil and paper to final tabulation of results obtained on the computing device. But often "time of solution" means machine time (interval the machine is devoted to the problem). It also has been used to denote operating time which is the time the machine actually operates in solving the problem and excludes setup time, if any, close-down time, and time for major maintenance work, if any. However, with regard to the basic objective it is over-all time—a useful but not precise interval—which is of interest.

Flexibility. In a digital machine flexibility means that the device or devices for carrying out arithmetic processes may be interconnected in arbitrary fashion, or ordered to perform operations in arbitrary sequence, so that there would be no limitation on the operation to be performed next on a given number in the machine. This flexibility is to be contrasted with a machine permanently "wired" so that it is restricted to variations of one problem.

In a more general sense flexibility denotes the existence in a machine of a device or devices for carrying out the usual arithmetic processes: addition, subtraction, multiplication, division, differentiation, and integration, as well as it denotes flexibility of potential arrangements. Digital machines such as the ENIAC are described as general purpose machines because they are both flexible as to arrangement, and have devices for all the arithmetic processes, including differentiation and integration, which are carried out by approximate methods so accurately that they suffice for most purposes.

Virtually all practical problems requiring numerical solutions come within the scope, but not necessarily the capacity, of the general purpose digital computers. There is a lower limit of complexity or quantity below which it does not pay to use a *large-scale* general-purpose computer, but this practical limitation should not be allowed to shadow the fundamental fact of the preceding sentence: virtually all practical problems requiring numerical solutions come within the scope, but not necessarily the capacity, of the general purpose computers.

Computers such as a-c calculating boards and differential analyzers are not general purpose devices, and despite the extension of their application to problems not originally contemplated when they first were devised, they remain substantially specialized and not flexible in the sense that term is used here.

Accuracy. Most continuous variable or analogue computers give results which may be accurate to three or four figures; the accuracy depends on the problem under solution, the part of the solution considered, and other factors, as well as on the device. Many results from differential analyzers are obtained in the form of curves. A-c calculating boards yield answers obtained by instrument readings. In contrast, numerical solutions accurate to five or more figures sometimes are

Figure 3. A decade counter

required in practice. It is desirable that new tables of mathematical quantities be given to several significant figures beyond current use to allow for future needs which usually are more exacting than existing ones. Computations which involve many differences of nearly equal numbers require many more significant figures in the numbers than will be obtained in the result. Numerous other examples of the desirability of high accuracy may be cited. For these reasons, and the added fact that to help justify its existence a large-scale general purpose computing device like the ENIAC definitely should increase accuracy over such "old-time" (in the era of large-scale computers) devices as the differential analyzers, most large-scale general-purpose digital computers deal with numbers which may seem outsize to engineers, but which actually are not. In the ENIAC, provision is made for using 10-digit numbers (as in the case of simple 10-column desk calculators) in almost all parts of the machine. This figure was chosen after a rough study of a particular differential equation which had to be solved many thousands of times during the war, and the choice was made to insure that 5-figure accuracy was found in the results. There are definite reasons why the accuracy of the solution of a differential equation (or other mathematical form) may decrease radically when handled by digital methods; a simple example not associated with any computing device or with a differential equation would be a column of 1,000 numbers to be added (it will be recalled that in the ENIAC many thousands of operations may take place in obtaining a solution). If each number is given to 11 significant figures, and the decimal point is at the same position in all of them, then disregarding the 11th figure in each case and carrying out the addition would result in an error of the order of magnitude of 500, the 5 appearing in the eighth digit position. While such an error in simple addition might not appear directly in the ENIAC because provision for round-off is made, nevertheless, problems such as differential equations can involve processes which result in decreased accuracy (assuming the machine operates correctly) because of the way the machine solves the problem.

The "floating decimal point" which may improve accuracy for a fixed size of number to be used in a machine or, conversely, may decrease the size of number required for a given accuracy is a feature not included in the ENIAC.

Speed. Setup time is the time required to arrange the interconnections for the particular problem at hand.

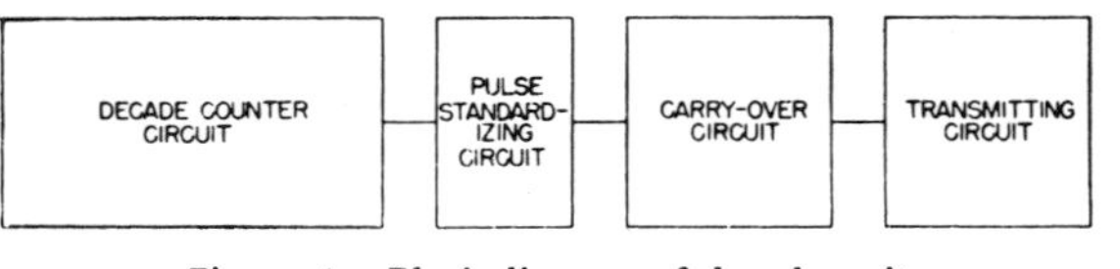

Figure 4. Block diagram of decade unit

In machines with automatic programming this is done by instructions given by the input device as the problem proceeds. In the ENIAC, interconnections (not including those which are the same for all problems and are built in) are made by hand, small patch cords being inserted at appropriate points in the machine and in the trunk system which extends around the front of the machine. The time required for this depends on the problem and the experience of the operator. It ranges from 30 minutes to a full day, and in this time the machine is not in operation, hence setup time is wasted so far as use of the machine is concerned.

The speed with which arithmetic operations are carried out is given by Table I, which shows the times required. This is an extremely important item, for it shows the great speed with which the ENIAC works in solving a problem. Take, for example, the time required to multiply two 10-digit numbers: it is approximately 3 milliseconds, or the speed is approximately 300 such multiplications per second. Now consider a problem requiring 1,000,000 multiplications. If there were no delays such as might be caused by slow-speed input and output equipment, this might be accomplished in an hour. Allowing 100 per cent leeway for all other operations except input and output (note that addition and subtraction are much faster than multiplication), this means that the problem might be solved in 2 hours, plus the time delay resulting from waiting for the input and output devices to complete their operations—if the latter operated very few times this would be negligible; if the problem required frequent operation of these devices this might amount to hours.

In connection with development work now under way on electronic large-scale general-purpose computing devices, increase in speed of arithmetic operation is not considered of major importance, although new machines may have speeds greater than those of the ENIAC by factors from slightly more than one to approximately ten.

It is interesting to note that the high speed of operation of the ENIAC makes less important the mathematical work which is often an intensive part of the preparation of a problem for solution. If the problem, for example, is to be hand-computed by a computing pool, it is often desirable and sometimes essential, that the solution be obtained by the most efficient method—quickly converging series, a special method of solution, maximum interval consistent with accuracy so that arithmetic work is minimized, and so forth. If this same problem were to be solved on the ENIAC, a saving of 50 per cent in solution time well might be negligible and the many days' work by high-quality personnel which often is put into such problems before any arithmetic work is done possibly might be cut down.

Reliability. There is no point in saying that a problem involving one million multiplications of 10-digit numbers can be carried out in two hours if the machine will have a breakdown before the problem is completed (although storage of results at regular intervals may save work up to the end of the interval preceding the trouble). The ENIAC was in numerous respects a pioneering device. Although several thousands of vacuum tubes have been used in a single network previously (this excludes systems such as the telephone system which uses many more tubes but in which a tube failure does not render the entire system inoperative) it is probable that no single device has had the 18,000 tubes which appear in the ENIAC. Since first use of the ENIAC much experience has been gained in operation and maintenance, and much of this bears directly on the question of reliability.

For 11 months the machine was in active use by Army Ordnance Ballistics Research Laboratory personnel at the Moore school. During this time a log book was kept listing all shutdowns and troubles encountered in the course of running problems. In this period of time the removal rate for tubes was one per 20 hours. In other words, 400 out of 18,000, or somewhat less than two per cent, of the tubes caused trouble during the course of the 8,000 hours of the study. It should be pointed out that the ENIAC was put into action immediately upon completion because of the urgency of some of the problems awaiting its operation. As a result there was no opportunity for a real "run in" period. This fact accounts for a rather high incidence of intermittent troubles resulting from bad soldered joints. There are some 500,000 soldered joints in the machine. (Besides its electronic circuits ENIAC includes a number of mechanical and electromechanical parts such as ventilating fans, protective relays, input and output relays, and punch card reader and punch. This equipment totals about ten per cent of the total components of the complete machine.) After the first few weeks of operation the practice of keeping power supplied continuously to the heaters of the vacuum tubes was followed as it was discovered that each shutdown resulted in two or three tubes failing. Power failures cannot be prevented

Table I. ENIAC Operating Speed

Operation	Time Required, Microseconds	Speed, Operations Per Second
Addition or subtraction	200	5,000
Multiplication of two 10-digit numbers	2,800	360
Calling up of the value of a function (from function table)	1,000	1,000

entirely and several occurred during the 11 months.

The method used for checking the results of a problem was generally as follows. Before and after each run of a problem test problems were run. In addition, many problems were checked within themselves, that is, certain terms or parts of these problems always must equal some constant such as one. Thus a check operation may be programmed comparing this part to one. The machine may be stopped automatically if this comparison is not right or the results may be printed for the operator to examine. Besides these precautions, each problem usually was run twice and the results compared.

Of the time that the ENIAC might have been actually computing, about 50 per cent was unproductive. Approximately half of this time was used for setting up the problem—connecting patch cords, setting switches, and checking these operations. This time is longer than is the rule now since operators and coders of problems were unfamiliar with the job and also because the problems solved initially were of extremely complex nature. The other half of this unused machine time was spent in maintenance and repairs.

A brief summary of the classes of failures requiring repairs follows:

1. Vacuum tube and electronic circuit component failure. These accounted for 20 per cent of the repair and maintenance time.

2. Circuit failures of mechanical nature such as bad solder joints, broken wires, short circuits. This accounted for 40 per cent of repair time.

3. Failures of the electromechanical input and output equipment accounting for 20 per cent of the repair time.

4. Failures of d-c power supply including rectifier tubes. Ten per cent of the repair time was used for these failures.

5. Failures of ventilating equipment and protective devices which accounted for the final ten per cent of the time.

It should be noted that these times include the time to find and repair the failure. In addition it can be said that the frequency of failures falling in category 2 was reduced greatly as time passed.

Capacity. The question of capacity of a large-scale machine, that is, the maximum "size" of problem it will handle, is essentially one of programming and memory in the case of the digital devices. With a machine such as a differential analyzer or an a-c board, it is a question of the amount of equipment (for example, the number of integrators or input tables in the differential analyzer). Restricting discussion to the large-scale general-purpose digital machines where automatic programming is incorporated, as in many of the new machines, but not in the ENIAC, the question reduces substantially to memory capacity.

To illustrate, in the ENIAC interconnections of the various units for a particular problem are by external connections, and when all of the places where a connection of a given type can be made are occupied, no further connections can be made to that unit. In contrast, a machine with automatic programming is arranged so that various units performing arithmetic operations can be called into use and after completing an operation can be freed so as to be available whenever next needed. This effectively eliminates the restriction on capacity resulting from programming. However it should not be thought that the ENIAC is limited unduly in this respect. The omission of automatic programming was for the sake of simplicity, but the extensive provision for interconnections has resulted in programming not being a serious limit on capacity.

For an appreciation of the reason that memory capacity is important, some consideration should be given to modern problems. To take one particular class, consider problems in electromagnetic field theory, hydrodynamics, or elasticity, which are expressed in terms of partial differential equations. Speaking broadly, these cannot be handled by differential analyzers which are intended primarily for ordinary differential equations. It was mentioned that Kron had used the General Electric Company's a-c board to solve some problems involving partial equations in each of these fields, but although this represented a distinct advance the method is quite limited. It is possible to solve a partial differential equation by a method similar to that used to solve an ordinary differential equation on a digital machine, that is, by replacing the "infinitesimal" differentials such as dt by finite but very small differences Δt. For partial equations the process is not so simple as there is more than one independent variable. In place of a "line" of computations followed by another using the results of the first line, and so on, a grid of results must be obtained, and these must be remembered in computing other results. Estimates by persons working in the fields usually place the desirable capacity of numbers to

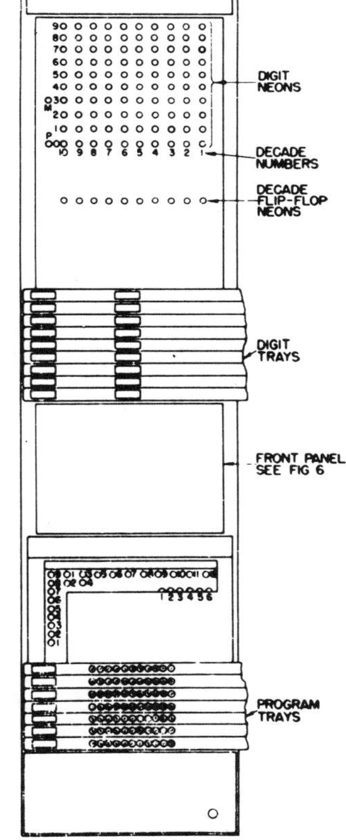

Figure 5. Front view of an accumulator

be remembered at between 1,000 and 5,000 (each number would have a certain number of digits—for example, 10 in the case of the ENIAC).

Present large-scale general-purpose digital machines could do this, but in the case of the ENIAC, for example, it would be necessary to record most of the 1,000 to 5,000 intermediate results by using the output device, and then to sort and reinsert each of these results at the right time by means of the input device. As previously mentioned, the input and output devices are slow-speed affairs, and the process would consume a great deal of time. Consequently it is reasonable to consider the capacity of the ENIAC limited in this respect. New development work in the field of the large-scale general-purpose electronic digital computers is devoted to a large extent to achieving adequate memory of the order of 1,000 to 5,000 10-digit numbers.

GENERAL DESCRIPTION OF THE ENIAC

In light of the preceding discussion the following characteristics of the ENIAC may be listed:

1. Large-scale.
2. General purpose.
3. Digital.
4. Electronic.
5. Uses 10-digit numbers.
6. Uses decimal system.
7. High speed.
8. Synchronous operation.
9. Some parallel operation possible.
10. Complete flexibility within limits of programming capacity.

In addition, it may be noted that the ENIAC consists of 40 panels erected along a U-shaped contour, plus d-c supplies for tube voltages, and so forth (Figure 2). The ENIAC has been housed in a room 30 by 50 feet. It contains approximately 18,000 vacuum tubes, and uses about 130 kw.

AN ACCUMULATOR

To describe in any detail all the sundry units in the ENIAC would require considerable space; in lieu of this the operation of one unit only—an accumulator—will be outlined, without going into circuit details.

Broadly speaking, the purpose of an accumulator is to perform additions and subtractions, to store a number, or transmit a number when called on, and to receive numbers for addition or subtraction or for storage. An electric device consisting of ten similar basic units of which all units except one were "off" or "normal" or otherwise distinguished, and one unit was "on" or in an "abnormal" state or otherwise oppositely distinguished would serve to form one column. Ten of these devices alongside one another could form the ten columns necessary to permit indication of a 10-digit number. The accumulators of the ENIAC use a "ring counter" for each column. The basic unit of the ring counter is a trigger circuit so arranged that nine of the trigger circuits are in a normal state and one in an abnormal state. Figure 3 shows a decade counter and Figure 4 a block diagram. The electric circuit includes a pulse-shaping circuit for assuring good wave form of the pulse and also includes carry-over circuits. Carry-over is required because after a counter reaches 9 the next pulse received by it should return it to 0 and add 1 in the next digit position. (Because the counter goes from 0, 1, 2, ... 9, and then back to 0, it is called a ring counter.) Carry-overs are of two types—when there is not already a 9 in the next digit position, in which instance one carry-over completes the work, and when the opposite is true. In the latter instance a further carry-over is required, and provision is made for this.

To eliminate the need for the counters to work in two directions, that is, from 0, 1, 2, ... to 9, and from 9, 8, 7, ... to 0, as would be required in ordinary addition and subtraction, the complements of negative numbers (with respect to 10^{10}) are used, and subtraction thus becomes a process in addition. However, as it is necessary to know whether a number is a simple positive number or a complement of a negative number, electrical means are provided for having an appropriate signal travel with the representative of a negative number. If P and M are used to denote respectively plus and minus quantities, then P 0,000,342,789 is the number +342,789 whereas M 9,999,452,111 represents −547,889 which is obtained by subtracting 9,999,452,111 from 10^{10}. If now the number +342,789 is in an accumulator, and the number −547,889 is sent to that accumulator to be combined with the former, the accumulator actually will receive the complement of the latter, and the operation will be

P 0,000,342,789 (in accumulator)
M 9,999,452,111 (sent to accumulator)
———————————
M 9,999,794,900 (result in accumulator)

There is a PM indicator in the accumulator which will indicate M for the result; this means that the answer is negative and that 9,999,794,900 is its complement. The true answer therefore is obtained by subtracting the latter from 10^{10} and is −205,100. Complements are obtained easily in the ENIAC, and the process of using them in place of negative numbers does not involve any great complexity.

An accumulator is not a counter, although each accumulator contains ten decade ring counters. A counter as its name implies "counts," and to get to any number such as 342,789 would go through the indications of all integers from 1 to that number. This would be a very long process. An accumulator, like a desk calculator, adds in all columns at the same time, and thus requires for its operation only the time for its counter unit (one in each digit position) to count from 0 to 9, plus time for causing carry-over, plus time for certain other processes. In the ENIAC the ring counters in each digit position are advanced by the reception of pulses. These pulses are 2 microseconds in duration, and follow one another at 10-microsecond intervals. Nine pulses are required to carry out additions and subtractions before carry-over, and 110 microseconds are needed for carry-over and other necessities. Thus one addition or subtraction requires 200 microseconds, which is the basic "addition time," and is the interval covered by the chart in Figure 1.

Figure 5 shows the front of an accumulator, and Figure 6 gives a drawing of the front control panel appearing approximately in the center of Figure 5. As the diagram of Figure 2 shows, there are 20 accumulators in the ENIAC. The external operation of an accumulator can be explained by noting the various uses of the parts appearing in Figure 6.

In the upper left-hand corner are receptacles marked interconnecting plus I_{L1} and I_{L2}. These are for interconnection to another accumulator to enable the pair of accumulators to handle a 20-digit number. Likewise at the upper right are receptacles I_{R1} and I_{R2} used for a similar purpose.

The digit input terminals are receptacles for numbers coming to the accumulator. The lettering α, β, γ, δ, ϵ indicates five separate receptacles, any one of which can be connected to any other unit of the machine by means of the digit trunks (transmission lines) which may be plugged into at each panel. There are 11 wires in each connection, ten to carry pulses corresponding respectively to the ten digits of a 10-digit number, and the 11th to carry the P or M indication. The α, β, γ, δ, ϵ receptacles allow five incoming interconnections, so that at different times during the solution of a problem the accumulator can receive numbers from the various other units to which connections are made.

The digit output terminals are the terminals through which the number in the accumulator at a given time may be sent out (A is for add output and through it goes the number in the accumulator, S is for subtract output and through it goes the complement of the number in the accumulator). Only one output receptacle is necessary, as this may be connected to as many digit trunks for transmission to other parts of the machine as required. This does not mean that these other units will receive all outputs of the accumulator under review. The pulses arrive at all these other units, but only in those which have received a program signal to accept them will the pulses (numbers) enter.

The panel marker which has the accumulator number has the on off switch for accumulator tube filament heaters, and shows the number of hours of operation of the tube heaters. This is desirable for maintenance. In the corresponding position on the right-hand side is the significant figure switch, which, by proper setting, results in the accumulator "clearing," that is, eliminating the number in it and indicating 0,000,000,000, except that a five is put in whatever digit position desired, if any. This is a common mathematical trick for rounding off numbers, and explains why certain simple round-off errors such as that given with an example of numerical addition earlier need not appear in the ENIAC. It does not do away however with round-off errors in general, and these are often of major importance. The selective clear switch, when in a proper position, enables the accumulator (and others with switches so set) to be cleared by a signal sent out to all from the initiating unit.

Omitting switches 1 to 4 for a moment, consider any one of the switches 5 to 12. Immediately beneath operation switch 5 for example is a repeat switch, and beneath this are shown terminals $5i$ and 5θ (i for input, θ instead of O for output because O resembles zero) all of which go with the operation switch 5. If operation switch 5 is set to α, and terminal $5i$ receives an appropriate program signal from elsewhere in the machine, then the accumulator will admit the number being sent to it on the α digit input terminal at the top of the panel, and this number will be added to whatever is then in the accumulator, of if nothing (0,000,000,000) is present the incoming number (and its sign indication P or M)

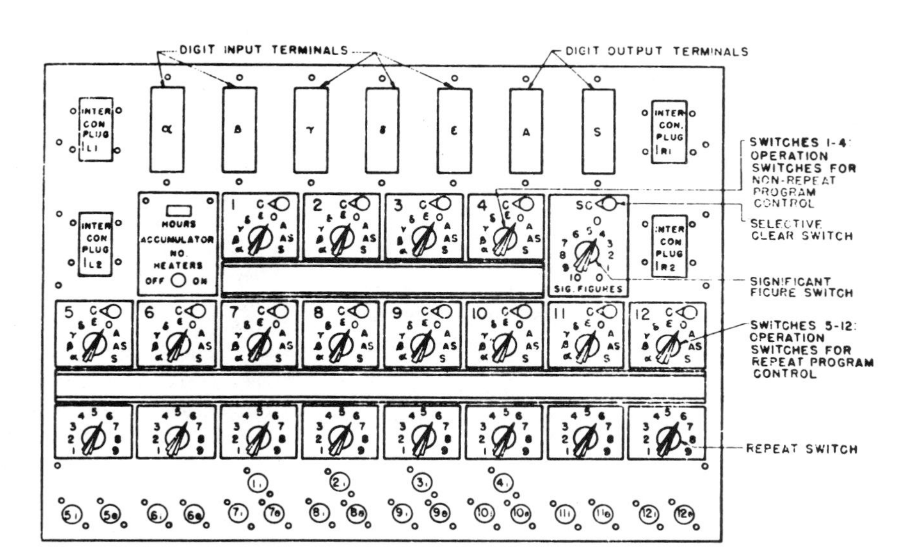

Figure 6. Accumulator front control panel

Terminals 1i, 2i, . . . 12i—Program input terminals
Terminals 5θ, 6θ, . . . 12θ—Program pulse output terminals

will be stored. If the repeat switch is set to any value such as six, the number will be received six times provided it is sent out from elsewhere in the machine at least six consecutive times. If it is sent out more than six times, the accumulator here under discussion will receive it only six times. This is a simple method of multiplying by small numbers, namely, adding the numbers together the appropriate number of times.

Consider now another switch, say operation switch 6. If it is set to A, its repeat switch to nine, and the program terminals 6θ connected to another unit of the machine (or to any of the i terminals of this accumulator) then the following will take place: The number in the machine will be sent out over the A digit output terminals (top right of figure) when an appropriate signal is received on $6i$; this will be repeated eight more times (total of nine times) because of the setting of the repeat switch; after this is done a program signal will be sent from 6θ to another unit to start operations there; if the clear correct switch is set to C the number in the accumulator will be cleared out of the accumulator in accordance with the setting of the significant figure switch but if the clear correct switch is set to 0, the number in the accumulator will be retained there.

Operation switches 1–4 are similar to those of 5 to 12, but have no associated repeat switches and no program output (θ) terminals.

This brief description of the external operation of an accumulator will tend to give an idea of the external operation of the other ENIAC units, and of the interconnections between units and the programming (except for use of the master programmer) which must be set up for a particular problem.

A common example of the use of accumulators as the only arithmetic units in a problem is that of generating the squares of all integers from 1 to 10,000 correct to 10 significant figures. Write

$$(n+1)^2=n^2+2n+1$$

which says that the square of an integer plus twice that number plus one is the square of the next higher integer. The work can be carried out thus: set the constant transmitter to supply the number one, and connect it to accumulators 1 and 2 using one of the α, β, γ, δ, or ϵ digital input terminals in each. Connect the A digit output terminal of number 1 to one of the unused α, β, γ, δ, or ϵ terminals of number 2; program so that number 1 transmits twice to number 2 and number 2 receives twice, and thereafter the constant transmitter supplies one to each accumulator. The machine then proceeds as follows, assuming the numbers one and one appear initially in both accumulators: number 1 transmits twice and number 2 receives, so that one is sent twice from number 1 to number 2, resulting in three appearing in number 2 and one in number 1; thereafter the constant transmitter sends one to each accumulator so that number 1 contains two, and number 2 contains four. This is the end of the first line, and if suitable programming were arranged the result could be recorded, and the second line started.

Number 1 now sends two twice to number 2, which brings the number in the latter up to eight. The constant transmitter now adds one to each accumulator's number, so that number 1 has three and number 2 has nine. This ends the second line. This process then is continued. It is so fast that if no time is taken out for recording it can compute in six seconds the squares of all integers from 1 to 10,000 correct to ten significant figures. (Actually the process can continue beyond 10,000 to the integer whose square last falls within 10^{10}, and could go farther if accumulators were connected for 20-digit operation.)

BRIEF HISTORY

While the ENIAC exists independent of its history, it is interesting to note that it was actual pressure of computing work that led to its inception. During the war the differential analyzer of the Moore school of electrical engineering of the University of Pennsylvania was used intensively for ballistic computations; the ballistic research laboratory of the Army Ordnance Department maintained a computing center of approximately 100 trained persons (college graduates) on the Pennsylvania campus in co-operation with the University, and under separate contracts the Moore school maintained other computing groups. In addition several hundred persons (Army employees) were given intensive 3-months' training courses in mathematics, plus approximately 100 specially selected members of the Women's Army Corps. The Ordnance Department had virtually duplicate facilities at Aberdeen, Md., both as to differential analyzer and number of computers. The total computing center was probably one of the largest, if not the largest, in the world.

Despite all this it was soon evident that if the work were to continue at its then rate of growth, and if personnel could not be obtained more easily than was possible in the early years of the war, the work would outrun the capacity of the computing center. It was in this atmosphere of pressure that development of the ENIAC was undertaken, and it was because of this pressure that there are deficiencies and omissions now apparent in the machine, which nevertheless retains its position as the first electronic (high-speed) large-scale general-purpose digital machine, and which is the forerunner of numerous others now under development. Although the ENIAC is a general purpose machine, its name (electronic numerical integrator and computer) reflects the preoccupation with numerical integration of differential equations which was such an important part of war computing work.

REFERENCE

1. The Automatic Sequence Controlled Calculator, **Howard H. Aiken, Grace M. Hopper.** *ELECTRICAL ENGINEERING*, volume 65, August–September, 1946, pages 384–91; October 1946, pages 449–54; November 1946, pages 522–8.

Chestnut Discusses the Convergence of Automatic Control and Electronics

IN the special 50th anniversary issue of the PROCEEDINGS OF THE IRE, Chestnut reviewed the history of automatic control and electronics, and concluded that they had developed relatively independently until about the time of World War II. Subsequently, "the intrinsic similarity between communication and control problems began to be established," resulting in a "blending of control, communication, computation, and instrumentation through the common concept of information as influenced greatly by the common medium of electronics." Chestnut suggested that this merger of heretofore disparate fields might be very useful in studying economic, social, and political systems. Historians of science and technology will recognize parallels between Chestnut's interpretation and the increasingly close linkages between science and technology during the past century. It is another example of the synergism that has occurred frequently in the history of electrical engineering.

Harold Chestnut (1917–) was born in New York and received the B.S. and M.S. degrees in electrical engineering from M.I.T. in 1939 and 1940, respectively. He became a test engineer with General Electric in 1940 and remained with G.E. in various capacities. He became a control systems engineer in 1956 and later a consultant in systems analysis. He was President of the IEEE in 1973. See *Engineers of Distinction*, p. 53, 1973 and *Who's Who in Engineering*, p. 1426, 1964.

Automatic Control and Electronics*

HAROLD CHESTNUT†, SENIOR MEMBER, IRE

Summary—The past fifty years have seen the automatic control and electronic fields come close together and form an effective means for increasing man's productivity and his ability to control energy and materials. By extending automatic control concepts to new processes, by developing more flexible controls capable of changing their characteristics to optimize performance of the process being controlled, and by increasing the capability of the sensing means in difficult environments, man will be able to make even more effective his ability to control automatically in the years ahead.

Electronics is increasingly able to provide physical means for providing the realization of automatic control principles and concepts. Increasing effort to achieve reliable electronic automatic control means must be continued in the years ahead to make possible the realization of the promised gains indicated by the automatic control theory. In addition, more use should be made of standardized design ranges of electrical and mechanical features so that all automatic control equipment can be made in less time and at a lower relative cost. The future appears bright for expanded use of automatic control and electronics as we look ahead for the next fifty years of the IRE.

IT IS SIGNIFICANT that on the occasion of the 50th Anniversary of the IRE, we think of Automatic Control and Electronics as being closely related. Today, automatic control has much of its technical structure firmly based on the same foundation of theory that was developed originally for electronic communications. A large proportion of automatic controls whether they be for the control of energy processes, materials processes, or information processes are electronic in nature. Further, the design, fabrication, and test of automatic controls are heavily dependent on the use of electronic computers both analog and digital.

In a somewhat parallel relationship, many of the applications of electronics are now considerably improved or enhanced by automatic control of such factors as voltage, frequency, or gain. Precision and high-speed fabrication methods for electronic devices and equipment are greatly facilitated by automatically controlled means of manufacture. In addition, many of the logic operations and other judgment criteria now being used more extensively for automatic control have served to make possible or increase the effectiveness of communication means such as automatic long-distance telephones and scatter transmission. Truly, the relationship between automatic control and electronics has developed into one that is most fruitful and mutually advantageous. Through use of some similar components these two fields have been able to capitalize to a certain extent on one another's practical developments, measurement equipment, and fabrication techniques.

* Received by the IRE, July 26, 1961.

† General Engineering Laboratory, General Electric Company, Schenectady, N. Y.

Developments During Past Fifty Years

At the same time of the inception of the IRE in 1912, however, neither electronics nor automatic control were terms that existed in the present sense of the words. It was essentially in the period of the 1920's and 1930's that these terms began to come into use, but the fields of engineering and science they represented tended to operate quite independently of each other.

Electronics in its early days was oriented largely in terms of its contribution to the higher frequencies of the communication and radio industries and operated in a physical environment that was in some respects more controlled and less rigorous than was that experienced by automatic control equipment used in heavy industry. However, as one reads the early IRE Proceedings in the 1913–1915 era, one has the feeling that environmental facts such as the vagaries of the transmission path for RF energy represented highly nonlinear and time-varying phenomena that are similar to those which challenge today's adaptive control.

Automatic control had its background in the speed regulation of steam engines and other power devices. As such, it was initially concerned with the control of high-energy processes of relatively slow operating speeds operating in strenuous environments which might experience high vibration and shock. Because of the high value of the equipment or material over which it exerted its influence, automatic control from its early days was forced to face high reliability requirements. With its lower signal levels and energy implications, electronics reliability requirements were initially less demanding but have grown to be very demanding because of the high military importance of electronics. Gradually, over the years electronics and automatic control have blended their interests and now share many common problems.

Tying Together of Automatic Control and Electronics

The 1930's saw the development by Nyquist,[1] Black,[2] and Bode[3,8] of the frequency response concept of feedback amplifiers for use for communication purposes. By

[1] H. Nyquist, "Regeneration theory," *Bell Sys. Tech. J.*, vol. 11, pp. 126–147; January, 1932.

[2] H. S. Black, "Stabilized feedback amplifiers," *Bell Sys. Tech. J.*, vol. 13, pp. 1–18; January, 1934.

[3] H. W. Bode, "Amplifiers," U. S. Patent No. 2,123,178; 1938.

Reprinted from *Proc. IRE*, vol. 50, pp. 787–792, May 1962.

the early 1940's the automatic control engineers spearheaded by Brown,[4,5] Hall[4,6] and others, began to embrace this method of analysis and design for feedback control system design. Wiener,[7] Shannon and Blackman,[8] and others during World War II tackled prediction and control problems in terms of communication and signal concepts and the intrinsic similarity between communication and control problems began to be established.

It was during the 1920's that the initial emphasis took place on making the electronic tubes then in use rugged enough for the industrial and military environments that would be encountered in many automatic control applications. The possibilities of using electronic tubes as part of the automatic controls for industry were attempted in a limited fashion prior to 1940. However, it was the military requirements of the early 1940's with the increased need for speed and accuracy that provided the greatest impetus for the wedding together of automatic control and electronic ideas.

Post War Developments

The emphasis on servomechanisms[9] and linear control theory during and immediately after World War II was followed by an effort to analyze nonlinear and sampled-data controls. The work of Kochenburger[10] on the describing function has been most helpful in understanding and designing nonlinear systems. In the field of sampled-data controls, Ragazzini,[11,12] Zadeh,[11] Franklin,[12] and later Jury[13] and Tou[14] have done much to clarify the basic analytical appreciation of this phenomenon so important to time-shared controls that are required by electronic digital computers. Truxal's[15] work in bringing together these various analytical methods into a clear perspective has been most helpful.

The widespread use of practical electronic computers, first analog in the 1940's and later digital in the 1950's, has greatly extended the control-systems engineer's analytical ability to understand and design automatic control systems. At the same time, some control systems were being built in which electronic computations were performed as part of the actual control process. The internal logic and control needs, such as floating decimal point and time sharing within some digital computers, have pointed the way to control principles that have been used for automatic controls quite apart from their computer contexts. In similar fashion to the electronic automatic gain control used since the early days of radio, the concept of variable gain or other adaptive controls has in the 1950's found its way into increasing control usage.[16]

Electronics in the form of radar or sonar had been developed during World War II to provide input, detection, and tracking signals for automatic search, ranging and tracking systems. As time has gone on, more extensive use is being found for electronic means of sensing in instrumentation systems both for display purposes and for inputs to automatic control systems. The high-speed, noncontacting aspect of electronic sensing as well as its capability of measuring certain quantities, as for example by X-ray or nuclear magnetic resonance, provide control with capabilities that it might otherwise not be able to attain. The recent development of analytical instruments making extensive use of electronics is providing still further tools with which to increase the capability of automatic control in the chemical and petroleum fields.

Present Status

Gradually, from this evolutionary blending of control, communication, computation, and instrumentation through the common concept of information as influenced greatly by the common medium of electronics, there is taking shape a new grouping of ideas and equipment that is variously called automatic control, instrumentation, or information processing. The basic building blocks of this new automatic control include the functions of sensing, converting, programming, communicating, regulating, computing, storing, actuating, and display as described in Table I.[17]

In the past there has been a considerable effort on individual systems designed to perform the automatic control functions indicated in Table I, and there will continue to be emphasis to improve such equipments.

[4] G. S. Brown and A. C. Hall, "Dynamic behavior and design of servomechanisms," *Trans. Am. Soc. Mech. Engrs.*, vol. 68, pp. 503–524; July, 1946.

[5] G. S. Brown and D. P. Campbell, "Principles of Servomechanisms," John Wiley and Sons, Inc., New York, N. Y.; 1948.

[6] A. C. Hall, "Application of circuit theory to the design of servomechanisms," *J. Franklin Inst.*, vol. 242, pp. 279–307; October, 1946.

[7] N. Wiener, "The Extrapolation, Interpolation, and Smoothing of Stationary Time Series with Engineering Applications," John Wiley and Sons, Inc., New York, N. Y.; 1949.

[8] R. B. Blackman, H. W. Bode, and C. E. Shannon, "Data-Smoothing and Prediction in Fire Control Systems," Research and Development Board, August, 1948, or see H. W. Bode and C. E. Shannon, "A simplified derivation of linear least square smoothing and prediction theory," PROC. IRE, vol. 38, pp. 417–425, April, 1950.

[9] H. M. James, N. B. Nichols, and R. S. Phillips, "Theory of Servomechanisms," McGraw-Hill Book Co., Inc., New York, N. Y.; 1947.

[10] R. J. Kochenburger, "Frequency response method of analyzing and synthesizing contactor servomechanisms," *Trans. AIEE*, vol. 69, pt. 1, pp. 270–294; 1950.

[11] J. R. Ragazzini and L. A. Zadeh, "The analysis of sampled data systems," *Trans. AIEE*, vol. 71, pt. 2, pp. 225–234; November, 1952.

[12] J. R. Ragazzini and G. F. Franklin, "Sampled-Data Control Systems," McGraw-Hill Book Co., Inc., New York, N. Y.; 1958.

[13] E. I. Jury, "Sampled-Data Control Systems," John Wiley and Sons, Inc., New York, N. Y.; 1958.

[14] J. T. Tou, "Digital and Sampled-Data Control Systems," McGraw-Hill Book Co., Inc., New York, N. Y.; 1959.

[15] J. G. Truxal, "Automatic Feedback Control System Synthesis," McGraw-Hill Book Co., Inc., New York, N. Y.; 1955.

[16] J. A. Aseltine, A. R. Mancini, and C. W. Sarture, "A survey of adaptive control systems," IRE TRANS. ON AUTOMATIC CONTROL, vol. AC-6, pp. 102–108; December, 1958.

[17] H. Chestnut and W. Mikelson, "The impact of information conversion on control," IRE TRANS. ON AUTOMATIC CONTROL, vol. AC-4, pp. 21–26; December, 1959.

TABLE I
TABLE OF DEFINITIONS OF BASIC AUTOMATIC CONTROL OR INFORMATION CONVERSION FUNCTIONS

Function	Definition or Explanation
Sensing	Generates primary data which describe phenomena or things.
Converting	Changes data from one form to another to facilitate its transmission, storage or manipulation.
Storing	Memorizes for short or long periods of time data, instructions, or programs.
Communicating	Transmits and receives data from one place to another.
Computing	Performs basic and more involved mathematical processes of comparing, adding, subtracting, multiplying, dividing, integrating, etc.
Programming	Schedules and directs an operation in accord with an over-all plan.
Regulating	Operates on final control elements of a process to maintain its controlled variable in accord with a reference quantity.
Actuating	Initiates, interrupts, or varies the transmission of power for purposes of controlling "energy conversion" or "materials" conversion processes.
Presenting	Displays data in a form useful for human intelligence.

At present, however, there is increasing emphasis on the understanding, design, and installation of increasingly complex systems that are a combination of a number of such equipments. The stability, performance, cost, reliability and maintainability of such systems are facets of concern for present-day automatic control activities. Such efforts represent one of the important directions in which automatic control work is proceeding.

National and International Cooperation

A recent interesting development in the organization of people and technical societies interested in automatic control has been the closer association of such individuals and groups on a national and an international basis. Because automatic control embraces such varied equipment as pneumatic, hydraulic, mechanical, electrical, as well as electronic devices and includes instruments, computers, regulators, and many other control means, the American Auotomatic Control Council (AACC) was formed in 1957 to represent the five major American technical societies interested in automatic control in the International Federation of Automatic Control (IFAC) which was also founded in 1957. In addition to the IRE, represented by the Professional Group on Automatic Control, AACC includes the American Institute of Electrical Engineers, the American Society of Mechanical Engineers, the American Institute of Chemical Engineers, and the Instrument Society of America. Through AACC, American control engineers are able to broaden their areas of interests to include a great number of fields of application of automatic control as well as to consider alternate physical means for accomplishing any given desired result. The benefits to be gained by having access to information on the developments in automatic control from all over the world are significant. Such increased technical contacts can do much to accelerate the future development of automatic control in the years ahead.

NEW AREAS OF CONTROL DEVELOPMENT

As we look to other ways in which automatic control developments are progressing, there are a number of encouraging activites taking place. Included in these are:

1) More extensive use of automatic control concepts in new fields where they have not been used to any appreciable degree.
2) Use of more flexible controls through the application of adaptive control concepts and including such means as self-learning or automatic optimizing.
3) Development of new and better sensors capable of measuring more quickly and accurately quantities that will be helpful to new controls.

New Fields

Some of the new fields where automatic control principles have not been used extensively, but appear to hold considerable promise, include the business and economic fields, the social, biological and political fields. Forrester[18] and others working in the field of industrial management have been able to develop dynamic models of individual businesses or industries and have shown the effects of feedback and various strategies on the stability and speed of response of such economic systems. The recent formation of many new countries and the development and growth of many existing ones are being considered from their possibilities of control even though the degree of automatic control is perhaps quite minor at present.

The possibility of using information, sensing, and control concepts to study social, biological, and even political situations is being tried to a limited extent. Data obtained here should be of value to automatic-control people in developing new general control concepts such as self-learning as well as for their value in developing control ideas for specific processes of these types. Uses of automatic control in the form of heart-pacers, synthetic limb controls, and medical operating aids are indicative of some of the directions which automatic control may develop. The use of over-all hospital information systems is another application of automatic control into the biological-social area.

Fig. 1 shows how information conversion is being

[18] J. W. Forrester, "The Impact of Feedback Control Concepts on the Management Sciences," The FIER Distinguished Lecture 1960, FIER, N. Y.; 1960.

used in conjunction with energy and/or materials processes to an increasing degree. Recent activities in the chemical and petroleum industry have been directed at the determination of the dynamic characteristics of the processes themselves and the most effective methods of controlling them. Significant cost reductions appear to be possible by more automatic *control of systems of such variables rather than of fixed control of the individual variables themselves.*

Experiences in the steel and electric utility industries are beginning to supply actual data from advanced automatic control systems involving control of materials, energy, and information as indicated in Fig. 2. In addition to the conventional automatic control shown by the solid lines, programmed inputs are supplied by the programmer, information about the actual performance of the process are recorded by the data logger, and modified inputs to the reference are determined by the computer operating on- or off-line. The need for extensive automatic control in such advanced energy processes as nuclear fission, magneto-fluid dynamics and power generation in space also represent new challenges to the control field.

An important aid in the application of automatic control in new fields will be the development of appropriate models for representing in a space-time sense the phenomenon taking place in the "process." These models must be quantitative to be most effective but the application of automatic control principles may make it possible for somewhat inexact or approximate models to be employed.

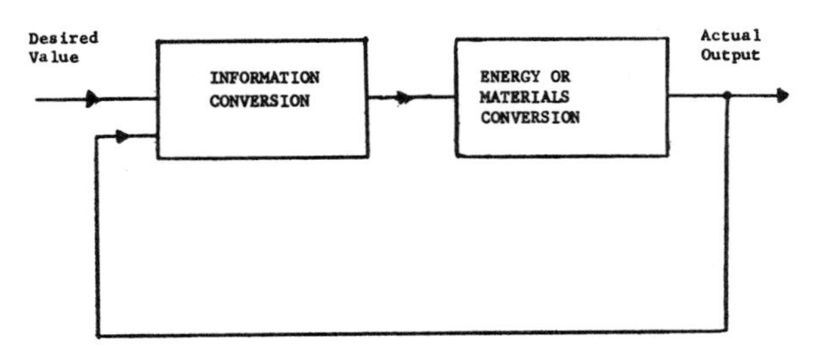

Fig. 1—Control system showing interrelationship of information, energy and materials conversion.

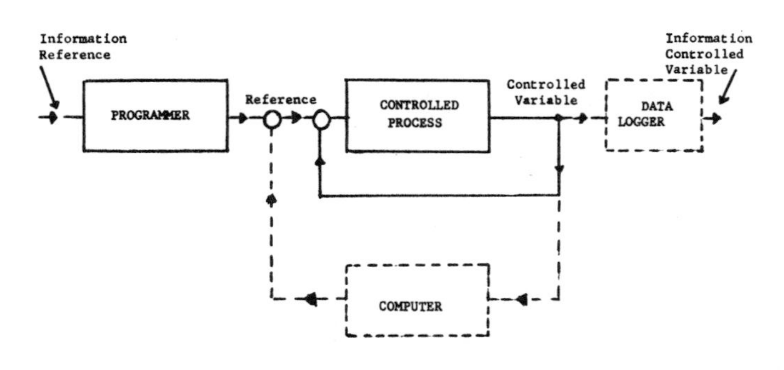

Fig. 2—Combined feedback control and information conversion system.

More Flexible Controls

The efforts at controlling such variable parameter processes as supersonic flight including that of space vehicles or combined materials and energy processes have highlighted the need for more flexible or adaptive controls. Fig. 3 serves to illustrate how changes in the character of the process being controlled can be used to provide an input to an adaptive control which in turn serves to modify the principal control of the process. Other forms of adaptive control use sensing and identification of the reference input or of the actuating error itself to provide a basis on which the adaptive control operates. Increasing sophistication, including extensive computation, is taking place in both the criteria being used as a basis for identifying the nature of the process as well as the nature of the changes that are made to take place in the principal control itself. Additional effort is required to provide a clear analytical understanding of the stability and performance phenomena of such highly nonlinear, time-varying, dynamic control systems.

Fig. 4 serves to illustrate the nature of the multivariable form of the adaptive control problem. By virtue of changing environments as well as the inherently nonlinear nature of the process itself, the control may be faced with the need for changing its logic and strategy in a fashion that can be only generally stated by the automatic control designer. Such controls need to possess a self-learning capability in which their characteristics are allowed fairly broad latitude if the process is so poorly known initially as to provide relatively little information to the designer. Based on quite general design criteria for acceptable performance, the control will use its memory as well as logic rules and endeavor to learn from its experience. By modifying its

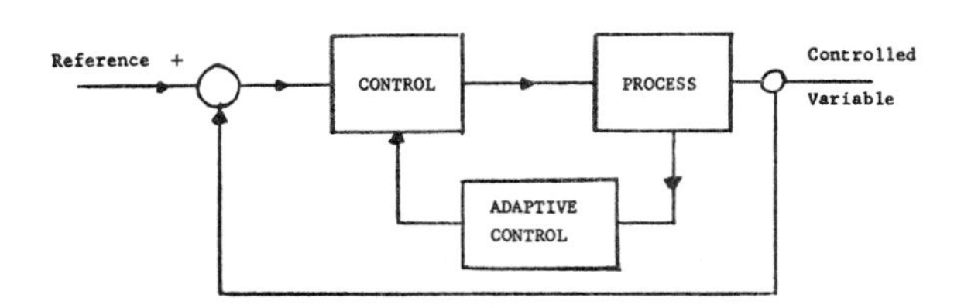

Fig. 3—Adaptive controls sought for processes which are not linear, constant, or fully known.

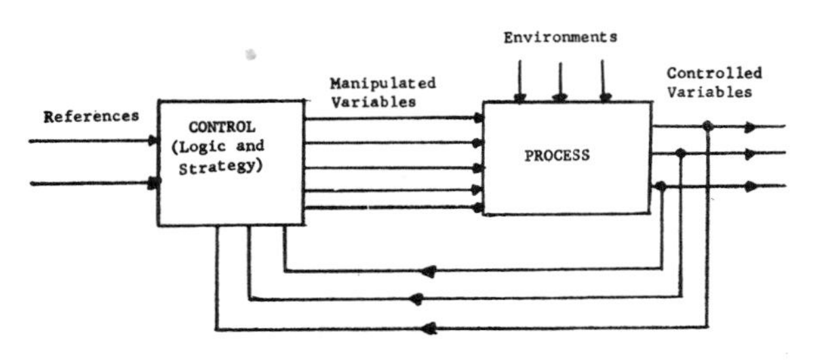

Fig. 4—More processes are multivariable with poorly controlled environments.

characteristics it will try to improve its performance in terms of some initially-agreed-upon objectives.

A variation of this problem is found in the information field itself where pattern or character recognition problems may possess a similar lack of definition for some information processes in their initial stages. Another form in which flexible adaptive control may prove increasingly effective is in the area of automatic design optimization. With engineering efforts representing a larger portion of the cost of producing new products and systems, more extensive computer methods for performing engineering evaluations are being employed. The expansion of control concepts of optimization more broadly to design[19] represents an attractive field for more flexible automatic-control concept application.

New Sensors

Paramount to the solution of any automatic control problem is the ability to sense the conditions that exist in the process. Although modern control theory is firmly based on the assumption that an adequate measure of the variables being controlled can be obtained, there are a number of materials and energy processes where sensors capable of making such measurements are not available. As Lord Kelvin has stated so well, it is necessary to measure things quantitatively in order to be able to understand and control them. Physical and chemical composition measurements in high and corrosive atmospheres are illustrative of specific problems in the sensor area. New analytical instruments capable of accurate and rapid measurements and suitable for on-line control applications will do much to further the usefulness of automatic control. Although creative thinking in conceiving new measurement concepts will be valuable, the development of new electronic components to operate in unfavorable environments such as these will also be very beneficial.

Associated Needs for Continued Control Progress

Accompanying the developments in the better understanding and broader application of automatic control, there exist associated needs for improved components and equipment to provide the means for accomplishing automatic control more reliably, quicker, and at lower relative cost. It is here that continued electronic development can be most effective.

Reliability[20]

The vast amount of work done in the past fifteen years on the reliability of electron tubes and their associated electronic components has resulted in marked electronic circuit improvement over the years and doubtless improvements will continue to be made. More extensive use of ceramic or relatively inert materials accompanied by strict observance of sound mechanical design principles have pointed the way to more reliable electronic and automatic control equipment. Conservative electrical designs in the sense of using components at operating conditions well below the design limits have likewise shown that improved reliability can be obtained.

The development of transistors and general solid-state electronics has provided smaller, lighter, and lower-power-consuming devices capable of high reliability that are finding increasing use in automatic control equipment. Certainly, the weight, size, and reliability needs in space vehicle control have emphasized the desirability of these miniature electronic devices. The possibility of micro-electronics for control in space appear to offer further advantages if the predicted reliability can be realized and adequate means for automatic controlled manufacture can be achieved at reasonable cost. Present thinking would seem to indicate however, that in the industrial area the need for extremely small size is not as important as high reliability and low cost.

Accompanying the efforts to improve the reliability of individual electronic elements or groups of elements, further emphasis is required in the judicious use of redundancy, fail-safe features, and/or other system means for providing over-all reliable operation. The large amount of electronic automatic control equipment currently being utilized in many industries and the large economic investment in processes being controlled automatically emphasize the present and future needs for automatic control of appreciably higher reliability.

Standardized Designs for Reduced Cost and Time

In many cases automatic control equipment has been designed from a custom-design point of view. Since the energy or materials portion of the system, *i.e.*, the controlled process, tended to be unique with each installation, the automatic control tended to be designed and tailored to the installation needs. This has resulted in the time required to produce automatic control being longer than if more standard control components were used. It is appropriate that increasing efforts be made to work toward a more standardized approach of automatic control system design in which appropriate voltage, frequency, impedance and power levels are established broadly across the control industry so that compatible components may be interconnected directly to a greater extent.

Compatible design should be directed toward achieving a greater measure of over-all electrical and mechanical uniformity for subassemblies and equipments so that shorter equipment delivery times can be realized and the cost for engineering and manufacturing per

[19] E. L. Harder, "Computers and automation," *Impact of Sci. on Soc.*, vol. 1, pp. 3–15; 1960.

[20] "Reliability of Military Electronic Equipment Report," Advisory Group on Reliability of Electronic Equipment (AGREE), U. S. Govt. Printing Office, Washington, D. C.; 1957.

individual equipment can be reduced. Automatic control and electronics are still changing at a sufficiently high rate that it does not appear appropriate that rigid standardization be introduced at this time. However, the rapid spread in the use of digital and logic control methods and the extensive use of transistor circuitry point up the desirability of having equipment with common characteristics for such electrical features as voltage level, impedances, and frequency in a number of ranges. By this means, fewer engineering designs and associated manufacturing fixtures will be required and more time and effort can be devoted to making sure the equipment is reliable and of suitable quality and performance. Industry efforts presently being undertaken in this direction should be strengthened and advanced more rapidly. With more compatible components the desirable objectives of reliable automatic control can be obtained more quickly and more cheaply.

Conclusions

The past 50 years have seen the automatic control and electronics fields come close together and form an effective means for increasing man's productivity and his ability to control energy and materials. By extending automatic control concepts to new processes, by developing more flexible controls capable of changing their characteristics to optimize the performance of the process being controlled, and by increasing the capability of the sensing means in difficult environments, man will be able to make even more effective his ability to control automatically in the years ahead.

Electronics is increasingly able to provide the physical means for the realization of automatic control principles and concepts. The increasing effort to achieve reliable electronic automatic control means must be continued in the years ahead to make possible the realization of the promised gains indicated by the automatic control theory. In addition, more use should be made of standardized design ranges of electrical and mechanical features so that automatic control equipment can be made in less time and at a lower relative cost. The future appears bright for expanded use of automatic control and electronics as we look ahead to the next 50 years of the IRE.

Fano on the Social Role of Computer Communications

In this provocative and thoughtful essay, Robert Fano addressed some fundamental issues involving technology, society, and human values. He pointed out that "radically different social consequences" could result from different ways of using computer-communications systems and that, in the absence of individual awareness and active concern, the impact "may turn out to be a nightmare." He suggested that computer-communications networks were tending to create a "new nervous system of society" that was just as necessary for the mass production of services as a transportation network had "proved to be essential to the mass production of goods." Fano stressed the need for universal access to the knowledge stored and communicated by the computer-communications network since "knowledge restricted to a segment of society can be transformed into power over the rest of society." See also Joseph Weizenbaum, "On the Impact of the Computer on Society," *Science*, pp. 609–614, May 12, 1972 and the paper by Daniel Noble on the new era of brain-extension systems in the PROCEEDINGS OF THE IEEE for September 1976.

Robert M. Fano (1917–   ) was born in Torino, Italy and studied engineering in the School of Engineering at Torino from 1935 to 1939 before coming to the U.S. He received B.S. and doctorate degrees from M.I.T. in 1941 and 1947, respectively. He remained at M.I.T. as a teacher and research engineer, and worked at the M.I.T. Radiation Lab and later at the Lincoln Laboratories. He was elected to the National Academy of Engineers in 1973. See *Engineers of Distinction*, p. 94, 1973 and *Who's Who in Engineering*, p. 756, 1959.

On the Social Role of Computer Communications

ROBERT M. FANO, FELLOW, IEEE

Invited Paper

Abstract—Computer-communication systems appear essential to meeting many needs in our society resulting from greater interdependence and complexity of operation and from rising expectations. There are, in principle, different ways of utilizing them for the same purposes, with radically different social consequences. The choice, which amounts in effect to a social decision, is, in practice, severely restricted by the computer hardware and software and by the communication facilities that happen to be available at the time. The paper discusses some of the pressures that lead to more widespread use of computers in the operation of society and illustrates how specific characteristics of computer-communication systems may influence social trends and, in particular, individual freedom.

THE "MARRIAGE" of computers and communication has been celebrated and consummated. By now the honeymoon is over and the two partners are beginning to face the realities of their interdependence.

Electrical communications systems have extended our senses by enabling us to reproduce at a distance printed text, then sound, and finally pictures. These capabilities have greatly facilitated the communication process (in a broad sense) on which the functioning of organized society depends. Distance no longer limits communication within society; instead, the limiting factor is the ability of people to control, comprehend, and utilize the information that is being transmitted. Now it is the human mind that needs to be extended. This is where computers come into the world of communication.

The utilization of computers has followed a very different path. They were originally developed to act as fast calculating machines, but soon thereafter their usefulness as data processors became evident. This led to their widespread use in automating clerical operations.

It was then discovered that they can act, through close man–machine interaction, as skillful assistants to people in a great variety of intellectual activities from text editing to engineering design and from browsing through very large files to formulating and solving very complex problems. It is now becoming apparent that computers could facilitate in a major way communication within society with respect to just those functions that have been performed in the past by people and that are now overwhelming them. Not only could information originating from different sources be merged, restructured, and selectively disseminated to meet individual needs, but people might be able to interact in real time through computers "in the presence of" information that can be consulted and amended by them, either jointly or independently. Furthermore, information stored in a computer may include procedures as well as data, and therefore it may represent dynamic models of situations that have in the past been difficult to record on a static medium, such as paper, for transmission to others.

Manuscript received October 12, 1971; revised June 19, 1972.

The author is with the Massachusetts Institute of Technology, Cambridge, Mass. 02139.

In order to exploit the computer's potential for facilitating communication within society, it is essential that computers be able to communicate easily with people and with other computers. In other words, it is essential that computers be embedded in a communication network. Of course, we cannot expect communication systems and computers in their present form to fit each other's needs and the objectives of the resulting computer-communication system. They will have to evolve in new directions, and so will have to evolve the legal, regulatory, and economic environment in which they exist. This evolution has just begun. Yet strong pressures on private and public organizations have led to the implementation of a multitude of computer-communication systems employing existing facilities, equipment, and techniques without much regard for their long-term social consequences. In effect, a nervous system of society is being developed piecemeal, on an *ad hoc* basis, to satisfy local needs. The task of understanding in depth the role of computer-communication systems in the operation of society, of evolving suitable policies in their respect, and of guiding their development from a social, economic, and technical standpoint, is a gigantic one. The following are examples of sociotechnical aspects of the task confronting us.

The Pressure of Numbers

A major force that has led to the use of computers in the operation of society is the growing volume of transactions of various types that must be handled. For instance, the Bank of America was led, in the middle of the 1950's, to pioneer in the use of computers by the realization that manual handling of checks would have required, in the foreseeable future, the entire adult population of California [1]. It is clear by now that, if computers were not available, the level of activity attainable in many segments of society would be strictly limited by the fraction of the population that could be devoted to the necessary bookkeeping tasks. It is not generally realized, however, that a similar limitation is beginning to emerge with respect to tasks of a substantially higher intellectual character, and that unless we succeed in circumventing it through proper exploitation of computers, the operation of society may well crack under the weight of its own complexity.

We hear many complaints about inadequacy of medical care, of education, and of many other services. These inadequacies are usually blamed on a variety of factors, such as lack of funds, incompetence, poor planning, or resistance to change. It may well be, however, that these factors are merely symptoms of a more fundamental limitation, namely, that society does not possess the human resources necessary to perform adequately all the necessary tasks, particularly in view of its growing complexity and the rising expectations on the part of the population. It is by no means obvious that a society can provide itself with all the specialized services that it needs according to present-day standards. Certainly, there

Reprinted from *Proc. IEEE*, vol. 60, pp. 1249–1253, Nov. 1972.

is some limit to the amount and quality of such services, although we do not know how our expectations compare with it. Can this limit be circumvented?

Storing Knowledge

Part of the problem that we face stems from the fact that providing specialized services requires a great deal of knowledge and experience that cannot be readily transferred from one person to another. A substantial fraction of a person's life is devoted to acquiring them, in spite of the fact that many other people already possess them. Could a good part of this knowledge and experience be stored in computer programs and made available on demand? It is, of course, true that every computer program stores some human knowledge. However, it is usually knowledge of very limited general value. Could truly significant knowledge be stored in a program so that a computer could perform tasks requiring real intellectual skills? This has been accomplished in a few instances.

A good example is symbolic integration, that is, nonnumeric evaluation of indefinite integrals. As a result of contributions by several people over more than a decade, there exists now a complex of programs that can perform symbolic integration with a skill comparable to that of a very competent mathematician [2]. It is important to note that symbolic integration usually involves many steps, each consisting of the application of one of several well-known techniques. However, there are no precise rules for deciding which technique should be used in each particular instance. One must resort to a trial-and-error process, guided by mathematical judgment built on experience. In other words, symbolic integration is a true intellectual skill that requires considerable effort to acquire, and that can be easily lost from lack of practice.

It is important to note that storing knowledge into a program does not interfere with people acquiring it. On the contrary, it gives them a choice not previously available. They can use the program directly to accomplish their goals, or they can learn from the program the techniques and skills that it employs. In fact, it would be easier to learn from a program than from a book, because the program could guide the learner in his work, correct his mistakes, and answer his questions, much as an instructor would. The potential is clearly there, but much remains to be done before a significant fraction of human knowledge could be stored in computer programs. In particular, we cannot realistically expect that the people that possess the knowledge will develop the appropriate programs from scratch. It ought to be possible for them to transfer their knowledge to a general-purpose program by making statements and answering questions more or less as they would with another person.

Mass Production of Services

Another part of the problem of increasing the quantity and quality of the services that society can provide itself with is best understood by analogy with the mass production of goods. The techniques of mass production have increased not only the quantity of goods available, but also their quality. Specifically, the quality of mass-produced goods is limited only by the total knowledge and capabilities of the society, while the quality of the goods produced by an artisan is strictly limited by his own knowledge and skill. With respect to services, we are still at the artisan stage. The quality of the services that we can obtain are still limited primarily by the knowledge, experience, and skill of the person that provides them, whether physician, teacher, or repairman. The average quality is far from satisfactory, and still the supply is inadequate to meet the needs of the entire population. What is needed, by analogy, is mass production of services, that is, a way of providing services that makes their quality consistent with the total knowledge and capabilities of the whole society, and their availability adequate to meet the needs of the entire population. For this purpose, information, instead of matter, must be transported and brought to bear on a specific situation, where and when the need arises. Thus a widespread and economic computer-communication network is clearly essential to the mass production of services, just as a widespread and economical transportation network has proved to be essential to the mass production of goods.

Information and Control

The services that are needed include access to information and advice on a great variety of matters that are essential to steering oneself safely and in a personally rewarding manner in a very complex environment which changes at a rapid rate. The fact that obtaining adequate information and advice is often prohibitively costly in time and money has consequences that transcend personal frustrations and injustices.

Control is a necessary function in organized society. It is most effectively exercised at the local level, where detailed information is readily available. However, it cannot be properly exercised out of the context of broader objectives and other activities with which coordination must be achieved. If knowledge of the context is not available locally, specific instructions must be issued from where the context is understood. Thus the present trend towards centralization of control can be seen as a direct consequence of the growth of interdependence and our inability as individuals to maintain a working knowledge of an increasingly complex environment. On the other hand, centralization of control runs into informational problems of its own. Specific instructions, to be effective, must be based on detailed local information. Collecting timely data about a broad spectrum of activities and generating from them detailed control information can quickly become an unmanageable task. Furthermore, information that can be readily acquired and utilized locally may be very difficult to represent for use elsewhere.

Are the informational problems of centralized control easier to solve than those of distributed control, or vice versa? The evidence emerging from the design of complex computer systems and communication networks seems to favor distributed control over central control. One reason for this is that less information needs to be explicitly stored. Distributed control is clearly more consistent with individual freedom and autonomy, but cannot survive without adequate information flow. Central control is undesirable from a personal standpoint, but its survival is much less dependent on information. Computer-communication systems will be required for either type of control, but their characteristics are bound to be quite different in the two cases. Thus if we follow the path that appears easier and safer in the short run, we may find ourselves in the future where we do not want to be, yet no longer able to change direction.

Computers and People

There is an important lesson that computer people had to learn during the last decade, at considerable cost in money, time, and frustration. The lesson is that it is inappropriate to design or evaluate computer equipment out of the context of the software that provides the interface to its users. The reason is simple. The equipment characteristics limit in a major way the interface characteristics that can be obtained. An extension of the same lesson remains to be learned, namely, that it is inappropriate, and in this case dangerous, to design or evaluate a computer system out of the context of the community of people that will be affected by its use, either directly or indirectly. The reason is equally simple. The characteristics of the computer system limit in a significant way the structure and mode of operation of the associated community of people and thereby influence the attitude and behavior of the individuals that comprise it. In turn, the characteristics of the community influence the evolution of the computer system itself. Because of the resulting feedback loop, the overall system consisting of hardware, software, and people may acquire unintended and possibly undesirable characteristics of its own.

In other words, when we employ computers in any task involving people, we make in effect a social decision, the consequences of which may be far reaching. The class of social decision that can, in fact, be implemented at a given time is largely defined by the specific characteristics of the available computer systems and the related facilities and techniques that are at our disposal. It may well be that all the options available at a particular time have undesirable social consequences. Yet the help of computers may be so urgently needed that disregarding such social consequences may be regarded as the lesser of two evils. The point is that, once the technology is specified, our freedom of choice in the utilization of computers may well be very limited. Our freedom of choice is instead widest and most effectively exercised ahead of time when we select the characteristics of the technology that we intend to make available in the future.

Man–Machine Interaction

A set of characteristics of computer systems that are very important with respect to potential social effects are those that influence the ease, flexibility, and intimacy of man–machine interaction from an intellectual as well as a physical standpoint. If the activities carried out by computers cannot be readily monitored and guided by people, and if human processing of information cannot be easily intermixed with computer processing, computers tend to become unchallengeable authorities with respect to the functions they perform. As a simple example, when data are stored on magnetic tape or on any other storage device within a computer system, they are removed from direct human inspection. If the computer system does not provide convenient means for browsing through them, it becomes difficult and costly to locate errors and correct them. As a result, the data and the conclusions reached from them become unchallengeable simply because it would be too time-consuming or costly to challenge them.

Man–machine interaction is of particular importance when computers are employed in any facet of the operation of society involving problem solving and decision making. It is didactically convenient, when explaining how a problem has been solved, to discuss first its formulation and then its solution. However, the separation into distinct phases does not correspond to what actually takes place. Instead, problem formulation and problem solution proceed concurrently most of the time, and this is so not by choice but by necessity. One cannot be sure that a problem has been properly formulated without exploring some of the consequences of the formulation. The fact that an important constraint has been forgotten often becomes evident through the examination of a solution that is logically correct but clearly unacceptable. Since computers are very powerful aids to problem solving, while problem formulation lies inherently in the human domain, the ability to establish close collaboration between man and computer becomes of crucial importance. If means for establishing this close collaboration are not available, some aspect of the formulation is very likely to be delegated by default to the computer. The delegation may take the form of forcing a problem into a mold in which it does not fit, or of accepting an inadequate solution because of time limitations or out of sheer frustration. If the problem is an important one, the consequences of delegating its formulation to computers may be quite serious and possibly tragic. The late Prof. Norbert Wiener used to warn us of this point way back in the late 1940's [3]. Computers, he used to say, are literal minded just like magics: they solve the problem that has been presented to them, not the one that ought to have been presented. Such warnings were usually followed by some tale of magics that made the point all too obvious.

Communication Through Computers

Another set of characteristics of computer-communication systems of considerable importance are those that facilitate people-to-people communication [4] over time and intellectual barriers as well as distance. These are the characteristics necessary to provide good coupling to people as members of an interacting community above and beyond good coupling to people as individuals. A system feature of major importance in this respect is the ability to share data and procedures under flexible and secure control. Not only is this ability essential to any computer-aided activity involving close collaboration between people, but it is bound to influence in a major way the structure and mode of operation of any community in which important functions are dependent on the use of computers.

If secure control over access to information cannot be exercised selectively within a computer system, it must be exercised externally by traditional means. This implies that all programs must be individually analyzed and approved before execution; checking the output would not be sufficient because improperly obtained information could be easily disguised. In practice, access to information and to computers would have to be severely limited, and control in individual organizations and in society as a whole would, by necessity, become increasingly centralized.

The sharing of information within a computer system presents difficult operational and semantic problems, even if the need for controls is disregarded. Information must be identified, retrieved from wherever it is stored, and used in a proper manner in conjunction with other information. If a computer system is to act as an information broker, it must be able to converse intelligently about the information it contains in its memory. While significant progress is being made in this direction, much remains to be done. In the meantime, we must be wary of plans to circumvent these difficulties through the imposition of rigid formats. Forcing situations into a preestablished mold often results in dangerous distortions of reality.

The protection of individual and organizational privacy is not the only objective that requires dependable means for insuring the integrity of the information stored in a computer system and for controlling its use. Controls are also needed to implement contractual agreements between developers of useful software and data bases and users of these facilities. Also, uncontrolled collection, storage, and dissemination of information lead necessarily to "pollution" of the information environment, a phenomenon which is already having disturbing if not serious consequences. Finally, public safety requires that secure control be exercised over information that could endanger the public, for instance, by causing panic or by enabling or fostering illegal activities.

The protection of individual privacy presents some special problems because of the large number of people involved and the proliferation of files containing information about individuals. The presence in a file of incorrect, incomplete, or misleading information is at least as dangerous to an individual as the possibility of unauthorized or otherwise improper disclosure of information about him. Thus it is not only necessary to protect personal information against malicious or accidental modifications; we must also make it possible for each individual to check the accuracy of information kept about him without infringing upon the privacy of others and without an unreasonable expenditure of time and effort on his part. It is difficult to conceive how this could be accomplished without providing direct access to the computer system in which the information is stored and without reducing through sharing the present proliferation of similar files.

The development of means for controlling the use of information within computer systems is still in its infancy. A few time-sharing systems exist in which access to individual files can be granted and withdrawn selectively for reading, writing, or executing, but the vast majority of computer systems do not permit even this basic form of control. Yet much more elaborate forms of control are needed. For instance, it is not sufficient to specify who may have access to a particular data file; it is also essential to place restrictions on the information that may be extracted from it. Furthermore, means are needed for controlling and auditing the controls themselves, according to externally established lines of authority and responsibility.

It was very encouraging at the 1972 Spring Joint Computer Conference to hear from the Chairman of the Board of IBM of the commitment IBM is making to data security in its systems. This is the first recognition on the part of a computer manufacturer that the problem of data security needs immediate attention. A similar commitment to action is also needed from the other segments of the computer-communication industry. The implementation of adequate facilities for controlled sharing of information presents very difficult problems even within the limited context of individual computer systems. These problems become substantially more difficult in the broader context of computer-communication networks, where programs executed in one computer may call on programs or data stored in other computers. The task confronting us is indeed very large and very complex, and the consequences of not devoting to it the necessary resources could be extremely serious.

Knowledge and Power

Computers provide access to knowledge, and knowledge restricted to a segment of society can be transformed into power over the rest of society. Thus unless computers are made truly accessible to the population at large, there will necessarily develop a dangerous power gap between those who have access to them and those who do not, and particularly between organizations, public or private, and individuals.

The ability to share knowledge through computer systems may well play in the future a role analogous to that of the printing press. Before the invention of the printing press, literacy was limited to a small segment of the population. This is not surprising because most people read what other people write. By now, it is generally accepted that literacy is a prerequisite to effective participation in the life of society. On the other hand, there is very little incentive today to learn how to exploit the capabilities of computers, because the data stored in computer systems and the programs necessary to extract relevant information from them are not readily accessible. Yet computer-communication networks are on their way to forming a new nervous system of society. How can individual citizens play their rightful role in society without having access to it?

It seems essential, therefore, that we redirect our efforts toward assisting individuals in their daily life and facilitating communication within society. This implies, for instance, that we must invest in the development of a computer-communication network capable of serving the public at large. It must be a single interconnected network in order to avoid introducing artificial barriers between people and between different aspects of their individual activities. This does not imply that the network must be owned and operated by a single organization, nor does it imply anything specific about the structure of the network and the distribution of processing and storage capacity. It does imply, however, that the network must look to each user as a single system, in the sense that he must be largely unaware of and unaffected by the internal structure and economic organization of the network.

In spite of recent efforts in the direction of general-purpose computer-communication networks, the computer and communication technologies and the legal and regulatory environment are largely ill suited to this objective. The trend is, instead, in the direction of dedicated special-purpose systems unrelated to and incompatible with one another. If this trend continues for much longer, it may become very difficult in practice to merge these special-purpose systems into a single general-purpose network. It is well to remember in this regard that, while wire telephony has evolved into a single general-purpose network, radio communication has evolved, instead, into an aggregate of special-purpose incompatible systems that would be very difficult, technically and politically, to merge into a single coherent network.

Making computer-based services available to the public implies much more than the development of the necessary hardware facilities. The task of making a computer-communication network useful to the public and also economically and intellectually accessible to the entire population is indeed a gigantic one that will require considerable effort over a long period of time. Unless we focus on this goal very soon, the growing gap of knowledge in society is likely to become so wide and the resulting gap of power so entrenched that it will be extremely difficult, if not impossible, to return to anything resembling a democratic society.

Where To?

This paper has attempted to illustrate two basic points. The first point is that the structural and operational complexity of modern society and the growing interdependence within it are straining our ability to comprehend and deal successfully with the multitude of problems that we face individually and collectively. The computer and communication technologies will not, by themselves, solve our problems, but it is very unlikely that we will be able to solve them without substantial help from them. In practice, their help is being increasingly sought in almost every aspect of the operation of society, and their utilization has already had a marked effect on many of them. This trend is bound to continue simply because there does not seem to be anywhere else we can turn for help in preventing society from collapsing under the weight of its own complexity.

The second point is that the growing utilization of computer and communication technologies will undoubtedly have a major influence on the operation of society, on our daily lives, and even on our values and our perception of the world. The character of these social effects, however, will depend largely on how we will choose to develop and utilize the computer and communication technologies. The choice, roughly speaking, is between automating those functions in society that people seem unable to continue to perform adequately and helping people to cope successfully with the growing magnitude and complexity of the tasks that confront them. The present trend is toward automation of functions in a way that lessens significantly human control over them by removing the pertinent information from easy access on the part of people. This trend is not the result of a conscious choice, but rather because it would be too inconvenient, or too uneconomical, or even impossible to do otherwise in view of the technology that happens to be available at this time. Continuation of this trend is very likely to lead to a society operated by a rigid bureaucracy whose power will stem from widespread surveillance and control over information, that is, to a society of the "1984" type. This is most likely to occur unintentionally as a result of the actions of many well-meaning people attempting to solve the problems they face in the best way they know how at the time. The opposite choice involves exploiting the computer and communication technologies to augment the intellectual capabilities of people and to facilitate communication between them, with the objective of enabling society as a whole to operate effectively at a much higher level of complexity. A major result of the power revolution has been the development of a great variety of devices that have augmented the power, the precision, and the skill of our muscles. The computer and communication technologies can, in a similar way, augment the power, the precision, and the skill of our mind. Moving in this direction, however, will require a conscious and determinate choice, because most short-range considerations, economical, technical, and political, seem to militate against it. In particular, we must consciously shape our technology in this direction and make sure that appropriate technical capabilities will be available when needed, rather than make do with whatever happens to be available as a result of short-range local considerations.

Even in the best of circumstances there are going to be many conflicts between new modes of operation in society and traditional habits of people. As frustrating as these conflicts may be to those of us who are fond of technical innovation, they are much more frustrating to the people at the receiving end, who do not wish to devote the intellectual and emotional energy required to change their outlooks and habits. Thus the concluding plea is for awareness of the individual, of his needs, his feelings, and particularly his pride as a human being. Without such an awareness the impact of computers and communication on society may turn out to be a nightmare.

References

[1] J. Weizenbaum, "The two cultures of the computer age," *Technol. Rev.*, p. 55, Apr. 1969.

[2] J. Moses, "Symbolic integration," Mass. Inst. Technol., Cambridge, Mass., Tech. Rep. MAC-TR-47 (thesis), Sept. 1967.

[3] N. Wiener, *The Human Use of Human Beings.* Boston, Mass.: Houghton Mifflin, 1950, p. 212.

[4] J. C. R. Licklider and R. W. Taylor, "The computer as a communication device," *Sci. Technol.*, no. 76, p. 21, Apr. 1968.

Author Index

Subject Index

C

F

G

H

I

J

K

L

S

T

U

V

W

X

Y

Editor's Biography

James E. Brittain (S'56–M'59) was born in North Carolina in 1931. During the Korean War, he was an Instructor in the U.S. Air Force in several radar and electronic countermeasures schools. He received a B.S. degree in electrical engineering from Clemson University, Clemson, SC, in 1957 and an M.S. degree in electrical engineering from the University of Tennessee, Knoxville, in 1958. After a year of research on antennas at the University of Tennessee Engineering Experiment Station, he worked as an Assistant Professor of Electrical Engineering at Clemson University until 1966, during which time he developed a strong interest in electrical history. Subsequently he received M.A. and Ph.D. degrees in the history of science and technology from Case Western Reserve University, Cleveland, OH, with both a master's thesis and doctoral dissertation being on the early history of electrical engineering.

Dr. Brittain was appointed Assistant Professor of the History of Science and Technology at the Georgia Institute of Technology, Atlanta, in 1969 and is now an Associate Professor at the same institution. He is also Associate Editor of *Technology and Culture*, and has been a member of the IEEE History Committee since 1972. He was the Guest Editor of a special bicentennial issue of the PROCEEDINGS OF THE IEEE devoted to the electrical history of America that appeared in September 1976. He has recently been appointed to the Editorial Advisory Board of the IEEE TRANSACTIONS ON EDUCATION, with responsibility for articles dealing with electrical engineering history. An upper level course in electrical history is among the courses that he teaches at Georgia Tech. He recently completed an inventory of historic engineering and industrial sites in Georgia for the *Historic American Engineering Record.*